Fractional Calculus and Waves in Linear Viscoelasticity

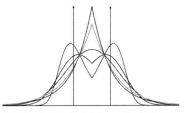

An Introduction to Mathematical Models

Fractional Calculus and Waves in Linear Viscoelasticity

Francesco Mainardi
University of Bologna, Italy

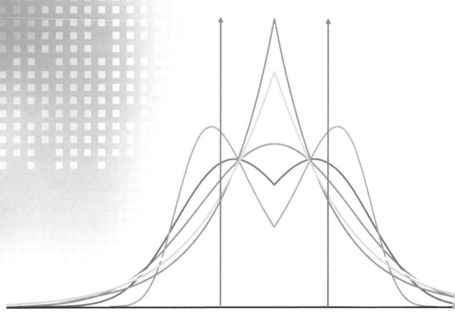

An Introduction to Mathematical Models

Imperial College Press

Published by

Imperial College Press
57 Shelton Street
Covent Garden
London WC2H 9HE

Distributed by

World Scientific Publishing Co. Pte. Ltd.
5 Toh Tuck Link, Singapore 596224
USA office: 27 Warren Street, Suite 401-402, Hackensack, NJ 07601
UK office: 57 Shelton Street, Covent Garden, London WC2H 9HE

British Library Cataloguing-in-Publication Data
A catalogue record for this book is available from the British Library.

FRACTIONAL CALCULUS AND WAVES IN LINEAR VISCOELASTICITY
An Introduction to Mathematical Models

Copyright © 2010 by Imperial College Press

All rights reserved. This book, or parts thereof, may not be reproduced in any form or by any means, electronic or mechanical, including photocopying, recording or any information storage and retrieval system now known or to be invented, without written permission from the Publisher.

For photocopying of material in this volume, please pay a copying fee through the Copyright Clearance Center, Inc., 222 Rosewood Drive, Danvers, MA 01923, USA. In this case permission to photocopy is not required from the publisher.

ISBN-13 978-1-84816-329-4
ISBN-10 1-84816-329-0

Printed in Singapore by B & JO Enterprise

To the memory of my parents, Enrico and Domenica

Preface

The aim of this monograph is essentially to investigate the connections among *fractional calculus*, *linear viscoelasticity* and *wave motion*. The treatment mainly reflects the research activity and style of the author in the related scientific areas during the last decades.

Fractional calculus, in allowing integrals and derivatives of any positive order (the term "fractional" is kept only for historical reasons), can be considered a branch of mathematical physics which deals with integro-differential equations, where integrals are of convolution type and exhibit weakly singular kernels of power law type.

Viscoelasticity is a property possessed by bodies which, when deformed, exhibit both viscous and elastic behaviour through simultaneous dissipation and storage of mechanical energy. It is known that viscosity refers mainly to fluids and elasticity mainly to solids, so we shall refer viscoelasticity to generic continuous media in the framework of a *linear* theory. As a matter of fact the linear theory of viscoelasticity seems to be the field where we find the most extensive applications of fractional calculus for a long time, even if often in an implicit way.

Wave motion is a wonderful world impossible to be precisely defined in a few words, so it is preferable to be guided in an intuitive way, as G.B. Whitham has pointed out. Wave motion is surely one of the most interesting and broadest scientific subjects that can be studied at any technical level. The restriction of wave propagation to linear viscoelastic media does not diminish the importance of this research area from mathematical and physical view points.

This book intends to show how fractional calculus provides a suitable (even if often empirical) method of describing dynamical properties of linear viscoelastic media including problems of wave propagation and diffusion. In all the applications the special transcendental functions are fundamental, in particular those of Mittag-Leffler and Wright type.

Here mathematics is emphasized for its own sake, but in the sense of a language for everyday use rather than as a body of theorems and proofs: unnecessary mathematical formalities are thus avoided. Emphasis is on problems and their solutions rather than on theorems and their proofs. So as not to bore a "practical" reader with too many mathematical details and functional spaces, we often skim over the regularity conditions that ensure the validity of the equations. A "rigorous" reader will be able to recognize these conditions, whereas a "practionist" reader will accept the equations for sufficiently well-behaved functions. Furthermore, for simplicity, the discussion is restricted to the scalar cases, i.e. one-dimensional problems.

The book is likely to be of interest to applied scientists and engineers. The presentation is intended to be self-contained but the level adopted supposes previous experience with the elementary aspects of mathematical analysis including the theory of integral transforms of Laplace and Fourier type.

By referring the reader to a number of appendices where some special functions used in the text are dealt with detail, the author intends to emphasize the mathematical and graphical aspects related to these functions.

Only seldom does the main text give references to the literature, the references are mainly deferred to notes sections at the end of chapters and appendices. The notes also provide some historical perspectives. The bibliography contains a remarkably large number of references to articles and books not mentioned in the text, since they have attracted the author's attention over the last decades and cover topics more or less related to this monograph. The interested reader could hopefully take advantage of this bibliography for enlarging and improving the scope of the monograph itself and developing new results.

This book is divided into six chapters and six appendices whose contents can be briefly summarized as follows. Since we have chosen to stress the importance of fractional calculus in modelling viscoelasticity, the first two chapters are devoted to providing an outline of the main notions in fractional calculus and linear viscoelasticity, respectively. The third chapter provides an analysis of the viscoelastic models based on constitutive equations containing integrals and derivatives of fractional order.

The remaining three chapters are devoted to wave propagation in linear viscoelastic media, so we can consider this chapter-set as a second part of the book. The fourth chapter deals with the general properties of dispersion and dissipation that characterize the wave propagation in linear viscoelastic media. In the fifth chapter we discuss asymptotic representations for viscoelastic waves generated by impact problems. In particular we deal with the techniques of wave-front expansions and saddle-point approximations. We then discuss the matching between the two above approximations carried out by the technique of rational Padè approximants. Noteworthy examples are illustrated with graphics. Finally, the sixth chapter deals with diffusion and wave-propagation problems solved with the techniques of fractional calculus. In particular, we discuss an important problem in material science: the propagation of pulses in viscoelastic solids exhibiting a constant quality factor. The tools of fractional calculus are successfully applied here because the phenomenon is shown to be governed by an evolution equation of fractional order in time.

The appendices are devoted to the special functions that play a role in the text. The most relevant formulas and plots are provided. We start in appendix A with the Eulerian functions. In appendices B, C and D we consider the Bessel, the Error and the Exponential Integral functions, respectively. Finally, in appendices E and F we analyse in detail the functions of Mittag-Leffler and Wright type, respectively. The applications of fractional calculus in diverse areas has considerably increased the importance of these functions, still ignored in most handbooks.

Francesco Mainardi
Bologna, December 2009

Acknowledgements

Over the years a large number of people have given advice and encouragement to the author; I wish to express my heartfelt thanks to all of them. In particular, I am indebted to the co-authors of several papers of mine without whom this book could not have been conceived. Of course, the responsibility of any possible mistake or misprint is solely that of the author.

Among my senior colleagues, I am grateful to Professor Michele Caputo and Professor Rudolf Gorenflo, for having provided useful advice in earlier and recent times, respectively. Prof. Caputo introduced me to the fractional calculus during my PhD thesis (1969–1971). Prof. Gorenflo has collaborated actively with me and with my students on several papers since 1995. It is my pleasure to enclose a photo showing the author between them, taken in Bologna, April 2002.

For a critical reading of some chapters I would like to thank my colleagues Virginia Kiryakova (Bulgarian Academy of Sciences), John W. Hanneken and B.N. Narahari Achar (University of Memphis, USA), Giorgio Spada (University of Urbino, Italy), José Carcione and Fabio Cavallini (OGS, Trieste, Italy).

Among my former students I would like to name (in alphabetical order): Gianni De Fabritiis, Enrico Grassi, Daniele Moretti, Antonio Mura, Gianni Pagnini, Paolo Paradisi, Daniele Piazza, Ombretta Pinazza, Donatella Tocci, Massimo Tomirotti, Giuliano Vitali and Alessandro Vivoli. In particular, I am grateful to Antonio Mura for help with graphics.

I am very obliged to the staff at the Imperial College Press, especially to Katie Lydon, Lizzie Bennett, Sarah Haynes and Rajesh Babu, for taking care of the preparation of this book.

Finally, I am grateful to my wife Giovanna and to our children Enrico and Elena for their understanding, support and encouragement.

F. Mainardi between R. Gorenflo (left) and M. Caputo (right)
Bologna, April 2002

Contents

Preface   vii

Acknowledgements   xi

List of Figures   xvii

1. Essentials of Fractional Calculus   1
   1.1 The fractional integral with support in $\mathbb{R}^+$ 2
   1.2 The fractional derivative with support in $\mathbb{R}^+$ 5
   1.3 Fractional relaxation equations in $\mathbb{R}^+$ 11
   1.4 Fractional integrals and derivatives with support in $\mathbb{R}$ 15
   1.5 Notes 17

2. Essentials of Linear Viscoelasticity   23
   2.1 Introduction...................... 23
   2.2 History in $\mathbb{R}^+$: the Laplace transform approach ... 26
   2.3 The four types of viscoelasticity 28
   2.4 The classical mechanical models 30
   2.5 The time - and frequency - spectral functions 41
   2.6 History in $\mathbb{R}$: the Fourier transform approach and the dynamic functions 45
   2.7 Storage and dissipation of energy: the loss tangent . 46
   2.8 The dynamic functions for the mechanical models.. 51
   2.9 Notes 54

3. Fractional Viscoelastic Models 57
   3.1 The fractional calculus in the mechanical models 57
       3.1.1 Power-Law creep and the Scott-Blair model 57
       3.1.2 The correspondence principle 59
       3.1.3 The fractional mechanical models 61
   3.2 Analysis of the fractional Zener model 63
       3.2.1 The material and the spectral functions 63
       3.2.2 Dissipation: theoretical considerations 66
       3.2.3 Dissipation: experimental checks 69
   3.3 The physical interpretation of the fractional Zener model via fractional diffusion 71
   3.4 Which type of fractional derivative? Caputo or Riemann-Liouville? 73
   3.5 Notes 74

4. Waves in Linear Viscoelastic Media: Dispersion and Dissipation 77
   4.1 Introduction 77
   4.2 Impact waves in linear viscoelasticity 78
       4.2.1 Statement of the problem by Laplace transforms 78
       4.2.2 The structure of wave equations in the space-time domain 82
       4.2.3 Evolution equations for the mechanical models 83
   4.3 Dispersion relation and complex refraction index 85
       4.3.1 Generalities 85
       4.3.2 Dispersion: phase velocity and group velocity 88
       4.3.3 Dissipation: the attenuation coefficient and the specific dissipation function 90
       4.3.4 Dispersion and attenuation for the Zener and the Maxwell models 91
       4.3.5 Dispersion and attenuation for the fractional Zener model 92
       4.3.6 The Klein-Gordon equation with dissipation 94

	4.4	The Brillouin signal velocity	98
	4.4.1	Generalities	98
	4.4.2	Signal velocity via steepest–descent path	100
	4.5	Notes	107

5. Waves in Linear Viscoelastic Media: Asymptotic Representations   109

	5.1	The regular wave–front expansion	109
	5.2	The singular wave–front expansion	116
	5.3	The saddle–point approximation	126
	5.3.1	Generalities	126
	5.3.2	The Lee-Kanter problem for the Maxwell model	127
	5.3.3	The Jeffreys problem for the Zener model	131
	5.4	The matching between the wave–front and the saddle–point approximations	133

6. Diffusion and Wave–Propagation via Fractional Calculus   137

	6.1	Introduction	137
	6.2	Derivation of the fundamental solutions	140
	6.3	Basic properties and plots of the Green functions	145
	6.4	The Signalling problem in a viscoelastic solid with a power-law creep	151
	6.5	Notes	153

Appendix A   The Eulerian Functions   155

A.1	The Gamma function: $\Gamma(z)$	155
A.2	The Beta function: $B(p,q)$	165
A.3	Logarithmic derivative of the Gamma function	169
A.4	The incomplete Gamma functions	171

Appendix B   The Bessel Functions   173

B.1	The standard Bessel functions	173
B.2	The modified Bessel functions	180
B.3	Integral representations and Laplace transforms	184
B.4	The Airy functions	187

Appendix C   The Error Functions · 191

- C.1 The two standard Error functions 191
- C.2 Laplace transform pairs 193
- C.3 Repeated integrals of the Error functions 195
- C.4 The Erfi function and the Dawson integral 197
- C.5 The Fresnel integrals 198

Appendix D   The Exponential Integral Functions · 203

- D.1 The classical Exponential integrals $\text{Ei}\,(z)$, $\mathcal{E}_1(z)$ . . . 203
- D.2 The modified Exponential integral $\text{Ein}\,(z)$ 204
- D.3 Asymptotics for the Exponential integrals 206
- D.4 Laplace transform pairs for Exponential integrals . . 207

Appendix E   The Mittag-Leffler Functions · 211

- E.1 The classical Mittag-Leffler function $E_\alpha(z)$ 211
- E.2 The Mittag-Leffler function with two parameters . . 216
- E.3 Other functions of the Mittag-Leffler type 220
- E.4 The Laplace transform pairs 222
- E.5 Derivatives of the Mittag-Leffler functions 227
- E.6 Summation and integration of Mittag-Leffler functions 228
- E.7 Applications of the Mittag-Leffler functions to the Abel integral equations 230
- E.8 Notes 232

Appendix F   The Wright Functions · 237

- F.1 The Wright function $W_{\lambda,\mu}(z)$ 237
- F.2 The auxiliary functions $F_\nu(z)$ and $M_\nu(z)$ in $\mathbb{C}$ 240
- F.3 The auxiliary functions $F_\nu(x)$ and $M_\nu(x)$ in $\mathbb{R}$ . . . 242
- F.4 The Laplace transform pairs 245
- F.5 The Wright M-functions in probability 250
- F.6 Notes 258

Bibliography   261

Index   343

List of Figures

1.1 Plots of $\psi_\alpha(t)$ with $\alpha = 1/4, 1/2, 3/4, 1$ versus t; top: linear scales ($0 \leq t \leq 5$); bottom: logarithmic scales ($10^{-2} \leq t \leq 10^2$). 13

1.2 Plots of $\phi_\alpha(t)$ with $\alpha = 1/4, 1/2, 3/4, 1$ versus t; top: linear scales ($0 \leq t \leq 5$); bottom: logarithmic scales ($10^{-2} \leq t \leq 10^2$). 14

2.1 The representations of the basic mechanical models: a) spring for Hooke, b) dashpot for Newton, c) spring and dashpot in parallel for Voigt, d) spring and dashpot in series for Maxwell. 31

2.2 The mechanical representations of the Zener [a), b)] and anti-Zener [c), d)] models: a) spring in series with Voigt, b) spring in parallel with Maxwell, c) dashpot in series with Voigt, d) dashpot in parallel with Maxwell. 34

2.3 The four types of canonic forms for the mechanical models: a) in creep representation; b) in relaxation representation. 35

2.4 The mechanical representations of the compound Voigt model (top) and compound Maxwell model (bottom). .. 36

2.5 The mechanical representations of the Burgers model: the creep representation (top), the relaxation representation (bottom). 39

2.6 Plots of the dynamic functions $G'(\omega)$, $G''(\omega)$ and loss tangent $\tan\delta(\omega)$ versus $\log\omega$ for the Zener model. 53

3.1 The Mittag-Leffler function $E_\nu(-t^\nu)$ versus t ($0 \le t \le 15$) for some rational values of ν, i.e. $\nu = 0.25$, 0.50, 0.75, 1. 61

3.2 The material functions $J(t)$ (top) and $G(t)$ (bottom) of the fractional Zener model versus t ($0 \le t \le 10$) for some rational values of ν, i.e. $\nu = 0.25$, 0.50, 0.75, 1. 64

3.3 The time–spectral function $\hat{R}_*(\tau)$ of the fractional Zener model versus τ ($0 \le \tau \le 2$) for some rational values of ν, i.e. $\nu = 0.25$, 0.50, 0.75, 0.90. 65

3.4 Plots of the loss tangent $\tan\delta(\omega)$ scaled with $\Delta/2$ against the logarithm of $\omega\bar\tau$, for some rational values of ν: $a)\,\nu = 1$, $b)\,\nu = 0.75$, $c)\,\nu = 0.50$, $d)\,\nu = 0.25$. 68

3.5 Plots of the loss tangent $\tan\delta(\omega)$ scaled with it maximum against the logarithm of $\omega\bar\tau$, for some rational values of ν: $a)\,\nu = 1$, $b)\,\nu = 0.75$, $c)\,\nu = 0.50$, $d)\,\nu = 0.25$. 69

3.6 Q^{-1} in brass: comparison between theoretical (continuous line) and experimental (dashed line) curves. 70

3.7 Q^{-1} in steel: comparison between theoretical (continuous line) and experimental (dashed line) curves. 71

4.1 Phase velocity V, group velocity U and attenuation coefficient δ versus frequency ω for a) Zener model, b) Maxwell model. 92

4.2 Phase velocity over a wide frequency range for some values of ν with $\tau_\epsilon = 10^{-3}\,s$ and a) $\gamma = 1.1$: 1) $\nu = 1$, 2) $\nu = 0.75$, 3) $\nu = 0.50$, 4) $\nu = 0.25$. b) $\gamma = 1.5$: 5) $\nu = 1$, 6) $\nu = 0.75$, 7) $\nu = 0.50$, 8) $\nu = 0.25$. 93

4.3 Attenuation coefficient over a wide frequency range for some values of ν with $\tau_\epsilon = 10^{-3}\,s$, $\gamma = 1.5$: 1) $\nu = 1$, 2) $\nu = 0.75$, 3) $\nu = 0.50$, 4) $\nu = 0.25$. 93

4.4 Dispersion and attenuation plots: $m = 0$ (left), $m = 1/\sqrt{2}$ (right). 97

4.5 Dispersion and attenuation plots: $m = \sqrt{2}$ (left), $m = \infty$ (right). 97

4.6 The evolution of the steepest–descent path $L(\theta)$: case (+). 104

4.7 The evolution of the steepest–descent path $L(\theta)$: case (−). 105

List of Figures

5.1 The pulse response for the Maxwell model depicted versus $t - x$ for some fixed values of x. 115

5.2 The pulse response for the Voigt model depicted versus t. 123

5.3 The pulse response for the Maxwell 1/2 model depicted versus $t - x$. 124

5.4 The evolution of the steepest-descent path $L(\theta)$ for the Maxwell model. 129

5.5 The Lee-Kanter pulse for the Maxwell model depicted versus x. 130

5.6 The position of the saddle points as a function of time elapsed from the wave front: 1) $1 < \theta < n_0$; 2) $\theta = n_0$; 3) $\theta > n_0$; where $n_0 = \sqrt{1.5}$. 132

5.7 The step-pulse response for the Zener (S.L.S.) model depicted versus x: for small times (left) and for large times (right). 135

5.8 The Lee-Kanter pulse response for the Zener (S.L.S.) model depicted versus x: for small times (left) and for large times (right). 135

6.1 The Cauchy problem for the time-fractional diffusion-wave equation: the fundamental solutions versus $|x|$ with a) $\nu = 1/4$, b) $\nu = 1/2$, c) $\nu = 3/4$. 146

6.2 The Signalling problem for the time-fractional diffusion-wave equation: the fundamental solutions versus t with a) $\nu = 1/4$, b) $\nu = 1/2$, c) $\nu = 3/4$. 147

6.3 Plots of the fundamental solution $\mathcal{G}_s(x,t;\nu)$ versus t at fixed $x = 1$ with $a = 1$, and $\nu = 1 - \epsilon$ ($\gamma = 2\epsilon$) in the cases: left $\epsilon = 0.01$, right $\epsilon = 0.001$. 153

A.1 Plots of $\Gamma(x)$ (continuous line) and $1/\Gamma(x)$ (dashed line) for $-4 < x \leq 4$. 158

A.2 Plots of $\Gamma(x)$ (continuous line) and $1/\Gamma(x)$ (dashed line) for $0 < x \leq 3$. 159

A.3 The left Hankel contour Ha_- (left); the right Hankel contour Ha_+ (right). 161

A.4 Plot of $I(\beta) := \Gamma(1 + 1/\beta)$ for $0 < \beta \leq 10$. 163

A.5 $\Gamma(x)$ (continuous line) compared with its first order Stirling approximation (dashed line). 164
A.6 Relative error of the first order Stirling approximation to $\Gamma(x)$ for $1 \leq x \leq 10$. 164
A.7 Plot of $\psi(x)$ for $-4 < x \leq 4$. 171

B.1 Plots of $J_\nu(x)$ with $\nu = 0, 1, 2, 3, 4$ for $0 \leq x \leq 10$. 178
B.2 Plots of $Y_\nu(x)$ with $\nu = 0, 1, 2, 3, 4$ for $0 \leq x \leq 10$. 178
B.3 Plots of $I_\nu(x)$, $K_\nu(x)$ with $\nu = 0, 1, 2$ for $0 \leq x \leq 5$. . . . 183
B.4 Plots of $e^{-x} I_\nu(x)$, $e^x K_\nu$ with $\nu = 0, 1, 2$ for $0 \leq x \leq 5$. . 184
B.5 Plots of $Ai(x)$ (continuous line) and its derivative $Ai'(x)$ (dotted line) for $-15 \leq x \leq 5$. 188
B.6 Plots of $Bi(x)$ (continuous line) and its derivative $Bi'(x)$ (dotted line) for $-15 \leq x \leq 5$. 189

C.1 Plots of $\text{erf}(x)$, $\text{erf}'(x)$ and $\text{erfc}(x)$ for $-2 \leq x \leq +2$. . . 193
C.2 Plot of the three sisters functions $\phi(a,t)$, $\psi(a,t)$, $\chi(a,t)$ with $a = 1$ for $0 \leq t \leq 5$. 195
C.3 Plot of the Dawson integral $\text{Daw}(x)$ for $0 \leq x \leq 5$. 198
C.4 Plots of the Fresnel integrals for $0 \leq x \leq 5$. 200
C.5 Plot of the Cornu spiral for $0 \leq x \leq 1$. 201

D.1 Plots of the functions $f_1(t)$, $f_2(t)$ and $f_3(t)$ for $0 \leq t \leq 10$. 208

F.1 Plots of the Wright type function $M_\nu(x)$ with $\nu = 0, 1/8, 1/4, 3/8, 1/2$ for $-5 \leq x \leq 5$; top: linear scale, bottom: logarithmic scale. 243
F.2 Plots of the Wright type function $M_\nu(x)$ with $\nu = 1/2, 5/8, 3/4, 1$ for $-5 \leq x \leq 5$: top: linear scale; bottom: logarithmic scale. 244
F.3 Comparison of the representations of $M_\nu(x)$ with $\nu = 1-\epsilon$ around the maximum $x \approx 1$ obtained by Pipkin's method (continuous line), 100 terms-series (dashed line) and the saddle-point method (dashed-dotted line). Left: $\epsilon = 0.01$; Right: $\epsilon = 0.001$. 245
F.4 The Feller-Takayasu diamond for Lévy stable densities. . 253

Chapter 1

Essentials of Fractional Calculus

In this chapter we introduce the linear operators of fractional integration and fractional differentiation in the framework of the so-called *fractional calculus*. Our approach is essentially based on an integral formulation of the fractional calculus acting on sufficiently well-behaved functions defined in $\mathbb{R}^+$ or in all of $\mathbb{R}$. Such an integral approach turns out to be the most convenient to be treated with the techniques of Laplace and Fourier transforms, respectively. We thus keep distinct the cases $\mathbb{R}^+$ and $\mathbb{R}$ denoting the corresponding formulations of fractional calculus by *Riemann–Liouville* or *Caputo* and *Liouville–Weyl*, respectively, from the names of their pioneers.

For historical and bibliographical notes we refer the interested reader to the the end of this chapter.

Our mathematical treatment is expected to be accessible to applied scientists, avoiding unproductive generalities and excessive mathematical rigour.

Remark : Here, and in all our following treatment, the integrals are intended in the *generalized* Riemann sense, so that any function is required to be *locally* absolutely integrable in $\mathbb{R}^+$. However, we will not bother to give descriptions of sets of admissible functions and will not hesitate, when necessary, to use *formal* expressions with generalized functions (*distributions*), which, as far as possible, will be re-interpreted in the framework of classical functions.

1.1 The fractional integral with support in $\mathbb{R}^+$

Let us consider *causal functions*, namely complex or real valued functions $f(t)$ of a real variable t that are vanishing for $t < 0$.

According to the Riemann–Liouville approach to fractional calculus the notion of fractional integral of order α ($\alpha > 0$) for a causal function $f(t)$, sufficiently well-behaved, is a natural analogue of the well-known formula (usually attributed to Cauchy), but probably due to Dirichlet, which reduces the calculation of the n–fold primitive of a function $f(t)$ to a single integral of convolution type.

In our notation, the Cauchy formula reads for $t > 0$:

$$_0I_t^n f(t) := f_n(t) = \frac{1}{(n-1)!} \int_0^t (t-\tau)^{n-1} f(\tau)\, d\tau, \quad n \in \mathbb{N}, \quad (1.1)$$

where $\mathbb{N}$ is the set of positive integers. From this definition we note that $f_n(t)$ vanishes at $t = 0$, jointly with its derivatives of order $1, 2, \ldots, n-1$.

In a natural way one is led to extend the above formula from positive integer values of the index to any positive real values by using the Gamma function. Indeed, noting that $(n-1)! = \Gamma(n)$, and introducing the arbitrary *positive* real number α, one defines the *Riemann–Liouville fractional integral* of order $\alpha > 0$:

$$_0I_t^\alpha f(t) := \frac{1}{\Gamma(\alpha)} \int_0^t (t-\tau)^{\alpha-1} f(\tau)\, d\tau, \quad t > 0, \; \alpha \in \mathbb{R}^+, \quad (1.2)$$

where $\mathbb{R}^+$ is the set of positive real numbers. For complementation we define $_0I_t^0 := I$ (Identity operator), i.e. we mean $_0I_t^0 f(t) = f(t)$.

Denoting by $\circ$ the composition between operators, we note the *semigroup property*

$$_0I_t^\alpha \circ {_0I_t^\beta} = {_0I_t^{\alpha+\beta}}, \quad \alpha, \beta \geq 0, \quad (1.3)$$

which implies the *commutative property* $_0I_t^\beta \circ {_0I_t^\alpha} = {_0I_t^\alpha} \circ {_0I_t^\beta}$. We also note the effect of our operators $_0I_t^\alpha$ on the power functions

$$_0I_t^\alpha t^\gamma = \frac{\Gamma(\gamma+1)}{\Gamma(\gamma+1+\alpha)} t^{\gamma+\alpha}, \quad \alpha \geq 0, \; \gamma > -1, \; t > 0. \quad (1.4)$$

The properties (1.3) and (1.4) are of course a natural generalization of those known when the order is a positive integer. The proofs are based on the properties of the two Eulerian integrals, i.e. the Gamma and Beta functions, see Appendix A,

$$\Gamma(z) := \int_0^\infty e^{-u}\, u^{z-1}\, du\,, \quad \mathrm{Re}\,\{z\} > 0\,, \tag{1.5}$$

$$B(p,q) := \int_0^1 (1-u)^{p-1}\, u^{q-1}\, du \;=\; \frac{\Gamma(p)\,\Gamma(q)}{\Gamma(p+q)}\,, \quad \mathrm{Re}\,\{p,q\} > 0\,. \tag{1.6}$$

For our purposes it is convenient to introduce the causal function

$$\Phi_\alpha(t) := \frac{t_+^{\alpha-1}}{\Gamma(\alpha)}\,, \quad \alpha > 0\,, \tag{1.7}$$

where the suffix $+$ is just denoting that the function is vanishing for $t < 0$ (as required by the definition of a causal function). We agree to denote this function as *Gel'fand-Shilov function* of order α to honour the authors who have treated it in their book [Gel'fand and Shilov (1964)]. Being $\alpha > 0$, this function turns out to be *locally* absolutely integrable in $\mathrm{I\!R}^+$.

Let us now recall the notion of *Laplace convolution*, i.e. the convolution integral with two causal functions, which reads in our notation

$$f(t) * g(t) := \int_0^t f(t-\tau)\, g(\tau)\, d\tau = g(t) * f(t)\,.$$

We note from (1.2) and (1.7) that the fractional integral of order $\alpha > 0$ can be considered as the Laplace convolution between $\Phi_\alpha(t)$ and $f(t)$, i.e.,

$$_0I_t^\alpha f(t) = \Phi_\alpha(t) * f(t)\,, \quad \alpha > 0\,. \tag{1.8}$$

Furthermore, based on the Eulerian integrals, one proves the *composition rule*

$$\Phi_\alpha(t) * \Phi_\beta(t) = \Phi_{\alpha+\beta}(t)\,, \quad \alpha,\,\beta > 0\,, \tag{1.9}$$

which can be used to re-obtain (1.3) and (1.4).

The Laplace transform for the fractional integral. Let us now introduce the Laplace transform of a generic function $f(t)$, locally absolutely integrable in $\mathbb{R}^+$, by the notation[1]

$$\mathcal{L}\left[f(t); s\right] := \int_0^\infty e^{-st} f(t)\, dt = \widetilde{f}(s)\,, \quad s \in \mathbb{C}\,.$$

By using the sign $\div$ to denote the juxtaposition of the function $f(t)$ with its Laplace transform $\widetilde{f}(s)$, a Laplace transform pair reads

$$f(t) \div \widetilde{f}(s)\,.$$

Then, for the convolution theorem of the Laplace transforms, see e.g. [Doetsch (1974)], we have the pair

$$f(t) * g(t) \div \widetilde{f}(s)\,\widetilde{g}(s)\,.$$

As a consequence of Eq. (1.8) and of the known Laplace transform pair

$$\Phi_\alpha(t) \div \frac{1}{s^\alpha}\,, \quad \alpha > 0\,,$$

we note the following formula for the Laplace transform of the fractional integral,

$$\,_0I_t^\alpha f(t) \div \frac{\widetilde{f}(s)}{s^\alpha}\,, \quad \alpha > 0\,, \qquad (1.10)$$

which is the straightforward generalization of the corresponding formula for the n-fold repeated integral (1.1) by replacing n with α.

---

[1] A sufficient condition of the existence of the Laplace transform is that the original function is of exponential type as $t \to \infty$. This means that some constant a_f exists such that the product $e^{-a_f t}|f(t)|$ is bounded for all t greater than some T. Then $\widetilde{f}(s)$ exists and is analytic in the half plane $\mathcal{R}e\,(s) > a_f$. If $f(t)$ is piecewise differentiable, then the inversion formula

$$f(t) = \mathcal{L}^{-1}\left[\widetilde{f}(s); t\right] = \frac{1}{2\pi i} \int_{\gamma - i\infty}^{\gamma + i\infty} e^{st}\,\widetilde{f}(s)\, ds\,, \quad \mathcal{R}e\,(s) = \gamma > a_f\,,$$

with $t > 0$, holds true at all points where $f(t)$ is continuous and the (complex) integral in it must be understood in the sense of the Cauchy principal value.

1.2 The fractional derivative with support in $\mathbb{R}^+$

After the notion of fractional integral, that of fractional derivative of order α ($\alpha > 0$) becomes a natural requirement and one is attempted to substitute α with $-\alpha$ in the above formulas. We note that for this generalization some care is required in the integration, and the theory of generalized functions would be invoked. However, we prefer to follow an approach that, avoiding the use of generalized functions as far is possible, is based on the following observation: the local operator of the standard derivative of order n ($n \in \mathbb{N}$) for a given t, $D_t^n := \dfrac{d^n}{dt^n}$ is just the left inverse (and not the right inverse) of the non-local operator of the n-fold integral $_aI_t^n$, having as a starting point any finite $a < t$. In fact, for any well-behaved function $f(t)$ ($t \in \mathbb{R}$), we recognize

$$D_t^n \circ {}_aI_t^n f(t) = f(t), \quad t > a, \qquad (1.11)$$

and

$$_aI_t^n \circ D_t^n f(t) = f(t) - \sum_{k=0}^{n-1} f^{(k)}(a^+) \frac{(t-a)^k}{k!}, \quad t > a. \qquad (1.12)$$

As a consequence, taking $a \equiv 0$, we require that $_0D_t^\alpha$ be defined as *left-inverse* to $_0I_t^\alpha$. For this purpose we first introduce the positive integer

$$m \in \mathbb{N} \quad \text{such that} \quad m - 1 < \alpha \le m,$$

and then we define the *Riemann-Liouville fractional derivative* of order $\alpha > 0$:

$$_0D_t^\alpha f(t) := D_t^m \circ {}_0I_t^{m-\alpha} f(t), \quad \text{with} \quad m - 1 < \alpha \le m, \qquad (1.13)$$

namely

$$_0D_t^\alpha f(t) := \begin{cases} \dfrac{1}{\Gamma(m-\alpha)} \dfrac{d^m}{dt^m} \displaystyle\int_0^t \dfrac{f(\tau)\,d\tau}{(t-\tau)^{\alpha+1-m}}, & m-1 < \alpha < m, \\[2ex] \dfrac{d^m}{dt^m} f(t), & \alpha = m. \end{cases}$$

$$(1.13a)$$

For complementation we define $_0D_t^0 = I$.

In analogy with the fractional integral, we have agreed to refer to this fractional derivative as the *Riemann-Liouville fractional derivative*.

We easily recognize, using the semigroup property (1.3),

$$_0D_t^\alpha \circ {_0I_t^\alpha} = D_t^m \circ {_0I_t^{m-\alpha}} \circ {_0I_t^\alpha} = D_t^m \circ {_0I_t^m} = I. \qquad (1.14)$$

Furthermore we obtain

$$_0D_t^\alpha \, t^\gamma = \frac{\Gamma(\gamma+1)}{\Gamma(\gamma+1-\alpha)} \, t^{\gamma-\alpha}, \quad \alpha > 0, \quad \gamma > -1, \quad t > 0. \qquad (1.15)$$

Of course, properties (1.14) and (1.15) are a natural generalization of those known when the order is a positive integer. Since in (1.15) the argument of the Gamma function in the denominator can be negative, we need to consider the analytical continuation of $\Gamma(z)$ in (1.5) into the left half-plane.

Note the remarkable fact that when α is not integer ($\alpha \notin \mathbb{N}$) the fractional derivative $_0D_t^\alpha f(t)$ is not zero for the constant function $f(t) \equiv 1$. In fact, Eq. (1.15) with $\gamma = 0$ gives

$$_0D_t^\alpha 1 = \frac{t^{-\alpha}}{\Gamma(1-\alpha)}, \quad \alpha \geq 0, \quad t > 0, \qquad (1.16)$$

which identically vanishes for $\alpha \in \mathbb{N}$, due to the poles of the Gamma function in the points $0, -1, -2, \ldots$.

By interchanging in (1.13) the processes of differentiation and integration we are led to the so-called *Caputo fractional derivative* of order $\alpha > 0$ defined as:

$$_0^*D_t^\alpha f(t) := {_0I_t^{m-\alpha}} \circ D_t^m f(t) \quad \text{with} \quad m-1 < \alpha \leq m, \qquad (1.17)$$

namely

$$_0^*D_t^\alpha f(t) := \begin{cases} \dfrac{1}{\Gamma(m-\alpha)} \displaystyle\int_0^t \dfrac{f^{(m)}(\tau)}{(t-\tau)^{\alpha+1-m}} \, d\tau, & m-1 < \alpha < m, \\[1em] \dfrac{d^m}{dt^m} f(t), & \alpha = m. \end{cases}$$

$$(1.17a)$$

To distinguish the Caputo derivative from the Riemann-Liouville derivative we decorate it with the additional apex $*$. For non-integer

α the definition (1.17) requires the absolute integrability of the derivative of order m. Whenever we use the operator ${}_0^*D_t^\alpha$ we (tacitly) assume that this condition is met.

We easily recognize that in general

$$
{}_0D_t^\alpha f(t) := D_t^m \circ {}_0I_t^{m-\alpha} f(t) \neq {}_0I_t^{m-\alpha} \circ D_t^m f(t) =: {}_0^*D_t^\alpha f(t), \quad (1.18)
$$

unless the function $f(t)$ along with its first $m-1$ derivatives vanishes at $t = 0^+$. In fact, assuming that the exchange of the m-derivative with the integral is legitimate, we have

$$
{}_0^*D_t^\alpha f(t) = {}_0D_t^\alpha f(t) - \sum_{k=0}^{m-1} f^{(k)}(0^+) \frac{t^{k-\alpha}}{\Gamma(k-\alpha+1)}, \quad (1.19)
$$

and therefore, recalling the fractional derivative of the power functions (1.15),

$$
{}_0^*D_t^\alpha f(t) = {}_0D_t^\alpha \left[f(t) - \sum_{k=0}^{m-1} f^{(k)}(0^+) \frac{t^k}{k!} \right]. \quad (1.20)
$$

In particular for $0 < \alpha < 1$ (i.e. $m=1$) we have

$$
{}_0^*D_t^\alpha f(t) = {}_0D_t^\alpha f(t) - f(0^+) \frac{t^{-\alpha}}{\Gamma(1-\alpha)} = {}_0D_t^\alpha \left[f(t) - f(0^+) \right].
$$

From Eq. (1.20) we recognize that the *Caputo* fractional derivative represents a sort of regularization in the time origin for the *Riemann-Liouville* fractional derivative. We also note that for its existence all the limiting values,

$$
f^{(k)}(0^+) := \lim_{t \to 0^+} D_t^k f(t), \; k = 0, 1, \ldots m-1,
$$

are required to be finite. In the special case $f^{(k)}(0^+) \equiv 0$, we recover the identity between the two fractional derivatives.

We now explore the most relevant differences between the two fractional derivatives. We first note from (1.15) that

$$
{}_0D_t^\alpha t^{\alpha-1} \equiv 0, \quad \alpha > 0, \quad t > 0, \quad (1.21)
$$

and, in view of (1.20),

$$
{}_0^*D_t^\alpha 1 \equiv 0, \quad \alpha > 0, \quad (1.22)
$$

in contrast with (1.16). More generally, from Eqs. (1.21) and (1.22) we thus recognize the following statements about functions which for $t > 0$ admit the same fractional derivative of order α (in the Riemann-Liouville or Caputo sense), with $m - 1 < \alpha \leq m$, $m \in \mathbb{N}$,

$$_0D_t^\alpha f(t) = {}_0D_t^\alpha g(t) \iff f(t) = g(t) + \sum_{j=1}^m c_j\, t^{\alpha-j}, \qquad (1.23)$$

$${}_0^*D_t^\alpha f(t) = {}_0^*D_t^\alpha g(t) \iff f(t) = g(t) + \sum_{j=1}^m c_j\, t^{m-j}, \qquad (1.24)$$

where the coefficients c_j are arbitrary constants. Incidentally, we note that (1.21) provides an instructive example for the fact that $_0D_t^\alpha$ is not right-inverse to $_0I_t^\alpha$, since for $t > 0$

$$_0I_t^\alpha \circ {}_0D_t^\alpha\, t^{\alpha-1} \equiv 0, \quad {}_0D_t^\alpha \circ {}_0I_t^\alpha\, t^{\alpha-1} = t^{\alpha-1}, \quad \alpha > 0. \qquad (1.25)$$

We observe the different behaviour of the two fractional derivatives in the Riemann-Liouville and Caputo at the end points of the interval $(m-1, m)$, namely, when the order is any positive integer, as it can be noted from their definitions (1.13), (1.17). For $\alpha \to m^-$ both derivatives reduce to D_t^m, as explicitly stated in Eqs. (1.13a), (1.17a), due to the fact that the operator $_0I_t^0 = I$ commutes with D_t^m. On the other hand, for $\alpha \to (m-1)^+$ we have:

$$\begin{cases} {}_0D_t^\alpha f(t) \to D_t^m \circ {}_0I_t^1\, f(t) = D_t^{m-1} f(t) = f^{(m-1)}(t), \\ {}_0^*D_t^\alpha f(t) \to {}_0I_t^1 \circ D_t^m\, f(t) = f^{(m-1)}(t) - f^{(m-1)}(0^+). \end{cases} \qquad (1.26)$$

As a consequence, roughly speaking, we can say that $_0D_t^\alpha$ is, with respect to its order α, an operator continuous at any positive integer, whereas $_0^*D_t^\alpha$ is an operator only left-continuous.

Furthermore, we observe that the semigroup property of the standard derivatives is not generally valid for both the fractional derivatives when the order is not integer.

The Laplace transform for the fractional derivatives. We point out the major usefulness of the Caputo fractional derivative in treating initial-value problems for physical and engineering applications where initial conditions are usually expressed in terms of integer-order derivatives. This can be easily seen using the Laplace transformation[2]. In fact, for the Caputo derivative of order α with $m - 1 < \alpha \leq m$, we have

$$\mathcal{L}\{{}^*_0D^\alpha_t f(t); s\} = s^\alpha \tilde{f}(s) - \sum_{k=0}^{m-1} s^{\alpha-1-k} f^{(k)}(0^+),$$

$$f^{(k)}(0^+) := \lim_{t \to 0^+} D^k_t f(t). \tag{1.27}$$

The corresponding rule for the Riemann-Liouville derivative of order α is

$$\mathcal{L}\{{}_0D^\alpha_t f(t); s\} = s^\alpha \tilde{f}(s) - \sum_{k=0}^{m-1} s^{m-1-k} g^{(k)}(0^+),$$

$$g^{(k)}(0^+) := \lim_{t \to 0^+} D^k_t g(t), \, g(t) := {}_0I^{m-\alpha}_t f(t). \tag{1.28}$$

Thus it is more cumbersome to use the rule (1.28) than (1.27). The rule (1.28) requires initial values concerning an extra function $g(t)$ related to the given $f(t)$ through a fractional integral. However, when all the limiting values $f^{(k)}(0^+)$ for $k = 0, 1, \ldots$ are finite and the order is not integer, we can prove that the corresponding $g^{(k)}(0^+)$ vanish so that the formula (1.28) simplifies into

$$\mathcal{L}\{{}_0D^\alpha_t f(t); s\} = s^\alpha \tilde{f}(s), \quad m - 1 < \alpha < m. \tag{1.29}$$

For this proof it is sufficient to apply the Laplace transform to Eq. (1.19), by recalling that

$$\mathcal{L}\{t^\alpha; s\} = \Gamma(\beta + 1)/s^{\alpha+1}, \quad \alpha > -1, \tag{1.30}$$

and then to compare (1.27) with (1.28).

It may be convenient to simply refer to the Riemann-Liouville derivative and to the Caputo derivative to as R–L and C derivatives, respectively.

---

[2]We recall that under suitable conditions the Laplace transform of the m-derivative of $f(t)$ is given by

$$\mathcal{L}\{D^m_t f(t); s\} = s^m \tilde{f}(s) - \sum_{k=0}^{m-1} f^{(k)}(0^+) s^{m-1-k}, \quad f^{(k)}(0^+) := \lim_{t \to 0^+} D^k_t f(t).$$

We now show how the standard definitions (1.13) and (1.17) for the R–L and C derivatives of order α of a function $f(t)$ ($t \in \mathbb{R}^+$) can be derived, at least *formally*, by the convolution of $\Phi_{-\alpha}(t)$ with $f(t)$, in a sort of analogy with (1.8) for the fractional integral. For this purpose we need to recall from the treatise on generalized functions [Gel'fand and Shilov (1964)] that (with proper interpretation of the quotient as a limit if $t = 0$)

$$\Phi_{-n}(t) := \frac{t_+^{-n-1}}{\Gamma(-n)} = \delta^{(n)}(t), \quad n = 0, 1, \ldots, \quad (1.31)$$

where $\delta^{(n)}(t)$ denotes the generalized derivative of order n of the Dirac delta distribution. Here, we assume that the reader has some minimal knowledge concerning these generalized functions, sufficient for handling classical problems in physics and engineering.

Equation (1.31) provides an interesting (not so well known) representation of $\delta^{(n)}(t)$, which is useful in our following treatment of fractional derivatives. In fact, we note that the derivative of order n of a causal function $f(t)$ can be obtained for $t > 0$ *formally* by the *(generalized) convolution* between Φ_{-n} and f,

$$\frac{d^n}{dt^n} f(t) = f^{(n)}(t) = \Phi_{-n}(t) * f(t) = \int_{0^-}^{t^+} f(\tau) \delta^{(n)}(t - \tau) \, d\tau, \quad (1.32)$$

based on the well-known property

$$\int_{0^-}^{t^+} f(\tau) \delta^{(n)}(\tau - t) \, d\tau = (-1)^n f^{(n)}(t), \quad (1.33)$$

where $\delta^{(n)}(t-\tau) = (-1)^n \delta^{(n)}(\tau-t)$. According to a usual convention, in (1.32) and (1.33) the limits of integration are extended to take into account for the possibility of impulse functions centred at the extremes. Then, a *formal definition of the fractional derivative* of positive order α could be

$$\Phi_{-\alpha} * f(t) = \frac{1}{\Gamma(-\alpha)} \int_{0^-}^{t^+} \frac{f(\tau)}{(t-\tau)^{1+\alpha}} \, d\tau, \quad \alpha \in \mathbb{R}^+.$$

The formal character is evident in that the kernel $\Phi_{-\alpha}(t)$ is not locally absolutely integrable and consequently the integral is in general divergent. In order to obtain a definition that is still valid for classical functions, we need to *regularize* the divergent integral in some way. For this purpose let us consider the integer $m \in \mathbb{N}$ such that $m-1 < \alpha < m$ and write $-\alpha = -m + (m-\alpha)$ or $-\alpha = (m-\alpha) - m$. We then obtain

$$\Phi_{-\alpha}(t) * f(t) = \Phi_{-m}(t) * \Phi_{m-\alpha}(t) * f(t) = D_t^m \circ {}_0I_t^{m-\alpha} f(t), \quad (1.34)$$

or

$$\Phi_{-\alpha}(t) * f(t) = \Phi_{m-\alpha}(t) * \Phi_{-m}(t) * f(t) = {}_0I_t^{m-\alpha} \circ D_t^m f(t). \quad (1.35)$$

As a consequence we derive two alternative definitions for the fractional derivative (1.34) and (1.35) corresponding to (1.13) and (1.17), respectively. The singular behaviour of $\Phi_{-m}(t)$ as a proper generalized (i.e. non-standard) function is reflected in the non-commutativity of convolution for $\Phi_{m-\alpha}(t)$ and $\Phi_m(t)$ in these formulas.

Remark : We recall an additional definition for the fractional derivative recently introduced by Hilfer for the order interval $0 < \alpha \leq 1$, see [Hilfer (2000b)], p. 113 and [Seybold and Hilfer (2005)], which interpolates the definitions (1.13) and (1.17). Like the two derivatives previously discussed, it is related to the Riemann-Liouville fractional integral. In our notation it reads

$${}_0D_t^{\alpha,\beta} := {}_0I_t^{\beta(1-\alpha)} \circ D_t^1 \circ {}_0I_t^{(1-\beta)(1-\alpha)}, \quad \begin{cases} 0 < \alpha \leq 1, \\ 0 < \beta \leq 1. \end{cases} \quad (1.36)$$

We call it the *Hilfer fractional derivative* of order α and type β. The Riemann-Liouville derivative of order α corresponds to the type $\beta = 0$, while the Caputo derivative to the type $\beta = 1$.

1.3 Fractional relaxation equations in $\mathbb{R}^+$

The different roles played by the R-L and C fractional derivatives are more clear when the fractional generalization of the first-order differential equation governing the exponential relaxation phenomena is considered. Recalling (in non-dimensional units) the initial value problem

$$\frac{du}{dt} = -u(t), \quad t \geq 0, \quad \text{with} \quad u(0^+) = 1, \quad (1.37)$$

whose solution is
$$u(t) = \exp(-t), \quad (1.38)$$
the following three alternatives with respect to the R-L and C fractional derivatives with $\alpha \in (0,1)$ are offered in the literature:
$$_0^*D_t^\alpha u(t) = -u(t), \quad t \geq 0, \quad \text{with} \quad u(0^+) = 1, \quad (1.39a)$$
$$_0D_t^\alpha u(t) = -u(t), \quad t \geq 0, \quad \text{with} \quad \lim_{t \to 0^+} {}_0I_t^{1-\alpha} u(t) = 1, \quad (1.39b)$$
$$\frac{du}{dt} = -{}_0D_t^{1-\alpha} u(t), \quad t \geq 0, \quad \text{with} \quad u(0^+) = 1. \quad (1.39c)$$

In analogy with the standard problem (1.37) we solve these three problems with the Laplace transform technique, using the rules (1.27), (1.28) and (1.29), respectively. Problems (a) and (c) are *equivalent* since the Laplace transform of the solution in both cases comes out to be

$$\widetilde{u}(s) = \frac{s^{\alpha-1}}{s^\alpha + 1}, \quad (1.40)$$

whereas in case (b) we get

$$\widetilde{u}(s) = \frac{1}{s^\alpha + 1} = 1 - s\frac{s^{\alpha-1}}{s^\alpha + 1}. \quad (1.41)$$

The Laplace transforms in (1.40) and (1.41) can be expressed in terms of functions of Mittag-Leffler type, of which we provide information in Appendix E. In fact, in virtue of the Laplace transform pairs (E.52) and (E.53), we have

$$\mathcal{L}\left[E_\alpha(-\lambda t^\alpha); s\right] = \frac{s^{\alpha-1}}{s^\alpha + \lambda}, \quad \mathcal{L}\{t^{\beta-1} E_{\alpha,\beta}(-\lambda t^\alpha); s\} = \frac{s^{\alpha-\beta}}{s^\alpha + \lambda}, \quad (1.42)$$

where

$$E_\alpha(-\lambda t^\alpha) := \sum_{n=0}^{\infty} \frac{(-\lambda t^\alpha)^n}{\Gamma(\alpha n + 1)}, \quad E_{\alpha,\beta}(-\lambda t^\alpha) := \sum_{n=0}^{\infty} \frac{(-\lambda t^\alpha)^n}{\Gamma(\alpha n + \beta)}, \quad (1.43)$$

with $\alpha, \beta \in \mathbb{R}^+$ and $\lambda \in \mathbb{R}$.

Then we obtain in the equivalent cases (a) and (c):
$$u(t) = \psi_\alpha(t) := E_\alpha(-t^\alpha), \quad t \geq 0, \quad 0 < \alpha < 1, \quad (1.44)$$

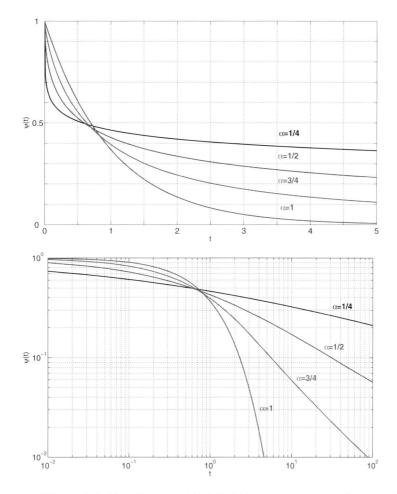

Fig. 1.1 Plots of $\psi_\alpha(t)$ with $\alpha = 1/4, 1/2, 3/4, 1$ versus t; top: linear scales ($0 \le t \le 5$); bottom: logarithmic scales ($10^{-2} \le t \le 10^2$).

and in case (b) :
$$\begin{aligned} u(t) = \phi_\alpha(t) &:= t^{-(1-\alpha)} E_{\alpha,\alpha}\left(-t^\alpha\right) \\ &:= -\frac{d}{dt} E_\alpha\left(-t^\alpha\right), \quad t \ge 0, \quad 0 < \alpha < 1. \end{aligned} \quad (1.45)$$

The plots of the solutions $\psi_\alpha(t)$ and $\phi_\alpha(t)$ are shown in Figs. 1.1 and 1.2, respectively, for some rational values of the parameter α, by adopting linear and logarithmic scales.

It is evident that for $\alpha \to 1^-$ the solutions of the three initial value problems reduce to the standard exponential function (1.38) since in all cases $\widetilde{u}(s) \to 1/(s+1)$. However, case (b) is of minor interest from a physical view-point since the corresponding solution (1.45) is infinite in the time-origin for $0 < \alpha < 1$.

Whereas for the equivalent cases (a) and (c) the corresponding solution shows a continuous transition to the exponential function for any $t \geq 0$ when $\alpha \to 1^-$, for the case (b) such continuity is lost.

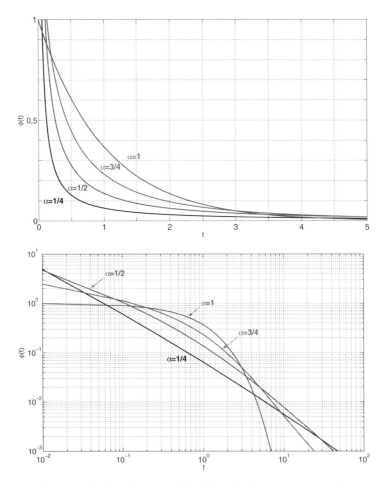

Fig. 1.2 Plots of $\phi_\alpha(t)$ with $\alpha = 1/4, 1/2, 3/4, 1$ versus t; top: linear scales ($0 \leq t \leq 5$); bottom: logarithmic scales ($10^{-2} \leq t \leq 10^2$).

It is worth noting the algebraic decay of $\psi_\alpha(t)$ and $\phi_\alpha(t)$ as $t \to \infty$:

$$\begin{cases} \psi_\alpha(t) \sim \dfrac{\sin(\alpha\pi)}{\pi} \dfrac{\Gamma(\alpha)}{t^\alpha}, \\[1em] \phi_\alpha(t) \sim \dfrac{\sin(\alpha\pi)}{\pi} \dfrac{\Gamma(\alpha+1)}{t^{(\alpha+1)}}, \end{cases} \quad t \to +\infty. \tag{1.46}$$

<u>Remark</u> : If we adopt the Hilfer intermediate derivative in fractional relaxation, that is

$$_0D_t^{\alpha,\beta} u(t) = -u(t),\ t \geq 0,\ \lim_{t\to 0^+}\ _0I_t^{(1-\alpha)(1-\beta)} u(t) = 1, \tag{1.47}$$

the Laplace transform of the solution turns out to be

$$\widetilde{u}(s) = \frac{s^{\beta(\alpha-1)}}{s^\alpha + 1}, \tag{1.48}$$

see [Hilfer (2000b)], so, in view of Eq. (1.43),

$$u(t) = H_{\alpha,\beta}(t) := t^{(1-\beta)(\alpha-1)}\, E_{\alpha,\alpha+\beta(1-\alpha)}(-t^\alpha),\ t \geq 0. \tag{1.49}$$

For plots of the Hilfer function $H_{\alpha,\beta}(t)$ we refer to [Seybold and Hilfer (2005)].

1.4 Fractional integrals and derivatives with support in $\mathbb{R}$

Choosing $-\infty$ as the lower limit in the fractional integral, we have the so-called *Liouville-Weyl fractional integral*. For any $\alpha > 0$ we write

$$_{-\infty}I_t^\alpha f(t) := \frac{1}{\Gamma(\alpha)} \int_{-\infty}^t (t-\tau)^{\alpha-1} f(\tau)\, d\tau, \quad \alpha \in \mathbb{R}^+, \tag{1.50}$$

and consequently, we define the *Liouville–Weyl fractional derivative* of order α as its left inverse operator:

$$_{-\infty}D_t^\alpha f(t) := D_t^m \circ\, _{-\infty}I_t^{m-\alpha} f(t),\ m-1 < \alpha \leq m, \tag{1.51}$$

with $m \in \mathbb{N}$, namely

$$_{-\infty}D_t^\alpha f(t) := \begin{cases} \dfrac{1}{\Gamma(m-\alpha)} \dfrac{d^m}{dt^m} \displaystyle\int_{-\infty}^t \dfrac{f(\tau)\, d\tau}{(t-\tau)^{\alpha+1-m}}, & m-1 < \alpha < m, \\[1em] \dfrac{d^m}{dt^m} f(t), & \alpha = m. \end{cases}$$

$$\tag{1.51a}$$

In this case, assuming $f(t)$ to vanish as $t \to -\infty$ along with its first $m-1$ derivatives, we have the identity
$$D_t^m \circ {}_{-\infty}I_t^{m-\alpha} f(t) = {}_{-\infty}I_t^{m-\alpha} \circ D_t^m f(t), \qquad (1.52)$$
in contrast with (1.18). While for the Riemann–Liouville fractional integral (1.2) a sufficient condition for the convergence of the integral is given by the asymptotic behaviour
$$f(t) = O\left(t^{\epsilon-1}\right), \quad \epsilon > 0, \ t \to 0^+, \qquad (1.53)$$
a corresponding sufficient condition for (1.50) to converge is
$$f(t) = O\left(|t|^{-\alpha-\epsilon}\right), \quad \epsilon > 0, \ t \to -\infty. \qquad (1.54)$$

Integrable functions satisfying the properties (1.53) and (1.54) are sometimes referred to as functions of *Riemann class* and *Liouville class*, respectively, see [Miller and Ross (1993)]. For example, power functions t^γ with $\gamma > -1$ and $t > 0$ (and hence also constants) are of Riemann class, while $|t|^{-\delta}$ with $\delta > \alpha > 0$ and $t < 0$ and $\exp(ct)$ with $c > 0$ are of Liouville class. For the above functions we obtain
$$\begin{cases} {}_{-\infty}I_t^\alpha |t|^{-\delta} = \dfrac{\Gamma(\delta-\alpha)}{\Gamma(\delta)} |t|^{-\delta+\alpha}, \\[2mm] {}_{-\infty}D_t^\alpha |t|^{-\delta} = \dfrac{\Gamma(\delta+\alpha)}{\Gamma(\delta)} |t|^{-\delta-\alpha}, \end{cases} \qquad (1.55)$$
and
$$\begin{cases} {}_{-\infty}I_t^\alpha \, e^{ct} = c^{-\alpha} \, e^{ct}, \\[2mm] {}_{-\infty}D_t^\alpha \, e^{ct} = c^\alpha \, e^{ct}. \end{cases} \qquad (1.56)$$

Causal functions can be considered in the above integrals with the due care. In fact, in view of the possible jump discontinuities of the integrands at $t = 0$, in this case it is worthwhile to write
$$\int_{-\infty}^t (\ldots) \, d\tau = \int_{0^-}^t (\ldots) \, d\tau.$$
As an example we consider for $0 < \alpha < 1$ the identity
$$\frac{1}{\Gamma(1-\alpha)} \int_{0^-}^t \frac{f'(\tau)}{(t-\tau)^\alpha} \, d\tau = \frac{f(0^+) \, t^{-\alpha}}{\Gamma(1-\alpha)} + \frac{1}{\Gamma(1-\alpha)} \int_0^t \frac{f'(\tau)}{(t-\tau)^\alpha} \, d\tau,$$
that is consistent with (1.19) for $m = 1$, that is
$$_0D_t^\alpha f(t) = \frac{f(0^+) \, t^{-\alpha}}{\Gamma(1-\alpha)} + {}_0^*D_t^\alpha f(t).$$

1.5 Notes

The fractional calculus may be considered an *old* and yet *novel* topic. It is an *old* topic because, starting from some speculations of G.W. Leibniz (1695, 1697) and L. Euler (1730), it has been developed progressively up to now. A list of mathematicians, who have provided important contributions up to the middle of the twentieth century, includes P.S. Laplace (1812), S.F. Lacroix (1819), J.B.J. Fourier (1822), N.H. Abel (1823–1826), I. Liouville (1832–1873), B. Riemann (1847), H. Holmgren (1865–1867), A.K. Grünwald (1867–1872), A.V. Letnikov (1868–1872), H. Laurent (1884), P.A. Nekrassov (1888), A. Krug (1890), I. Hadamard (1892), O. Heaviside (1892–1912), S. Pincherle (1902), G.H. Hardy and I.E. Littlewood (1917–1928), H. Weyl (1917), P. Lévy (1923), A. Marchaud (1927), H.T. Davis (1924–1936), E.L. Post (1930), A. Zygmund (1935–1945), E.R. Love (1938–1996), A. Erdélyi (1939–1965), H. Kober (1940), D.V. Widder (1941), M. Riesz (1949), W. Feller (1952).

However, it may be considered a *novel* topic as well. Only since the Seventies has it been the object of specialized conferences and treatises. For the first conference the merit is due to B. Ross who, shortly after his Ph.D. dissertation on fractional calculus, organized the *First Conference on Fractional Calculus and its Applications* at the University of New Haven in June 1974, and edited the proceedings, see [Ross (1975a)]. For the first monograph the merit is ascribed to K.B. Oldham and I. Spanier, see [Oldham and Spanier (1974)] who, after a joint collaboration begun in 1968, published a book devoted to fractional calculus in 1974.

Nowadays, the series of texts devoted to fractional calculus and its applications includes over ten titles, including (alphabetically ordered by the first author) [Kilbas *et al.* (2006); Kiryakova (1994); Miller and Ross (1993); Magin (2006); Nishimoto (1991); Oldham and Spanier (1974); Podlubny (1999); Rubin (1996); Samko *et al.* (1993); West *et al.* (2003); Zaslavsky (2005)]. This list is expected to grow up in the forthcoming years. We also cite three books (still) in Russian: [Nakhushev (2003); Pskhu (2005); Uchaikin (2008)].

Furthermore, we call attention to some treatises which contain a detailed analysis of some mathematical aspects and/or physical applications of fractional calculus, although without explicit mention in their titles, see e.g. [Babenko (1986); Caputo (1969); Davis (1936); Dzherbashyan (1966); Dzherbashyan (1993); Erdélyi et al. (1953-1954); Gel'fand and Shilov (1964); Gorenflo and Vessella (1991)].

In recent years considerable interest in fractional calculus has been stimulated by the applications it finds in different areas of applied sciences like physics and engineering, possibly including fractal phenomena. In this respect A. Carpinteri and F. Mainardi have edited a collection of lecture notes entitled *Fractals and Fractional Calculus in Continuum Mechanics* [Carpinteri and Mainardi (1997)], whereas Hilfer has edited a book devoted to the applications in physics [Hilfer (2000a)]. In these books the mathematical theory of fractional calculus was reviewed by [Gorenflo and Mainardi (1997)] and by [Butzer and Westphal (2000)].

Now there are more books of proceedings and special issues of journals published that refer to the applications of fractional calculus in several scientific areas including special functions, control theory, chemical physics, stochastic processes, anomalous diffusion, rheology. Among the special issues which appeared in the last decade we mention: *Signal Processing*, Vol. 83, No. 11 (2003) and Vol. 86, No. 10 (2006); *Nonlinear Dynamics*, Vol. 29, No. 1-4 (2002) and Vol. 38, No. 1-4 (2004); *Journal of Vibration and Control*, Vol. 13, No. 9-10 (2007) and Vol. 14, No. 1-4 (2008); *Physica Scripta*, Vol. T136, October (2009), We also mention the electronic proceedings [Matignon and Montseny (1998)] and the recent books, edited by [Le Méhauté et al. (2005)], [Sabatier et al. (2007)], [Klages et al. (2008)], [Mathai and Haubold (2008)], which contain selected and improved papers presented at conferences and advanced schools, concerning various applications of fractional calculus.

Already since several years, there exist two international journals devoted almost exclusively to the subject of fractional calculus: *Journal of Fractional Calculus* (Editor-in-Chief: K. Nishimoto, Japan) started in 1992, and *Fractional Calculus and Applied Analysis* (Managing Editor: V. Kiryakova, Bulgaria) started in 1998, see http://www.diogenes.bg/fcaa/. Furthermore, web–sites devoted to fractional calculus have been set up, among which we call special attention to http://www.fracalmo.org whose name is originated by FRActional CALculus MOdelling. This web–site was set up in December 2000 by the initiative of the author and some colleagues: it contains interesting news and web–links.

Quite recently the new journal *Fractional Dynamic Systems* (http://fds.ele-math.com/) has been announced to start in 2010.

The reader interested in the history of fractional calculus is referred to Ross' bibliographies in [Oldham and Spanier (1974)], [Ross (1975b); (1977)] and to the historical notes contained in the textbooks and reviews already cited.

Let us recall that exhaustive tables of fractional integrals are available in the second volume of the Bateman Project devoted to Integral Transforms [Erdélyi *et al.* (1953-1954)], in Chapter XIII.

It is worthwhile and interesting to say here something about the commonly used naming for the types of fractional integrals and derivatives that have been discussed in this chapter. Usually names are given to honour the scientists who provided the main contributions, but not necessarily to those who first introduced the corresponding notions. Surely Liouville and then Riemann (as a student!) contributed significantly towards fractional integration and differentiation, but their notions have a history. As a matter of fact it was Abel who, in his 1823 paper [Abel (1823)], solved his celebrated integral equation by using fractional integration and differentiation of order 1/2. Three years later Abel considered the generalization to any order $\alpha \in (0,1)$ in [Abel (1826)]. So Abel, using the operators that nowadays are ascribed to Riemann and Liouville, preceded these eminent mathematicians by at least ten years. Because Riemann, like Abel, worked on the positive real semi-axis $\mathbb{R}^+$, whereas Liouville and later Weyl mainly on all of $\mathbb{R}$, we would use the names of Abel-Riemann and Liouville-Weyl for the fractional integrals with support in $\mathbb{R}^+$ and $\mathbb{R}$, respectively. However, whereas for $\mathbb{R}$ we keep the names of Liouville-Weyl, for $\mathbb{R}^+$, in order to be consistent with the existing literature, we agree to use the names of Riemann-Liouville, even if this is an injustice towards Abel.

In IR we have not discussed the approach investigated and used in several papers by Beyer, Kempfle and Schaefer, that is appropriate for causal processes not starting at a finite instant of time, see e.g. [Beyer and Kempfle (1994); Beyer and Kempfle (1995); Kempfle (1998); Kempfle and Schäfer (1999); Kempfle and Schäfer (2000); Kempfle et al. (2002a); Kempfle et al. (2002b)]. They define the time-fractional derivative on the whole real line as a pseudo-differential operator via its Fourier symbol.

In this book, special attention is devoted to an alternative form of fractional derivative (where the orders of fractional integration and ordinary differentiation are interchanged) that nowadays is known as the *Caputo derivative*. As a matter of fact, such a form is found in a paper by Liouville himself as noted by Butzer and Westphal [Butzer and Westphal (2000)] but Liouville, not recognizing its role, disregarded this notion. As far as we know, up to to the middle of the tuentieth century most authors did not take notice of the difference between the two forms and of the possible use of the alternative form. Even in the classical book on Differential and Integral Calculus by the eminent mathematician R. Courant, the two forms of the fractional derivative were considered as equivalent, see [Courant (1936)], pp. 339-341. As shown in Eqs. (1.19) and (1.20) the alternative form (denoted with the additional apex $*$) can be considered as a regularization of the *Riemann–Liouville* derivative which identically vanishes when applied to a constant. Only in the late sixties was the relevance of the alternative form recognized. In fact, in [Dzherbashyan and Nersesyan (1968)] and then in [Kochubei (1989); Kochubei (1990)] the authors used the alternative form as given by (1.19) in dealing with Cauchy problems for differential equations of fractional order. Formerly, Caputo, see [Caputo (1967); Caputo (1969)] introduced this form as given by Eq. (1.17) proving the corresponding rule in the Laplace transform domain, see Eq. (1.27). With his derivative Caputo was thus able to generalize the rule for the Laplace transform of a derivative of integer order and to solve some problems in Seismology in a proper way. Soon later, this derivative was adopted by [Caputo and Mainardi (1971a); (1971b)] in the framework of the theory of *Linear Viscoelasticity*.

Since the seventies a number of authors have re-discovered and used the alternative form, recognizing its major usefulness for solving physical problems with standard initial conditions. Although several papers by different authors appeared where the alternative derivative was adopted, it was only in the late nineties, with the tutorial paper [Gorenflo and Mainardi (1997)] and the book [Podlubny (1999)], that such form was popularized. In these references the Caputo form was named the *Caputo fractional derivative*, a term now universally accepted in the literature. The reader, however, is alerted that in a very few papers the Caputo derivative is referred to as the Caputo–Dzherbashyan derivative. Note also the transliteration as Djrbashyan.

As a relevant topic, let us now consider the question of notation. Following [Gorenflo and Mainardi (1997)] the present author opposes to the use of the notation $_0D_t^{-\alpha}$ for denoting the fractional integral; it is misleading, even if it is used in such distinguished treatises as [Oldham and Spanier (1974); Miller and Ross (1993); Podlubny (1999)]. It is well known that derivation and integration operators are not inverse to each other, even if their order is integer, and therefore such indiscriminate use of symbols, present only in the framework of the fractional calculus, appears unjustified. Furthermore, we have to keep in mind that for fractional order the derivative is yet an *integral* operator, so that, perhaps, it would be less disturbing to denote our $_0D_t^\alpha$ as $_0I_t^{-\alpha}$, than our $_0I_t^\alpha$ as $_0D_t^{-\alpha}$.

The notation adopted in this book is a modification of that introduced in a systematic way by [Gorenflo and Mainardi (1997)] in their CISM lectures, that, in its turn, was partly based on the book on Abel Integral Equations [Gorenflo and Vessella (1991)] and on the article [Gorenflo and Rutman (1994)].

As far as the Mittag-Leffler function is concerned, we refer the reader to Appendix E for more details, along with historical notes therein.

Chapter 2
Essentials of Linear Viscoelasticity

In this chapter the fundamentals of the linear theory of viscoelasticity are presented in the one-dimensional case. The classical approaches based on integral and differential constitutive equations are reviewed. The application of the Laplace transform leads to the so-called *material functions* (or step responses) and their (continuous and discrete) time spectra related to the creep and relaxation tests. The application of the Fourier transform leads to the so-called *dynamic functions* (or harmonic responses) related to the storage and dissipation of energy.

2.1 Introduction

We denote the stress by $\sigma = \sigma(x,t)$ and the strain by $\epsilon = \epsilon(x,t)$ where x and t are the space and time variables, respectively. For the sake of convenience, both stress and strain are intended to be normalized, i.e. scaled with respect to a suitable reference state $\{\sigma_*, \epsilon_*\}$.

At sufficiently small (theoretically infinitesimal) strains, the behaviour of a viscoelastic body is well described by the linear theory of viscoelasticity. According to this theory, the body may be considered as a linear system with the stress (or strain) as the excitation function (input) and the strain (or stress) as the response function (output).

To derive the most general stress–strain relations, also referred as the *constitutive equations*, two fundamental hypotheses are required: (i) invariance for time translation and (ii) causality; the former means that a time shift in the input results in an equal shift in the output, the latter that the output for any instant t_1 depends on the values of the input only for $t \leq t_1$. Furthermore, in this respect, the response functions to an excitation expressed by the unit step function $\Theta(t)$, known as Heaviside function defined as

$$\Theta(t) = \begin{cases} 0 & \text{if } t < 0, \\ 1 & \text{if } t > 0, \end{cases}$$

are known to play a fundamental role both from a mathematical and physical point of view.

The creep test and the relaxation test. We denote by $J(t)$ the strain response to the unit step of stress, according to the *creep test*

$$\sigma(t) = \Theta(t) \implies \epsilon(t) = J(t), \qquad (2.1a)$$

and by $G(t)$ the stress response to a unit step of strain, according to the *relaxation test*

$$\epsilon(t) = \Theta(t) \implies \sigma(t) = G(t). \qquad (2.1b)$$

The functions $J(t)$ and $G(t)$ are usually referred as the *creep compliance* and *relaxation modulus* respectively, or, simply, the *material functions* of the viscoelastic body. In view of the causality requirement, both functions are *causal*, i.e. vanishing for $t < 0$. Implicitly, we assume that all our causal functions, including $J(t)$ and $G(t)$, are intended from now on to be multiplied by the Heaviside function $\Theta(t)$.

The limiting values of the material functions for $t \to 0^+$ and $t \to +\infty$ are related to the instantaneous (or glass) and equilibrium behaviours of the viscoelastic body, respectively. As a consequence, it is usual to set

$$\begin{cases} J_g := J(0^+) & \text{glass compliance}, \\ J_e := J(+\infty) & \text{equilibrium compliance}; \end{cases} \qquad (2.2a)$$

and

$$\begin{cases} G_g := G(0^+) & \text{glass modulus}, \\ G_e := G(+\infty) & \text{equilibrium modulus}. \end{cases} \qquad (2.2b)$$

From experimental evidence, both the material functions are non-negative. Furthermore, for $0 < t < +\infty$, $J(t)$ turns out to be a non-decreasing function, whereas $G(t)$ a non-increasing function. Assuming that $J(t)$ is a *differentiable, increasing function of time*, we write

$$t \in \mathbb{R}^+, \quad \frac{dJ}{dt} > 0 \implies 0 \leq J(0^+) < J(t) < J(+\infty) \leq +\infty. \quad (2.3a)$$

Similarly, assuming that $G(t)$ is a *differentiable, decreasing function of time*, we write

$$t \in \mathbb{R}^+, \quad \frac{dG}{dt} < 0 \implies +\infty \geq G(0^+) > G(t) > G(+\infty) \geq 0. \quad (2.3b)$$

The above characteristics of monotonicity of $J(t)$ and $G(t)$ are related respectively to the physical phenomena of *strain creep* and *stress relaxation*, which indeed are experimentally observed. Later on, we shall outline more restrictive mathematical conditions that the material functions must usually satisfy to agree with the most common experimental observations.

The creep representation and the relaxation representation.
Hereafter, by using the Boltzmann superposition principle, we are going to show that the general stress – strain relation is expressed in terms of one material function [$J(t)$ or $G(t)$] through a linear hereditary integral of Stieltjes type. Choosing the *creep representation*, we obtain

$$\epsilon(t) = \int_{-\infty}^{t} J(t-\tau)\, d\sigma(\tau). \quad (2.4a)$$

Similarly, in the *relaxation representation*, we have

$$\sigma(t) = \int_{-\infty}^{t} G(t-\tau)\, d\epsilon(\tau). \quad (2.4b)$$

In fact, since the responses are to be invariant for time translation, we note that in $J(t)$ and $G(t)$, t is the time *lag* since application of stress or strain. In other words, an input $\sigma(t) = \sigma_1 \Theta(t-\tau_1)$ [$\epsilon(t) = \epsilon_1 \Theta(t-\tau_1)$] would be accompanied by an output $\epsilon(t) = \sigma_1 J(t-\tau_1)$ [$\sigma(t) = \epsilon_1 G(t-\tau_1)$]. As a consequence, a series of N stress steps $\Delta\sigma_n = \sigma_{n+1} - \sigma_n$ ($n = 1, 2, \ldots, N$) added consecutively at times

$$\tau_N > \tau_{N-1} > \cdots > \tau_1 > -\infty,$$

will induce the total strain according to
$$\sigma(t) = \sum_{n=1}^{N} \Delta\sigma_n \, \Theta(t - \tau_n) \implies \epsilon(t) = \sum_{n=1}^{N} \Delta\sigma_n \, J(t - \tau_n).$$
We can approximate arbitrarily well any physically realizable stress history by a step history involving an arbitrarily large number of arbitrarily small steps. By passing to the limit in the sums above, we obtain the strain and stress responses to arbitrary stress and strain histories according to Eqs. (2.4a) and (2.4b), respectively. In fact
$$\sigma(t) = \int_{-\infty}^{t} \Theta(t-\tau) \, d\sigma(\tau) = \int_{-\infty}^{t} d\sigma(\tau) \implies \epsilon(t) = \int_{-\infty}^{t} J(t-\tau) \, d\sigma(\tau),$$
and
$$\epsilon(t) = \int_{-\infty}^{t} \Theta(t-\tau) \, d\epsilon(\tau) = \int_{-\infty}^{t} d\epsilon(\tau) \implies \sigma(t) = \int_{-\infty}^{t} G(t-\tau) \, d\epsilon(\tau).$$
Wherever the stress [strain] history $\sigma(t)$ [$\epsilon(t)$] is differentiable, by $d\sigma(\tau)$ [$d\epsilon(\tau)$] we mean $\dot\sigma(\tau) \, d\tau$ [$\dot\epsilon(\tau) \, d\tau$], where we have denoted by a superposed dot the derivative with respect to the variable τ. If $\sigma(t)$ [$\epsilon(t)$] has a jump discontinuity at a certain time τ_0, the corresponding contribution is intended to be $\Delta\sigma_0 \, J(t - \tau_0)$ [$\Delta\epsilon_0 \, G(t - \tau_0)$].

All the above relations are thus a consequence of the Boltzmann superposition principle, which states that in linear viscoelastic systems the total response to a stress [strain] history is equivalent (in some way) to the sum of the responses to a sequence of incremental stress [strain] histories.

2.2   History in $\mathbb{R}^+$: the Laplace transform approach

Usually, the viscoelastic body is quiescent for all times prior to some starting instant that we assume as $t = 0$; in this case, under the hypotheses of causal histories, differentiable for $t \in \mathbb{R}^+$, the creep and relaxation representations (2.4a) and (2.4b) reduce to
$$\epsilon(t) = \int_{0^-}^{t} J(t - \tau) \, d\sigma(\tau) = \sigma(0^+) \, J(t) + \int_{0}^{t} J(t - \tau) \, \dot\sigma(\tau) \, d\tau, \quad (2.5a)$$
$$\sigma(t) = \int_{0^-}^{t} G(t - \tau) \, d\epsilon(\tau) = \epsilon(0^+) \, G(t) + \int_{0}^{t} G(t - \tau) \, \dot\epsilon(\tau) \, d\tau. \quad (2.5b)$$

Unless and until we find it makes any sense to do otherwise, we implicitly restrict our attention to causal histories.

Another form of the constitutive equations can be obtained from Eqs. (2.5a) and (2.5b) by integrating by parts. We thus have

$$\epsilon(t) = J_g\,\sigma(t) + \int_0^t \dot{J}(t-\tau)\,\sigma(\tau)\,d\tau\,, \qquad (2.6a)$$

and, if $G_g < \infty$,

$$\sigma(t) = G_g\,\epsilon(t) + \int_0^t \dot{G}(t-\tau)\,\epsilon(\tau)\,d\tau\,. \qquad (2.6b)$$

The causal functions $\dot{J}(t)$ and $\dot{G}(t)$ are referred as the *rate of creep (compliance)* and the *rate of relaxation (modulus)*, respectively; they play the role of *memory functions* in the constitutive equations (2.6a) and (2.6b). If $J_g > 0$ or $G_g > 0$ it may be convenient to consider the non-dimensional form of the memory functions obtained by normalizing them to the glass values[1].

The integrals from 0 to t in the R.H.S of Eqs. (2.5a) and (2.5b) and (2.6a) and (2.6b) can be re-written using the convolution form and then dealt with the technique of the Laplace transforms, according to the notation introduced in Chapter 1,

$$f(t) * g(t) \div \widetilde{f}(s)\,\widetilde{g}(s)\,.$$

Then, we show that application of the Laplace transform to Eqs. (2.5a) and (2.5b) and (2.6a) and (2.6b) yields

$$\widetilde{\epsilon}(s) = s\,\widetilde{J}(s)\,\widetilde{\sigma}(s)\,, \qquad (2.7a)$$

$$\widetilde{\sigma}(s) = s\,\widetilde{G}(s)\,\widetilde{\epsilon}(s)\,. \qquad (2.7b)$$

This means that *the use of Laplace transforms allow us to write the creep and relaxation representations in a unique form, proper for each of them.*

In fact, Eq. (2.7a) is deduced from (2.5a) or (2.6a) according to

$$\widetilde{\epsilon}(s) = \sigma(0^+)\,\widetilde{J}(s) + \widetilde{J}(s)\,[s\,\widetilde{\sigma}(s) - \sigma(0^+)] = J_g\,\widetilde{\sigma}(s) + [s\,\widetilde{J}(s) - J_g]\,\widetilde{\sigma}(s)\,,$$

---

[1] See later in Chapters 4 and 5 when we will use the non-dimensional memory functions

$$\Psi(t) := \frac{1}{J_g}\frac{dJ}{dt}\,, \quad \Phi(t) := \frac{1}{G_g}\frac{dG}{dt}\,.$$

and, similarly, Eq. (2.7b) is deduced from (2.5b) or (2.6b) according to

$$\widetilde{\sigma}(s) = \epsilon(0^+)\,\widetilde{G}(s) + \widetilde{G}(s)\,[s\,\widetilde{\epsilon}(s) - \epsilon(0^+)] = G_g\,\widetilde{\epsilon}(s) + [s\,\widetilde{G}(s) - G_g]\,\widetilde{\epsilon}(s)\,.$$

We notice that (2.7b) is valid also if $G_g = \infty$, provided that we use a more general approach to the Laplace transform, based on the theory of generalized functions, see e.g. [Doetsch (1974); Ghizzetti and Ossicini (1971); Zemanian (1972)].

2.3 The four types of viscoelasticity

Since the creep and relaxation integral formulations must agree with each other, there must be a one-to-one correspondence between the relaxation modulus and the creep compliance. The basic relation between $J(t)$ and $G(t)$ is found noticing the following *reciprocity relation* in the Laplace domain, deduced from Eqs. (2.7a) (2.7b),

$$s\,\widetilde{J}(s) = \frac{1}{s\,\widetilde{G}(s)} \iff \widetilde{J}(s)\,\widetilde{G}(s) = \frac{1}{s^2}\,. \qquad (2.8)$$

Indeed, inverting the R.H.S. of (2.8), we obtain

$$J(t) * G(t) := \int_0^t J(t-\tau)\,G(\tau)\,d\tau = t\,. \qquad (2.9)$$

We can also obtain (2.8) noticing that, if the strain causal history is $J(t)$, then the stress response is $\Theta(t)$, the unit step function, so Eqs. (2.4a) and (2.5a) give

$$\Theta(t) = \int_{0^-}^t G(t-\tau)\,dJ(\tau) = J_g\,G(t) + \int_0^t G(t-\tau)\,\dot{J}(\tau)\,d\tau\,. \qquad (2.10)$$

Then, applying the Laplace transform to (2.10) yields

$$\frac{1}{s} = J_g\,\widetilde{G}(s) + \widetilde{G}(s)\left[s\widetilde{J}(s) - J_g\right]\,.$$

Following [Pipkin (1986)], Eq. (2.10) allows us to obtain some notable relations in the time domain (inequalities and integral equations) concerning the material functions. Taking it for granted that, for $0 \leq t \leq +\infty$, $J(t)$ is non-negative and increasing, and $G(t)$ is non-negative decreasing, Eq. (2.10) yields

$$1 = \int_{0^-}^t G(t-\tau)\,dJ(\tau) \geq G(t) \int_{0^-}^t dJ(\tau) = G(t)\,J(t)\,,$$

namely

$$J(t)\,G(t) \leq 1\,. \tag{2.11}$$

We also note that if $J_g \neq 0$, we can rearrange (2.10) as a Volterra integral equation of the second kind, treating $G(t)$ as the unknown and $J(t)$ as the known function,

$$G(t) = J_g^{-1} - J_g^{-1} \int_0^t \dot{J}(t-\tau)\,G(\tau)\,d\tau\,. \tag{2.12a}$$

Similarly, if $G(t)$ is given and $G_g \neq \infty$, the equation for $J(t)$ is

$$J(t) = G_g^{-1} - G_g^{-1} \int_0^t \dot{G}(t-\tau)\,J(\tau)\,d\tau\,. \tag{2.12b}$$

Pipkin has also pointed out the following inequalities

$$G(t) \int_0^t J(\tau)\,d\tau \leq t \leq J(t) \int_0^t G(\tau)\,d\tau\,. \tag{2.13}$$

One of these inequalities (L.H.S.) is not as close as (2.11); the other (R.H.S.) gives new information. Furthermore, using with the due care the limiting theorems for the Laplace transform

$$f(0^+) = \lim_{s \to \infty} s\widetilde{f}(s)\,, \quad f(+\infty) = \lim_{s \to 0} s\widetilde{f}(s)\,,$$

we can deduce from the L.H.S of (2.8) that

$$J_g = \frac{1}{G_g}\,, \quad J_e = \frac{1}{G_e}\,, \tag{2.14}$$

with the convention that 0 and $+\infty$ are reciprocal to each other.

The remarkable relations allow us to classify the viscoelastic bodies according to their instantaneous and equilibrium responses. In fact, from Eqs. (2.2), (2.3) and (2.14) we easily recognize four possibilities for the limiting values of the creep compliance and relaxation modulus, as listed in Table 2.1.

Table 2.1 The four types of viscoelasticity.

Type	J_g	J_e	G_g	G_e
I	> 0	$< \infty$	$< \infty$	> 0
II	> 0	$= \infty$	$< \infty$	$= 0$
III	$= 0$	$< \infty$	$= \infty$	> 0
IV	$= 0$	$= \infty$	$= \infty$	$= 0$

We note that the viscoelastic bodies of type I exhibit both instantaneous and equilibrium elasticity, so their behaviour appears close to the purely elastic one for sufficiently short and long times. The bodies of type II and IV exhibit a complete stress relaxation (at constant strain) since $G_e = 0$ and an infinite strain creep (at constant stress) since $J_e = \infty$, so they do not present equilibrium elasticity. Finally, the bodies of type III and IV do not present instantaneous elasticity since $J_g = 0$ ($G_g = \infty$).

Other properties will be pointed out later on.

2.4 The classical mechanical models

To get some feeling for linear viscoelastic behaviour, it is useful to consider the simpler behaviour of analog *mechanical models*. They are constructed from linear springs and dashpots, disposed singly and in branches of two (in series or in parallel) as it is shown in Fig. 2.1.

As analog of stress and strain, we use the total extending force and the total extension, respectively. We note that when two elements are combined in series [in parallel], their compliances [moduli] are additive. This can be stated as a combination rule: *creep compliances add in series, while relaxation moduli add in parallel.*

The important role in the literature of the mechanical models is justified by the historical development. In fact, the early theories were established with the aid of these models, which are still helpful to visualize properties and laws of the general theory, using the combination rule.

Now, it is worthwhile to consider the simple models of Fig. 2.1 by providing their governing stress–strain relations along with the related material functions.

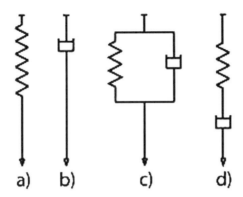

Fig. 2.1 The representations of the basic mechanical models: a) spring for Hooke, b) dashpot for Newton, c) spring and dashpot in parallel for Voigt, d) spring and dashpot in series for Maxwell.

The Hooke model. The spring a) in Fig. 2.1 is the elastic (or storage) element, as for it the force is proportional to the extension; it represents a perfect elastic body obeying the Hooke law. This model is thus referred to as the *Hooke model*. If we denote by m the pertinent elastic modulus we have

$$Hooke\ model\ :\ \sigma(t) = m\,\epsilon(t)\,, \qquad (2.15a)$$

so

$$\begin{cases} J(t) = 1/m\,, \\ G(t) = m\,. \end{cases} \qquad (2.15b)$$

In this case we have no creep and no relaxation so the creep compliance and the relaxation modulus are constant functions: $J(t) \equiv J_g \equiv J_e = 1/m$; $G(t) \equiv G_g \equiv G_e = 1/m$.

The Newton model. The dashpot b) in Fig. 2.1 is the viscous (or dissipative) element, the force being proportional to rate of extension; it represents a perfectly viscous body obeying the Newton law. This model is thus referred to as the *Newton model*. Denoting by b_1 the pertinent viscosity coefficient, we have

$$Newton\ model\ :\ \sigma(t) = b_1 \frac{d\epsilon}{dt} \qquad (2.16a)$$

so
$$\begin{cases} J(t) = \dfrac{t}{b_1}, \\ G(t) = b_1\,\delta(t). \end{cases} \qquad (2.16b)$$

In this case we have a linear creep $J(t) = J_+ t$ and instantaneous relaxation $G(t) = G_-\,\delta(t)$ with $G_- = 1/J_+ = b_1$.

We note that the Hooke and Newton models represent the limiting cases of viscoelastic bodies of type I and IV, respectively.

The Voigt model. A branch constituted by a spring in parallel with a dashpot is known as the *Voigt model*, c) in Fig. 2.1. We have

$$Voigt\ model\ :\ \sigma(t) = m\,\epsilon(t) + b_1 \frac{d\epsilon}{dt}, \qquad (2.17a)$$

so

$$\begin{cases} J(t) = J_1\left(1 - e^{-t/\tau_\epsilon}\right), & J_1 = \dfrac{1}{m},\ \tau_\epsilon = \dfrac{b_1}{m}, \\ G(t) = G_e + G_-\,\delta(t), & G_e = m,\ G_- = b_1, \end{cases} \qquad (2.17b)$$

where τ_ϵ is referred to as the *retardation time*.

The Maxwell model. A branch constituted by a spring in series with a dashpot is known as the *Maxwell model*, d) in Fig. 2.1. We have

$$Maxwell\ model\ :\ \sigma(t) + a_1 \frac{d\sigma}{dt} = b_1 \frac{d\epsilon}{dt}, \qquad (2.18a)$$

so

$$\begin{cases} J(t) = J_g + J_+ t, & J_g = \dfrac{a_1}{b_1},\ J_+ = \dfrac{1}{b_1}, \\ G(t) = G_1\,e^{-t/\tau_\sigma}, & G_1 = \dfrac{b_1}{a_1},\ \tau_\sigma = a_1, \end{cases} \qquad (2.18b)$$

where τ_σ is is referred to as the *the relaxation time*.

The Voigt and the Maxwell models are thus the simplest viscoelastic bodies of type III and II, respectively. The Voigt model exhibits an exponential (reversible) strain creep but no stress relaxation; it is also referred as the retardation element. The Maxwell model exhibits an exponential (reversible) stress relaxation and a linear (non reversible) strain creep; it is also referred to as the relaxation element.

Based on the combination rule introduced above, we can continue the previous procedure in order to construct the simplest models of type I and IV that require three parameters.

The Zener model. The simplest viscoelastic body of type I is obtained by adding a spring either in series to a Voigt model or in parallel to a Maxwell model, respectively. In this way, according to the combination rule, we add a positive constant both to the Voigt-like creep compliance and to the Maxwell-like relaxation modulus so that $J_g > 0$ and $G_e > 0$. Such a model was considered by Zener [Zener (1948)] with the denomination of *Standard Linear Solid* (*S.L.S.*) and will be referred here also as the *Zener model*. We have

$$Zener\ model\ :\ \left[1 + a_1 \frac{d}{dt}\right] \sigma(t) = \left[m + b_1 \frac{d}{dt}\right] \epsilon(t), \quad (2.19a)$$

so

$$\begin{cases} J(t) = J_g + J_1\left(1 - e^{-t/\tau_\epsilon}\right), & J_g = \frac{a_1}{b_1},\ J_1 = \frac{1}{m} - \frac{a_1}{b_1},\ \tau_\epsilon = \frac{b_1}{m}, \\ G(t) = G_e + G_1\, e^{-t/\tau_\sigma}, & G_e = m,\ G_1 = \frac{b_1}{a_1} - m,\ \tau_\sigma = a_1. \end{cases}$$

$$(2.19b)$$

We point out the condition $0 < m < b_1/a_1$ in order J_1, G_1 be positive and hence $0 < J_g < J_e < \infty$ and $0 < G_e < G_g < \infty$. As a consequence, we note that, for the *S.L.S.* model, the retardation time must be greater than the relaxation time, i.e. $0 < \tau_\sigma < \tau_\epsilon < \infty$.

The anti-Zener model. The simplest viscoelastic body of type IV requires three parameters, i.e. a_1, b_1, b_2; it is obtained by adding a dashpot either in series to a Voigt model or in parallel to a Maxwell model (Fig. 2.1c and Fig. 2.1d, respectively). According to the combination rule, we add a linear term to the Voigt-like creep compliance and a delta impulsive term to the Maxwell-like relaxation modulus so that $J_e = \infty$ and $G_g = \infty$. We may refer to this model as the *anti-Zener model*. We have

$$anti-Zener\ model\ :\ \left[1 + a_1 \frac{d}{dt}\right] \sigma(t) = \left[b_1 \frac{d}{dt} + b_2 \frac{d^2}{dt^2}\right] \epsilon(t), \quad (2.20a)$$

so

$$\begin{cases} J(t) = J_+ t + J_1\left(1 - e^{-t/\tau_\epsilon}\right), & J_+ = \frac{1}{b_1},\ J_1 = \frac{a_1}{b_1} - \frac{b_2}{b_1^2},\ \tau_\epsilon = \frac{b_2}{b_1}, \\ G(t) = G_-\, \delta(t) + G_1\, e^{-t/\tau_\sigma}, & G_- = \frac{b_2}{a_1},\ G_1 = \frac{b_1}{a_1} - \frac{b_2}{a_1^2},\ \tau_\sigma = a_1. \end{cases}$$

$$(2.20b)$$

We point out the condition $0 < b_2/b_1 < a_1$ in order J_1, G_1 to be positive. As a consequence we note that, for the *anti-Zener* model, the relaxation time must be greater than the retardation time, i.e. $0 < \tau_\epsilon < \tau_\sigma < \infty$, on the contrary of the Zener $(S.L.S.)$ model.

In Fig. 2.2 we exhibit the mechanical representations of the Zener model [a), b)] and the anti-Zener model [c), d)]. Because of their main characteristics, these models can be referred as the *three-element elastic model* and the *three-element viscous model*, respectively.

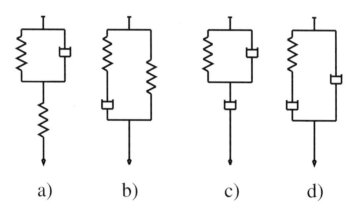

a)   b)   c)   d)

Fig. 2.2 The mechanical representations of the Zener [a), b)] and anti-Zener [c), d)] models: a) spring in series with Voigt, b) spring in parallel with Maxwell, c) dashpot in series with Voigt, d) dashpot in parallel with Maxwell.

By using the combination rule, general mechanical models can obtained whose material functions turn out to be of the type

$$\begin{cases} J(t) = J_g + \sum_n J_n \left(1 - e^{-t/\tau_{\epsilon,n}}\right) + J_+ t, \\ G(t) = G_e + \sum_n G_n\, e^{-t/\tau_{\sigma,n}} + G_-\, \delta(t), \end{cases} \quad (2.21)$$

where all the coefficients are non negative. We note that the four types of viscoelasticity of Table 2.1 are obtained from Eqs. (2.21) by taking into account that

$$\begin{cases} J_e < \infty \iff J_+ = 0, \quad J_e = \infty \iff J_+ \neq 0, \\ G_g < \infty \iff G_- = 0, \quad G_g = \infty \iff G_- \neq 0. \end{cases} \quad (2.22)$$

The canonic forms. In Fig. 2.3, following [Gross (1953)], we exhibit the general mechanical representations of Eqs. (2.21) in terms of springs and dashpots (illustrated here by boxes), so summarizing the four *canonic forms*.

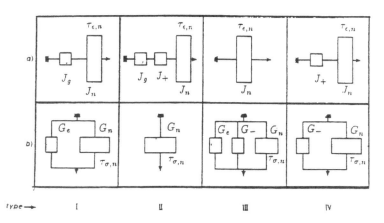

Fig. 2.3 The four types of canonic forms for the mechanical models: a) in creep representation; b) in relaxation representation.

The reader must note that in Fig. 2.3 the boxes denoted by J_g, G_e represent springs, those denoted by J_+, G_- represent dashpots and those denoted by $\{J_n, \tau_{\epsilon,n}\}$ and by $\{G_n, \tau_{\sigma,n}\}$ represent a sequence of Voigt models connected in series (*compound Voigt model*) and a sequence of Maxwell models connected in parallel (*compound Maxwell model*), respectively. The compound Voigt and Maxwell models are represented in Fig. 2.4.

As a matter of fact, each of the two representations can assume one of the *four canonic forms*, which are obtained by cutting out one, both, or none of the two single elements which have appeared besides the branches. Each of these four forms corresponds to each of the four types of linear viscoelastic behaviour (indicated in Table 2.1).

We recall that these material functions $J(t)$ and $G(t)$ are interrelated because of the *reciprocity relation* (2.8) in the Laplace domain.

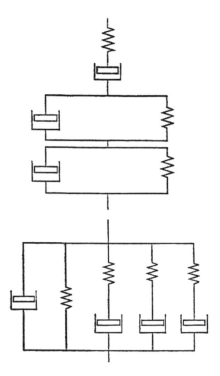

Fig. 2.4 The mechanical representations of the compound Voigt model (top) and compound Maxwell model (bottom).

Appealing to the theory of Laplace transforms, we get from (2.21)

$$\begin{cases} s\,\widetilde{J}(s) = J_g + \sum_n \dfrac{J_n}{1 + s\,\tau_{\epsilon,n}} + \dfrac{J_+}{s}, \\ s\,\widetilde{G}(s) = G_e + \sum_n \dfrac{G_n\,(s\,\tau_{\sigma,n})}{1 + s\,\tau_{\sigma,n}} + G_-\,s. \end{cases} \qquad (2.23)$$

The second equality can be re-written as

$$s\,\widetilde{G}(s) = (G_e + \beta) - \sum_n \dfrac{G_n}{1 + s\,\tau_{\sigma,n}} + G_-\,s, \text{ with } \beta := \sum_n G_n.$$

Therefore, as a consequence of (2.23), $s\,\widetilde{J}(s)$ and $s\,\widetilde{G}(s)$ turn out to be *rational* functions in $\mathbb{C}$ with simple poles and zeros on the negative real axis $Re[s] < 0$ and, possibly, with a simple pole or with a simple zero at $s = 0$, respectively. As a consequence, see e.g.

[Bland (1960)], the above functions can be written as

$$s\,\tilde{J}(s) = \frac{1}{s\,\tilde{G}(s)} = \frac{P(s)}{Q(s)}, \text{ where } \begin{cases} P(s) = 1 + \sum_{k=1}^{p} a_k\,s^k, \\ Q(s) = m + \sum_{k=1}^{q} b_k\,s^k, \end{cases} \quad (2.24)$$

where the orders of the polynomials are equal ($q = p$) or differ of unity ($q = p + 1$) and the zeros are alternating on the negative real axis. The least zero in absolute magnitude is a zero of $Q(s)$. The ratio of any coefficient in $P(s)$ to any coefficient in $Q(s)$ is positive. The four types of viscoelasticity then correspond to whether the least zero is ($J_+ \neq 0$) or is not ($J_+ = 0$) equal to zero and to whether the greatest zero in absolute magnitude is a zero of $P(s)$ ($J_g \neq 0$) or a zero of $Q(s)$ ($J_g = 0$). We also point out that the polynomials at the numerator and denominator are *Hurwitz polynomials*, in that they have no zeros for $Re[s] > 0$, with $m \geq 0$ and $q = p$ or $q = p + 1$. Furthermore, the resulting rational functions $s\,\tilde{J}(s)$, $s\,\tilde{G}(s)$ turn out to be *positive real functions* in $\mathbb{C}$, namely they assume positive real values for $s \in \mathbb{R}^+$.

The operator equation. According to the classical theory of viscoelasticity (see e.g. [Alfrey (1948); Gross (1953)]), the above properties mean that the stress–strain relation must be a linear differential equation with constant (positive) coefficients of the following form

$$\left[1 + \sum_{k=1}^{p} a_k \frac{d^k}{dt^k}\right] \sigma(t) = \left[m + \sum_{k=1}^{q} b_k \frac{d^k}{dt^k}\right] \epsilon(t). \quad (2.25)$$

Eq. (2.25) is referred to as the *operator equation* of the mechanical models, of which we have investigated the most simple cases illustrated in Figs. 2.1, 2.2. Of course, the constants m, a_k, b_k are expected to be subjected to proper restrictions in order to meet the physical requirements of realizability. For further details we refer the interested reader to [Hanyga (2005a); (2005b); (2005c)].

In Table 2.2 we summarize the four cases, which are expected to occur in the *operator equation* (2.25), corresponding to the four types of viscoelasticity.

Table 2.2 The four cases of the operator equation.

Type	Order	m	J_g	G_e	J_+	G_-
I	$q = p$	> 0	a_p/b_p	m	0	0
II	$q = p$	$= 0$	a_p/b_p	0	$1/b_1$	0
III	$q = p+1$	> 0	0	m	0	b_q/a_p
IV	$q = p+1$	$= 0$	0	0	$1/b_1$	b_q/a_p

We recognize that for $p = 1$ Eq. (2.25) includes the operator equations for the classical models with two parameters: Voigt and Maxwell; and with three parameters: Zener and anti-Zener. In fact, we recover the Voigt model (type III) for $m > 0$ and $p = 0, q = 1$, the Maxwell model (type II) for $m = 0$ and $p = q = 1$, the Zener model (type I) for $m > 0$ and $p = q = 1$, and the anti-Zener model (type IV) for $m = 0$ and $p = 1, q = 2$.

The Burgers model. With four parameters we can construct two models, the former with $m = 0$ and $p = q = 2$, the latter with $m > 0$ and $p = 1, q = 2$, referred in [Bland (1960)] to as four-element models of the first kind and of the second kind, respectively.

We restrict our attention to the former model, known as *Burgers model*, because it has found numerous applications, specially in geosciences, see e.g. [Klausner (1991); Carcione et al. (2006)]. We note that such a model is obtained by adding a dashpot or a spring to the representations of the Zener or of the anti-Zener model, respectively. Assuming the creep representation the dashpot or the spring is added in series, so the Burgers model results in a series combination of a Maxwell element with a Voigt element. Assuming the relaxation representation, the dashpot or the spring is added in parallel, so the Burgers model results in two Maxwell elements disposed in parallel. We refer the reader to Fig. 2.5 for the two mechanical representations of the Burgers model.

According to our general classification, the Burgers model is thus a four-element model of type II, defined by the four parameters $\{a_1, a_2, b_1, b_2\}$.

We have

$$Burgers\ model: \left[1 + a_1\frac{d}{dt} + a_2\frac{d^2}{dt^2}\right]\sigma(t) = \left[b_1\frac{d}{dt} + b_2\frac{d^2}{dt^2}\right]\epsilon(t), \quad (2.26a)$$

so

$$\begin{cases} J(t) = J_g + J_+ t + J_1\left(1 - e^{-t/\tau_\epsilon}\right), \\ G(t) = G_1\,e^{-t/\tau_{\sigma,1}} + G_2\,e^{-t/\tau_{\sigma,2}}. \end{cases} \quad (2.26b)$$

We leave to the reader to express as an exercise the physical quantities J_g, J_+, τ_ϵ and G_1, $\tau_{\sigma,1}$, G_2, $\tau_{\sigma,2}$, in terms of the four parameters $\{a_1, a_2, b_1, b_2\}$ in the operator equation (2.26a).

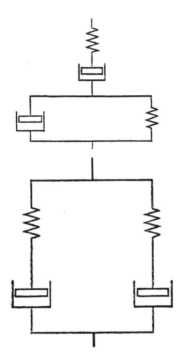

Fig. 2.5 The mechanical representations of the Burgers model: the creep representation (top), the relaxation representation (bottom).

Remark on the initial conditions :

We note that the initial conditions at $t = 0^+$ for the stress $\sigma(t)$ and strain $\epsilon(t)$,

$$\{\sigma^{(h)}(0^+),\ h = 0, 1, \ldots p-1\},\ \{\epsilon^{(k)}(0^+),\ k = 0, 1, \ldots q-1\},$$

do not appear in the operator equation, but they are required to be compatible with the integral equations (2.5a) and (2.5b) and consequently with the corresponding Laplace transforms provided by Eqs. (2.7a) and (2.7b). Since the above equations do not contain the initial conditions, some compatibility conditions at $t = 0^+$ must be *implicitly* required both for stress and strain. In other words, the equivalence between the integral equations (2.5a) and (2.5b), and the differential operator equation (2.25), implies that when we apply the Laplace transform to both sides of Eq. (2.25) the contributions from the initial conditions do not appear, namely they are vanishing or cancel in pair-balance. This can be easily checked for the simplest classical models described by Eqs. (2.17)–(2.20). For simple examples, let us consider the Voigt model for which $p = 0$, $q = 1$ and $m > 0$, see Eq. (2.17a), and the Maxwell model for which $p = q = 1$ and $m = 0$, see Eq. (2.18a).

For the Voigt model we get

$$s\widetilde{\sigma}(s) = m\widetilde{\epsilon}(s) + b_1\left[s\widetilde{\epsilon}(s) - \epsilon(0^+)\right],$$

so, for any causal stress and strain histories, it would be

$$s\widetilde{J}(s) = \frac{1}{m + b_1 s} \iff \epsilon(0^+) = 0. \qquad (2.27a)$$

We note that the condition $\epsilon(0^+) = 0$ is surely satisfied for any reasonable stress history since $J_g = 0$, but is not valid for any reasonable strain history; in fact, if we consider the relaxation test for which $\epsilon(t) = \Theta(t)$ we have $\epsilon(0^+) = 1$. This fact may be understood recalling that for the Voigt model we have $J_g = 0$ and $G_g = \infty$ (due to the delta contribution in the relaxation modulus).

For the Maxwell model we get

$$\widetilde{\sigma}(s) + a_1\left[s\widetilde{\sigma}(s) - \sigma(0^+)\right] = b_1\left[s\widetilde{\epsilon}(s) - \epsilon(0^+)\right],$$

so, for any causal stress and strain histories it would be

$$s\widetilde{J}(s) = \frac{a_1}{b_1} + \frac{1}{b_1 s} \iff a_1\sigma(0^+) = b_1\epsilon(0^+). \qquad (2.27b)$$

We now note that the condition $a_1\sigma(0^+) = b_1\epsilon(0^+)$ is surely satisfied for any causal history, both in stress and in strain. This fact may be understood recalling that, for the Maxwell model, we have $J_g > 0$ and $G_g = 1/J_g > 0$.

Then we can generalize the above considerations stating that the compatibility relations of the initial conditions are valid for all the four types of viscoelasticity, as far as the creep representation is considered. When the relaxation representation is considered, caution is required for the types III and IV, for which, for correctness, we would use the generalized theory of integral transforms suitable just for dealing with generalized functions.

2.5 The time - and frequency - spectral functions

From the previous analysis of the classical mechanical models in terms of a finite number of basic elements, one is led to consider two *discrete* distributions of characteristic times (the *retardation* and the *relaxation* times), as it has been stated in Eq. (2.21). However, in more general cases, it is natural to presume the presence of *continuous* distributions, so that, for a viscoelastic body, the material functions turn out to be of the following form

$$\begin{cases} J(t) = J_g + a \int_0^\infty R_\epsilon(\tau) \left(1 - e^{-t/\tau}\right) d\tau + J_+ t, \\ G(t) = G_e + b \int_0^\infty R_\sigma(\tau) e^{-t/\tau} d\tau + G_- \delta(t), \end{cases} \quad (2.28)$$

where all the coefficients and functions are non-negative. The function $R_\epsilon(\tau)$ is referred to as the *retardation spectrum* while $R_\sigma(\tau)$ as the *relaxation spectrum*. For the sake of convenience we shall omit the suffix to denote any one of the two spectra; we shall refer to $R(\tau)$ as the *time–spectral function*. in $\mathbb{R}^+$, with the supplementary normalization condition $\int_0^\infty R(\tau)\, d\tau = 1$ We require $R(\tau)$ be locally summable if the integral in $\mathbb{R}^+$ is convergent.

The discrete distributions of the classical mechanical models, see Eqs. (2.21), can be easily recovered from Eqs. (2.28). In fact, assuming $a \neq 0$, $b \neq 0$, we get after a proper use of the *delta-Dirac generalized functions*

$$\begin{cases} R_\epsilon(\tau) = \dfrac{1}{\alpha} \sum_n J_n\, \delta(\tau - \tau_{\epsilon,n})\,, & a = \sum_n J_n\,, \\ R_\sigma(\tau) = \dfrac{1}{\beta} \sum_n G_n\, \delta(\tau - \tau_{\sigma,n})\,, & b = \sum_n G_n\,. \end{cases} \quad (2.29)$$

We now devote particular attention to the time-dependent contributions to the material functions (2.28) which are provided by the continuous or discrete spectra using for them the notation

$$\begin{cases} J_\tau(t) := a \displaystyle\int_0^\infty R_\epsilon(\tau)\left(1 - e^{-t/\tau}\right) d\tau\,, \\ G_\tau(t) := b \displaystyle\int_0^\infty R_\sigma(\tau)\, e^{-t/\tau}\, d\tau\,. \end{cases} \quad (2.30)$$

We recognize that $J_\tau(t)$ (that we refer as the *creep function with spectrum*) is a non-decreasing, non-negative function in $\mathbb{R}^+$ with limiting values $J_\tau(0^+) = 0$, $J_\tau(+\infty) = a$ or ∞, whereas $G_\tau(t)$ (that we refer as the *relaxation function with spectrum*) is a non-increasing, non-negative function in $\mathbb{R}^+$ with limiting values $G_\tau(0^+) = b$ or ∞, $G_\tau(+\infty) = 0$. More precisely, in view of the spectral representations (2.30), we have

$$\begin{cases} J_\tau(t) \ge 0\,, \quad (-1)^n \dfrac{d^n J_\tau}{dt^n} \le 0\,, \\ \\ G_\tau(t) \ge 0\,, \quad (-1)^n \dfrac{d^n G_\tau}{dt^n} \ge 0\,. \end{cases} \quad t \ge 0\,, \quad n = 1, 2, \ldots\,. \quad (2.31)$$

Using a proper terminology of mathematical analysis, see e.g. [Berg and Forst (1975); Feller (1971); Gripenberg *et al.* (1990)], $G_\tau(t)$ is a *completely monotonic function* whereas $J_\tau(t)$ is a *Bernstein function*, since it is a non-negative function with a completely monotonic derivative. These properties have been investigated by several authors, including [Molinari (1975)], [Del Piero and Deseri (1995)] and recently, in a detailed way, by [Hanyga (2005a); Hanyga (2005b); Hanyga (2005c)].

The determination of the *time–spectral functions* starting from the knowledge of the creep and relaxation functions is a problem that can be formally solved through the Titchmarsh inversion formula of the Laplace transform theory according to [Gross (1953)]. For this purpose let us recall the Gross method of *Laplace integral pairs*, which is based on the introduction of the *frequency–spectral functions* $S_\epsilon(\gamma)$ and $S_\sigma(\gamma)$ defined as

$$S_\epsilon(\gamma) := a\,\frac{R_\epsilon(1/\gamma)}{\gamma^2}\,, \quad S_\sigma(\gamma) := b\,\frac{R_\sigma(1/\gamma)}{\gamma^2}\,, \qquad (2.32)$$

where $\gamma = 1/\tau$ denotes a retardation or relaxation frequency. We note that with the above choice it turns out

$$a\,R_\epsilon(\tau)\,d\tau = S_\epsilon(\gamma)\,d\gamma\,, \quad b\,R_\sigma(\tau)\,d\tau = S_\sigma(\gamma)\,d\gamma\,. \qquad (2.33)$$

Differentiating (2.30) with respect to time yields

$$\begin{cases} \dot{J}_\tau(t) = a\int_0^\infty \dfrac{R_\epsilon(\tau)}{\tau}\,e^{-t/\tau}\,d\tau = \int_0^\infty \gamma\,S_\epsilon(\gamma)\,e^{-t\gamma}\,d\gamma\,, \\[1em] -\dot{G}_\tau(t) = b\int_0^\infty \dfrac{R_\sigma(\tau)}{\tau}\,e^{-t/\tau}\,d\tau = \int_0^\infty \gamma\,S_\sigma(\gamma)\,e^{-t\gamma}\,d\gamma\,. \end{cases} \qquad (2.34)$$

We recognize that $\gamma\,S_\epsilon(\gamma)$ and $\gamma\,S_\sigma(\gamma)$ turn out to be the inverse Laplace transforms of $\dot{J}_\tau(t)$ and $-\dot{G}_\tau(t)$, respectively, where t is now considered the Laplace transform variable instead of the usual s. Adopting the usual notation for the Laplace transform pairs, we thus write

$$\begin{cases} \gamma\,S_\epsilon(\gamma) = a\,\dfrac{R_\epsilon(1/\gamma)}{\gamma} \div \dot{J}_\tau(t)\,, \\[1em] -\gamma\,S_\sigma(\gamma) = b\,\dfrac{R_\sigma(1/\gamma)}{\gamma} \div \dot{G}_\tau(t)\,. \end{cases} \qquad (2.35)$$

Consequently, when the *creep* and *relaxation* functions are given as analytical expressions, the corresponding frequency distributions can be derived by standard methods for the inversion of Laplace transforms; then, by using Eq. (2.32), the time–spectral functions can be easily derived.

Incidentally, we note that in the expressions defining the time and frequency spectra, often $d(\log\tau)$ and $d(\log\gamma)$ rather than $d\tau$ and $d\gamma$ are involved in the integrals. This choice changes the scaling of the above spectra in order to better deal with phenomena occurring on several time (or frequency) scales. In fact, introducing the new variables $u = \log\tau$ and $v = \log\gamma$, where $-\infty < u, v < +\infty$, the new spectra are related to the old ones as it follows

$$\hat{R}(u)\,du = R(\tau)\,\tau\,d\tau\,, \quad \hat{S}(v)\,dv = S(\gamma)\,\gamma\,d\gamma\,. \qquad (2.36)$$

Example of time and frequency spectra. As an example of spectrum determination, we now consider the creep function

$$J_\tau(t) = a\,\mathrm{Ein}\,(t/\tau_0)\,, \quad a > 0\,, \quad \tau_0 > 0\,, \tag{2.37}$$

where Ein denotes the modified exponential integral function, defined in the complex plane as an entire function whose integral and series representations read

$$\mathrm{Ein}\,(z) = \int_0^z \frac{1-\mathrm{e}^{-\zeta}}{\zeta}\,d\zeta = -\sum_{n=1}^{\infty}(-1)^n\frac{z^n}{n\,n!}\,, \quad z \in \mathbb{C}\,. \tag{2.38}$$

For more details, see Appendix D. As a consequence we get

$$\frac{dJ_\tau}{dt}(t) = a\,\frac{1-\mathrm{e}^{-\gamma_0 t}}{t}\,, \quad \gamma_0 = \frac{1}{\tau_0}\,. \tag{2.39}$$

By inspection of a table of Laplace transform pairs we get the inversion and, using (2.35), the following time and frequency–spectra

$$R_\epsilon(\tau) = \begin{cases} 0\,, & 0 < \tau < \tau_0\,, \\ 1/\tau\,, & \tau_0 < \tau < \infty\,; \end{cases} \tag{2.40a}$$

$$S_\epsilon(\gamma) = \begin{cases} a/\gamma\,, & 0 < \gamma < \gamma_0\,, \\ 0\,, & \gamma_0 < \gamma < \infty\,. \end{cases} \tag{2.40b}$$

Plotted against $\log\tau$ and $\log\gamma$ the above spectra are step-wise distributions.

Stieltjes transforms. To conclude this section, following [Gross (1953)], we look for the relationship between the Laplace transform of the creep/relaxation function and the corresponding time or frequency spectral function. If we choose the frequency spectral function, we expect that a sort of iterated Laplace transform be involved in the required relationship, in view of the above results. In fact, applying the Laplace transform to Eq. (2.34) we obtain

$$\begin{cases} s\widetilde{J_\tau}(s) = \displaystyle\int_0^\infty \frac{\gamma\,S_\epsilon(\gamma)}{s+\gamma}\,d\gamma\,, \\[2mm] s\widetilde{G_\tau}(s) = -\displaystyle\int_0^\infty \frac{\gamma\,S_\sigma(\gamma)}{s+\gamma}\,d\gamma + G_\tau(0^+)\,. \end{cases} \tag{2.41}$$

Introducing the function

$$\widetilde{L}(s) = \int_0^\infty \frac{\gamma S(\gamma)}{s+\gamma}\, d\gamma, \qquad (2.42)$$

where the suffix ϵ or σ is understood, we recognize that $\widetilde{L}(s)$ is the Stieltjes transform of $\gamma S(\gamma)$. The inversion of the Stieltjes transform may be carried out by Titchmarsh's formula,

$$\gamma S(\gamma) = \frac{1}{\pi} Im\left\{\widetilde{L}\left(\gamma e^{-i\pi}\right)\right\} = \frac{1}{\pi} \lim_{\delta \to 0} Im\left\{\widetilde{L}(-\gamma - i\delta)\right\}. \qquad (2.43)$$

Consequently, when the Laplace transforms of the creep and relaxation functions are given as analytical expressions, the corresponding frequency distributions can be derived by standard methods for the inversion of Stieltjes transforms; then, by using Eq. (2.32) the time–spectral functions can be easily derived.

2.6 History in IR: the Fourier transform approach and the dynamic functions

In addition to the unit step (that is acting for $t \geq 0$), another widely used form of excitation in viscoelasticity is the *harmonic* or *sinusoidal* excitation that is acting for all of IR since it is considered (ideally) applied since $t = -\infty$. The corresponding responses, which are usually referred to as the *dynamic functions*, provide, together with the material functions previously investigated, a complete description of the viscoelastic behaviour. In fact, according to [Findley et al. (1976)], creep and relaxation experiments provide information starting from a lower limit of time which is approximatively of $10\,s$, while dynamic experiments with sinusoidal excitations may provide data from about $10^{-8}\,s$ to about $10^3\,s$. Thus there is an overlapping region ($10\,s - 10^3\,s$) where data can be obtained from both types of experiments. Furthermore, the dynamic experiments provide information about storage and dissipation of the mechanical energy, as we shall see later.

In the following the basic concepts related to sinusoidal excitations is introduced. It is convenient to use the complex notation for sinusoidal functions, i.e. the excitations in stress and strain in non-dimensional form are written as

$$\sigma(t;\omega) = e^{i\omega t}, \; \epsilon(t;\omega) = e^{i\omega t}, \; \omega > 0, \; -\infty < t < +\infty, \quad (2.44)$$

where ω denotes the angular frequency ($f = \omega/2\pi$ is the cyclic frequency and $T = 1/f$ is the period). Of course, in Eq. (2.44) we understand to take the real or imaginary part of the exponential in view of the Euler formula $e^{\pm i\omega t} = \cos \omega t \pm i \sin \omega t$.

For histories of type (2.44), the integral stress–strain relations (2.4) can be used to provide the corresponding response functions. We obtain, after an obvious change of variable in the integrals,

$$\sigma(t) = e^{i\omega t} \implies \epsilon(t) = J^*(\omega) e^{i\omega t}, \; J^*(\omega) := i\omega \widehat{J}(\omega), \quad (2.45a)$$

and

$$\epsilon(t) = e^{i\omega t} \implies \sigma(t) = G^*(\omega) e^{i\omega t}, \; G^*(\omega) := i\omega \widehat{G}(\omega), \quad (2.45b)$$

where $\widehat{J}(\omega) = \int_0^\infty J(t) e^{-i\omega t} dt$ and $\widehat{G}(\omega) = \int_0^\infty G(t) e^{-i\omega t} dt$.

The functions $J^*(\omega)$ and $G^*(\omega)$ are usually referred as the *complex compliance* and *complex modulus*, respectively, or, simply, the *dynamic functions* of the viscoelastic body. They are related with the Fourier transforms of the causal functions $J(t)$ and $G(t)$ and therefore can be expressed in terms of their Laplace transforms as follows

$$J^*(\omega) = s \widetilde{J}(s)\Big|_{s=i\omega}, \quad G^*(\omega) = s \widetilde{G}(s)\Big|_{s=i\omega}, \quad (2.46)$$

so that, in agreement with the reciprocity relation (2.8),

$$J^*(\omega) G^*(\omega) = 1. \quad (2.47)$$

2.7 Storage and dissipation of energy: the loss tangent

Introducing the *phase shift* $\delta(\omega)$ between the sinusoidal excitation and the sinusoidal response in Eqs. (2.45a) and (2.45b), we can write

$$J^*(\omega) = J'(\omega) - iJ''(\omega) = |J^*(\omega)| e^{-i\delta(\omega)}, \quad (2.48a)$$

and

$$G^*(\omega) = G'(\omega) + iG''(\omega) = |G^*(\omega)| e^{+i\delta(\omega)}. \quad (2.48b)$$

As a consequence of energy considerations, recalled hereafter by following [Tschoegel (1989)], it turns out that $\delta(\omega)$ must be positive (in particular, $0 < \delta(\omega) < \pi/2$) as well as the quantities $J'(\omega)$, $J''(\omega)$ and $G'(\omega)$, $G''(\omega)$ entering Eqs. (2.48a) and (2.48b). Usually, J' and G' are called the *storage compliance* and the *storage modulus*, respectively, while J'' and G'' are called the *loss compliance* and the *loss modulus*, respectively; as we shall see, the above attributes connote something to do with energy storage and loss. Furthermore,

$$\tan \delta(\omega) = \frac{J''(\omega)}{J'(\omega)} = \frac{G''(\omega)}{G'(\omega)} \qquad (2.49)$$

is referred to as the *loss tangent*, a quantity that summarizes the damping ability of a viscoelastic body, as we will show explicitly below.

During the deformation of a viscoelastic body, part of the total work of deformation is dissipated as heat through viscous losses, but the remainder of the deformation-energy is stored elastically. It is frequently of interest to determine, for a given sample of material in a given mode of deformation, the total work of deformation as well as the amount of energy stored and the amount dissipated. Similarly, one may wish to know the rate at which the energy of deformation is absorbed by the material or the rate at which it is stored or dissipated.

The rate at which energy is absorbed per unit volume of a viscoelastic material during deformation is equal to the *stress power*, i.e. the rate at which work is performed. The stress power at time t is

$$\dot{W}(t) = \sigma(t)\,\dot{\epsilon}(t), \qquad (2.50)$$

i.e. it is the product of the instantaneous stress and rate of strain. The electrical analog of (2.50) is the well-known relation which states that the electrical power equals the product of instantaneous voltage and current. The total work of deformation or, in other words, the mechanical energy absorbed per unit volume of material in the deformation from the initial time t_0 up to the current time t, results in

$$W(t) = \int_{t_0}^{t} \dot{W}(\tau)\,d\tau = \int_{t_0}^{t} \sigma(t)\,\dot{\epsilon}(\tau)\,d\tau. \qquad (2.51)$$

Assuming the possibility of computing separately the energy stored, $W_s(t)$, and the energy dissipated, $W_d(t)$, we can write

$$W(t) = W_s(t) + W_d(t), \qquad \dot{W}(t) = \dot{W}_s(t) + \dot{W}_d(t). \tag{2.52}$$

Please note that all energy or work terms and their derivatives will henceforth refer to unit volume of the material even when this is not explicitly stated.

Elastically stored energy is potential energy. Energy can also be stored inertially as kinetic energy. Such energy storage may be encountered in fast loading experiments, e.g. in response to impulsive excitation, or in wave propagation at high frequency. In the linear theory of viscoelastic behaviour, however, inertial energy storage plays no role.

How much of the total energy is stored and how much is dissipated, i.e. the precise form of (2.52), depends, of course, on the nature of the material on the one hand, and on the type of deformation on the other. The combination of stored and dissipated energy is conveniently based on the representation of linear viscoelastic behaviour by models (the classical mechanical models) in that, by definition, the energy is dissipated uniquely in the dashpots and stored uniquely in the springs.

For the rate of energy dissipation we get

$$\dot{W}_d(t) = \sum_n \sigma_{dn}(t) \dot{\epsilon}_{dn}(t) = \sum_n \eta_n(t) \left[\dot{\epsilon}_{dn}(t)\right]^2, \tag{2.53}$$

where $\sigma_{dn}(t)$ and $\dot{\epsilon}_{dn}(t)$ are the stress and the rate of strain, respectively, in the n-th dashpot, which are related by the equality $\sigma_{dn}(t) = \eta_n \dot{\epsilon}_{dn}(t)$ with η_n denoting the coefficient of viscosity.

For the energy storage we get

$$\begin{aligned} W_s(t) &= \sum_n \int_{t_0}^{t} \sigma_{sn}(\tau) \dot{\epsilon}_{sn}(\tau) \, d\tau \\ &= \sum_n G_n \int_{\epsilon_{sn}(0)}^{\epsilon_{sn}(t)} \epsilon_{sn}(\tau) \, d\epsilon_{sn}(\tau) \\ &= \frac{1}{2} \sum_n G_n \left[\epsilon_{sn}(t)\right]^2, \end{aligned} \tag{2.54}$$

where $\sigma_{sn}(t)$ and $\epsilon_{sn}(t)$ are the stress and the strain, respectively, in the n-th spring, which are related by the equality $\sigma_{sn}(t) = G_n \epsilon_{sn}(t)$ with G_n denoting the elastic modulus.

Equations (2.53) and (2.54) are the basic relations for determining energy storage and dissipation, respectively, during a particular deformation. They are given meaning by finding the stresses, strains, rates of strain in the springs and dashpots of mechanical models in the given mode of deformation. The nature of the material is reflected in the distribution of the parameters G_n and η_n. Examples have been given by [Tschoegel (1989)], to which the interested reader is referred. In the absence of appropriate spring-dashpot models we may still think of energy-storing and energy-dissipating mechanisms but without identifying them with mechanical models, and modify the arguments as needed.

Let us now compute the total energy $W(t)$ and its rate $\dot{W}(t)$ for sinusoidal excitations, and possibly determine the corresponding contributions due to the storing and dissipating mechanisms [Tschoegel (1989)]. Taking the imaginary parts in (2.45b) we have

$$\epsilon(t) = \sin \omega t \implies \sigma(t) = G'(\omega) \sin \omega t + G''(\omega) \cos \omega t, \qquad (2.55)$$

where the terms on the R.H.S. represent, respectively, the components of the stress which are in phase and out of phase with the strain.

Since the rate of strain is $\omega \cos \omega t$, Eqs. (2.50) and (2.55) lead to

$$\dot{W}(t) = \frac{\omega}{2} \left[G'(\omega) \sin 2\omega t + G''(\omega)(1 + \cos 2\omega t) \right]. \qquad (2.56)$$

Integration of (2.56), subject to the initial condition $W(0) = 0$, yields

$$W(t) = \frac{1}{4} \left[G'(\omega)(1 - \cos 2\omega t) + G''(\omega)(2\omega t + \sin 2\omega t) \right]. \qquad (2.57)$$

In general, all storing mechanisms are not in phase as well as all dissipating mechanisms, so that in Eqs. (2.56) and (2.57) we cannot recognize the partial contributions to the storage and dissipation of energy. Only if *phase coherence* is assumed among the energy storing mechanisms on the one hand and the energy dissipating mechanisms on the other, we can easily separate the energy stored from that dissipated. We get

$$\dot{W}_s^c(t) = \frac{\omega}{2} G'(\omega) \sin 2\omega t, \quad \dot{W}_d^c(t) = \frac{\omega}{2} G''(\omega)(1 + \cos 2\omega t), \qquad (2.58)$$

hence
$$W_s^c(t) = \frac{G'(\omega)}{4}(1-\cos 2\omega t), \quad W_d^c(t) = \frac{G''(\omega)}{4}(2\omega t + \sin 2\omega t), \quad (2.59)$$

where the superscript c points out the hypothesis of coherence.

For the stored energy, a useful parameter is the *average* taken over a full cycle of the excitation. We find from (2.59)

$$\langle W_s(\omega) \rangle := \frac{1}{T} \int_t^{t+T} W_s^c(\tau) \, d\tau$$
$$= \frac{\omega \, G'(\omega)}{8\pi} \int_0^{2\pi/\omega} (1 - \cos 2\omega \tau) \, d\tau = \frac{G'(\omega)}{4}, \quad (2.60)$$

which is one half of the maximum coherently storable energy.

For the dissipated energy we consider the amount of energy that would be dissipated coherently over a full cycle of the excitation. We find from (2.56)

$$\Delta W_d(\omega) := \int_t^{t+T} \dot{W}_d^c(\tau) \, d\tau$$
$$= \frac{\omega \, G''(\omega)}{2} \int_0^{2\pi/\omega} (1 + \cos 2\omega \tau) \, d\tau = \pi \, G''(\omega). \quad (2.61)$$

We recognize that Eqs. (2.60) and (2.61) justify the names of $G'(\omega)$ and $G''(\omega)$ as *storage* and *loss modulus*, respectively.

Usually the dissipation in a viscoelastic medium is measured by introducing the so-called *specific dissipation function*, or *internal friction*, defined as

$$Q^{-1}(\omega) = \frac{1}{2\pi} \frac{\Delta W_d}{W_s^*}, \quad (2.62)$$

where ΔW_d is the amount of energy dissipated coherently in one cycle and W_s^* is the peak energy stored coherently during the cycle. It is worthwhile to note that Q^{-1} denotes the reciprocal of the so-called *quality factor*, that is denoted by Q in electrical engineering, see e.g. [Knopoff (1956)]. From Eqs. (2.49) and (2.60) and (2.62) it turns out that

$$Q^{-1}(\omega) = \tan \delta(\omega). \quad (2.63)$$

This equation shows that the damping ability of a linear viscoelastic body is dependent only on the tangent of the phase angle, namely the *loss tangent* introduced in Eq. (2.49), that is a function of frequency and is a measure of a physical property, but is independent of the stress and strain amplitudes.

2.8 The dynamic functions for the mechanical models

Let us conclude this chapter with the evaluation of the dynamic functions (complex moduli or complex compliances) for the classical mechanical models as it can be derived from their expressions according to Eq. (2.49), with special emphasis to their loss tangent.

For convenience, let us consider the Zener model, that contains as limiting cases the Voigt and Maxwell models, whereas we leave as an exercise the evaluation of the dynamic functions for the anti-Zener and Burgers models.

For this purpose we consider the dynamic functions, namely the complex compliance $J^*(\omega)$ and the complex modulus $G^*(\omega)$, for the Zener model, that can be derived from the Laplace transforms of the corresponding material functions $J(t)$ and $G(t)$ according to Eqs. (2.46). Using Eqs. (2.19a) and (2.19b) we get

$$J^*(\omega) = s\,\widetilde{J}(s)\Big|_{s=i\omega} = J_g + J_1 \frac{1}{1+s\tau_\epsilon}\Big|_{s=i\omega}, \qquad (2.64)$$

$$G^*(\omega) = s\,\widetilde{G}(s)\Big|_{s=i\omega} = G_e + G_1 \frac{s\tau_\sigma}{1+s\tau_\sigma}\Big|_{s=i\omega}. \qquad (2.65)$$

Then we get:

$$J^*(\omega) = J'(\omega) - J''(\omega)\,,\quad \begin{cases} J'(\omega) = J_g + J_1 \dfrac{1}{1+\omega^2\tau_\epsilon^2}\,, \\ J''(\omega) = J_1 \dfrac{\omega\tau_\epsilon}{1+\omega^2\tau_\epsilon^2}\,; \end{cases} \qquad (2.66)$$

$$G^*(\omega) = G'(\omega) + G''(\omega)\,,\quad \begin{cases} G'(\omega) = G_e + G_1 \dfrac{\omega\tau_\sigma}{1+\omega^2\tau_\sigma^2}\,, \\ G''(\omega) = G_1 \dfrac{\omega^2\tau_\sigma^2}{1+\omega^2\tau_\sigma^2}\,. \end{cases} \qquad (2.67)$$

Taking into account the definitions in (2.19b) that provide the interrelations among the constants in Eqs. (2.64) and (2.67), we find it convenient to introduce a new characteristic time

$$\tau := \sqrt{\tau_\sigma \tau_\epsilon}, \tag{2.68}$$

and

$$\Delta := \frac{\tau_\epsilon - \tau_\sigma}{\tau} = \begin{cases} \dfrac{J_e - J_g}{\sqrt{J_g J_e}}, \\[2mm] \dfrac{G_g - G_e}{\sqrt{G_g G_e}}. \end{cases} \tag{2.69}$$

Then, after simple algebraic manipulations, the loss tangent for the Zener model turns out to be

$$Zener\ model\ :\ \tan\delta(\omega) = \frac{J''(\omega)}{J'(\omega)} = \frac{G''(\omega)}{G'(\omega)} = \Delta \frac{\omega\tau}{1 + (\omega\tau)^2}. \tag{2.70}$$

We easily recognize that the loss tangent for the Zener model attains its maximum value $\Delta/2$ for $\omega = 1/\tau$.

It is instructive to adopt another notation in order to provide alternative expressions (consistent with the results by [Caputo and Mainardi (1971b)]), by introducing the characteristic frequencies related to the retardation and relaxation times:

$$\begin{cases} \alpha := 1/\tau_\epsilon = m/b_1, \\ \beta := 1/\tau_\sigma = 1/a_1, \end{cases} \text{with } 0 < \alpha < \beta < \infty. \tag{2.71}$$

As a consequence the constitutive equations (2.19a) and (2.19b) for the Zener model read

$$\left[1 + \frac{1}{\beta}\frac{d}{dt}\right]\sigma(t) = m\left[1 + \frac{1}{\alpha}\frac{d}{dt}\right]\epsilon(t), \quad m = G_e = G_g\frac{\alpha}{\beta}. \tag{2.72}$$

Then, limiting ourselves to consider the complex modulus, this reads

$$G^*(\omega) = G_e \frac{1 + i\omega/\alpha}{1 + i\omega/\beta} = G_g \frac{\alpha + i\omega}{\beta + i\omega}, \tag{2.73}$$

henceforth

$$G^*(\omega) = G'(\omega) + G''(\omega), \quad \begin{cases} G'(\omega) = G_g \dfrac{\omega^2 + \alpha\beta}{\omega^2 + \beta^2}, \\[2mm] G''(\omega) = G_g \dfrac{\omega(\beta - \alpha)}{\omega^2 + \beta^2}. \end{cases} \tag{2.74}$$

Finally, the loss tangent turns out to be

$$Zener\ model: \quad \tan\delta(\omega) = \frac{G''(\omega)}{G'(\omega)} = (\beta - \alpha)\frac{\omega}{\omega^2 + \alpha\beta}. \quad (2.75)$$

Now the loss tangent attains its maximum value $(\beta-\alpha)/(2\sqrt{\alpha\beta})$ for $\omega = \sqrt{\alpha\beta}$, a result consistent with that obtained with the previous notation.

It is instructive to plot in Fig. 2.6 the dynamic functions $G'(\omega)$, $G''(\omega)$ and the loss tangent $\tan\delta(\omega)$ versus ω for the Zener model. For convenience we use non-dimensional units and we adopt for ω a logarithmic scale from 10^{-2} to 10^2. We take $\alpha = 1/2$, $\beta = 2$ so $\alpha\beta \equiv 1$, and $G_g = 1$ so $G_e = \alpha\beta = 1/4$.

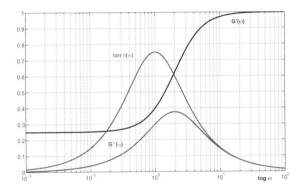

Fig. 2.6 Plots of the dynamic functions $G'(\omega)$, $G''(\omega)$ and loss tangent $\tan\delta(\omega)$ versus $\log\omega$ for the Zener model.

As expected, from Eq. (2.75) we easily recover the expressions of the loss tangent for the limiting cases of the Hooke, Newton, Voigt and Maxwell models. We obtain:

$$Hooke\ model\ (\alpha = \beta = 0): \quad \tan\delta(\omega) = 0, \quad (2.76)$$

$$Newton\ model\ (0 = \alpha < \beta = \infty): \quad \tan\delta(\omega) = \infty, \quad (2.77)$$

$$Voigt\ model\ (0 < \alpha < \beta = \infty): \quad \tan\delta(\omega) = \frac{\omega}{\alpha} = \omega\tau_\epsilon, \quad (2.78)$$

$$Maxwell\ model\ (0 = \alpha < \beta < \infty): \quad \tan\delta(\omega) = \frac{\beta}{\omega} = \frac{1}{\omega\tau_\sigma}. \quad (2.79)$$

We recover that the Hooke model exhibits only energy storage whereas the Newton model, only energy dissipation. The Voigt and Maxwell models exhibit both storage and dissipation of energy, in such a way that their loss tangent turns out to be directly proportional and inversely proportional to the frequency, respectively. As a consequence, with respect to the loss tangent, the Zener model exhibits characteristics common to the Voigt and Maxell models in the extremal frequency regions: precisely, its loss tangent is increasing for very low frequencies (like for the Voigt model), is decreasing for very high frequencies (like for the Maxwell model), and attains its (finite) maximum value within an intermediate frequency range.

2.9 Notes

The approach to linear viscoelasticity based on memory functions (the "hereditary" approach) was started by V. Volterra, e.g. [Volterra (1913); Volterra (1928); Volterra (1959)] and pursued in Italy by a number of mathematicians, including: Cisotti, Giorgi, Graffi, Tricomi, Benvenuti, Fichera, Caputo, Fabrizio and Morro.

Many results of the Italian school along with the recent theoretical achievements of the "hereditary" approach are well considered in the book [Fabrizio and Morro (1992)] and in the papers [Deseri et al. (2006)], [Fabrizio et al. (2009)].

Our presentation is mostly based on our past review papers [Caputo and Mainardi (1971b); Mainardi (2002a)] and on classical books [Bland (1960); Gross (1953); Pipkin (1986); Tschoegel (1989)].

For the topic of *realizability of the viscoelastic models* and for the related concept of complete monotonicity the reader is referred to the papers by A. Hanyga, see e.g. [Hanyga (2005a); Hanyga (2005b); Hanyga (2005c)] and the references therein.

We have not considered (in the present edition) the topic of *ladder networks*: the interested reader is invited to consult the excellent treatise [Tschoegel (1989)] and the references therein. We note that in the literature of ladder networks, the pioneering contributions by the late Ellis Strick, Professor of Geophysics at the University of Pittsburgh, are unfortunately not mentioned: these contributions turn out to be hidden in his unpublished lecture notes [Strick (1976)].

To conclude, applications of the linear theory of viscoelasticity appear in several fields of material sciences such as chemistry (e.g. [Doi and Edwards (1986); Ferry (1980)], seismology (e.g. [Aki and Richards (1980); Carcione (2007)]), soil mechanics (e.g. [Klausner (1991)]), arterial rheology (e.g. [Craiem et al. (2008)]), food rheology (e.g. [Rao and Steffe (1992)]), to mention just a few. Because papers are spread out in a large number of journals, any reference list cannot be exhaustive.

Chapter 3

Fractional Viscoelastic Models

Linear viscoelasticity is certainly the field of the most extensive applications of fractional calculus, in view of its ability to model hereditary phenomena with long memory.

Our analysis, based on the classical linear theory of viscoelasticity recalled in Chapter 2, will start from the power law creep to justify the introduction of the operators of fractional calculus into the stress-strain relationship. So doing, we will arrive at the fractional generalization of the classical mechanical models through a correspondence principle. We will devote particular attention to the generalization of the Zener model (Standard Linear Solid) of which we will provide a physical interpretation.

We will also consider the effects of the initial conditions in properly choosing the mathematical definition for the fractional derivatives that are expected to replace the ordinary derivatives in the classical models.

3.1 The fractional calculus in the mechanical models

3.1.1 *Power-Law creep and the Scott-Blair model*

Let us consider the viscoelastic solid with *creep compliance*,

$$J(t) = \frac{a}{\Gamma(1+\nu)} t^\nu, \quad a > 0, \quad 0 < \nu < 1, \tag{3.1}$$

where the coefficient in front of the power-law function has been introduced for later convenience. Such creep behaviour is found to be of great interest in a number of creep experiments; usually it is referred to as the *power-law creep*. This law is compatible with the mathematical theory presented in Section 2.5, in that there exists a corresponding non-negative *retardation spectrum* (in time and frequency). In fact, by using the method of Laplace integral pairs and the reflection formula for the Gamma function,

$$\Gamma(\nu)\,\Gamma(1-\nu) = \frac{\pi}{\sin \pi \nu}\,,$$

we find

$$R_\epsilon(\tau) = \frac{\sin \pi \nu}{\pi}\,\frac{1}{\tau^{1-\nu}} \iff S_\epsilon(\gamma) = a\,\frac{\sin \pi \nu}{\pi}\,\frac{1}{\gamma^{1+\nu}}\,. \qquad (3.2)$$

In virtue of the *reciprocity relationship* (2.8) in the Laplace domain we can find for such viscoelastic solid its *relaxation modulus*, and then the corresponding *relaxation spectrum*. After simple manipulations we get

$$G(t) = \frac{b}{\Gamma(1-\nu)}\,t^{-\nu}\,, \quad b = \frac{1}{a} > 0\,, \qquad (3.3)$$

and

$$R_\sigma(\tau) = \frac{\sin \pi \nu}{\pi}\,\frac{1}{\tau^{1+\nu}} \iff S_\sigma(\gamma) = b\,\frac{\sin \pi \nu}{\pi}\,\frac{1}{\gamma^{1-\nu}}\,. \qquad (3.4)$$

For our viscoelastic solid exhibiting power-law creep, the stress-strain relationship in the *creep representation* can be easily obtained by inserting the creep law (3.1) into the integral in (2.4a). We get:

$$\epsilon(t) = \frac{a}{\Gamma(1+\nu)}\int_{-\infty}^{t}(t-\tau)^\nu\,d\sigma\,. \qquad (3.5)$$

Writing $d\sigma = \dot{\sigma}(t)\,dt$ and integrating by parts, we finally have

$$\epsilon(t) = \frac{a}{\Gamma(1+\nu)}\int_{-\infty}^{t}(t-\tau)^{\nu-1}\,\sigma(\tau)\,d\tau = a \cdot {}_{-\infty}I_t^\nu\,[\sigma(t)]\,, \qquad (3.6)$$

where ${}_{-\infty}I_t^\nu$ denotes the *fractional integral* of order ν with starting point $-\infty$, the so-called Liouville-Weyl integral introduced in Section 1.3.

In the *relaxation representation* the stress-strain relationship is now obtained from (2.4b) and (3.3). Writing $d\epsilon = \dot{\epsilon}(t)\,dt$, we get

$$\sigma(t) = \frac{b}{\Gamma(1-\nu)}\int_{-\infty}^{t}(t-\tau)^{-\nu}\,\dot{\epsilon}(\tau)\,d\tau = b \cdot {}_{-\infty}D_t^\nu\,[\epsilon(t)]\,, \qquad (3.7)$$

where

$$_{-\infty}D_t^\nu := {_{-\infty}I_t^{1-\nu}} \circ D_t = D_t \circ {_{-\infty}I_t^{1-\nu}}, \text{ with } D_t := \frac{d}{dt}, \quad (3.8)$$

denotes the *fractional derivative* of order ν with starting point $-\infty$, the so-called Liouville-Weyl derivative introduced in Section 1.4.

From now on we will consider *causal histories*, so the starting point in Eqs. (3.5)-(3.8) is 0 instead of $-\infty$. This implies that the Liouville-Weyl integral and the Liouville-Weil derivative must be replaced by the Riemann-Liouville integral $_0I_t^\nu$, introduced in Section 1.1, and by the Riemann-Liouville (R-L) or by the Caputo (C) derivative, introduced in Section 1.2, denoted respectively by $_0D_t^\nu$ and $_0^*D_t^\nu$. Later, in Section 2.5, we will show the equivalence between the two types of fractional derivatives as far as we remain in the framework of our constitutive equations and our preference for the use of fractional derivative in the Caputo sense. Thus, for causal histories, we write

$$\epsilon(t) = a \cdot {_0I_t^\nu} \left[\sigma(t)\right], \quad (3.9)$$

$$\sigma(t) = b \cdot {_0D_t^\nu} \epsilon(t) = b \cdot {_0^*D_t^\nu} \left[\epsilon(t)\right], \quad (3.10)$$

where $ab = 1$.

Some authors, e.g. [Bland (1960)], refer to Eq. (3.10) (with the R-L derivative) as the *Scott-Blair stress-strain law*. Indeed Scott-Blair was the scientist who, in the middle of the past century, proposed such a constitutive equation to characterize a viscoelastic material whose mechanical properties are intermediate between those of a pure elastic solid (Hooke model) and a pure viscous fluid (Newton model).

3.1.2 The correspondence principle

The use of fractional calculus in linear viscoelasticity leads us to generalize the classical mechanical models, in that the basic Newton element (dashpot) is substituted by the more general Scott-Blair element (of order ν), sometimes referred to as *pot*. In fact, we can construct the class of these generalized models from Hooke and Scott-Blair elements, disposed singly and in branches of two (in series or in parallel).

The material functions are obtained using the combination rule; their determination is made easy if we take into account the following *correspondence principle* between the classical and fractional mechanical models, as introduced in [Caputo and Mainardi (1971b)], that is empirically justified. Taking $0 < \nu \leq 1$, such a correspondence principle can be formally stated by the following three equations where Laplace transform pairs are outlined:

$$\delta(t) \div 1 \Rightarrow \frac{t^{-\nu}}{\Gamma(1-\nu)} \div s^{1-\nu}, \qquad (3.11)$$

$$t \div \frac{1}{s^2} \Rightarrow \frac{t^{\nu}}{\Gamma(1+\nu)} \div \frac{1}{s^{\nu+1}}, \qquad (3.12)$$

$$e^{-t/\tau} \div \frac{1}{s+1/\tau} \Rightarrow E_\nu[-(t/\tau)^\nu] \div \frac{s^{\nu-1}}{s^\nu + (1/\tau)^\nu}, \qquad (3.13)$$

where $\tau > 0$ and E_ν denotes the Mittag-Leffler function of order ν.

In Fig. 3.1, we display plots of the function $E_\nu(-t^\nu)$ versus t for some (rational) values of ν.

Referring the reader to Appendix E for more details on this function, here we recall its asymptotic representations for small and large times,

$$E_\nu(-t^\nu) \sim 1 - \frac{t^\nu}{\Gamma(1+\nu)}, \quad t \to 0^+; \qquad (3.14)$$

$$E_\nu(-t^\nu) \sim \frac{t^{-\nu}}{\Gamma(1-\nu)}, \quad t \to +\infty. \qquad (3.15)$$

We easily recognize that, compared to the exponential obtained for $\nu = 1$, the *fractional* relaxation function $E_\nu(-t^\nu)$ exhibits a very different behaviour. In fact, for $0 < \nu < 1$, as shown in Eqs. (3.14) and (3.15) our function exhibits for small times a much faster decay (the derivative tends to $-\infty$ in comparison with -1), and for large times a much slower decay (algebraic decay in comparison with exponential decay).

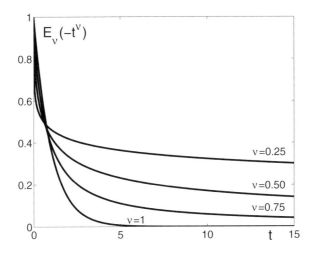

Fig. 3.1 The Mittag-Leffler function $E_\nu(-t^\nu)$ versus t ($0 \le t \le 15$) for some rational values of ν, i.e. $\nu = 0.25,\, 0.50,\, 0.75,\, 1$.

3.1.3 The fractional mechanical models

We now consider the fractional generalizations of the Newton, Voigt, Maxwell, Zener and anti-Zener models. For this purpose it is sufficient to replace the derivative of order 1 with the fractional derivative of order $\nu \in (0,1)$ (in the R-L or C sense) in their constitutive equations (2.16a)-(2.20a) and then make use of the correspondence principle stated by Eqs. (3.11)-(3.13). We then obtain the following stress-strain relationships and corresponding material functions:

$$fractional\ Newton\ (Scott-Blair)\ model : \sigma(t) = b_1 \frac{d^\nu \epsilon}{dt^\nu}, \quad (3.16a)$$

$$\begin{cases} J(t) = \dfrac{t^\nu}{b_1\, \Gamma(1+\nu)}, \\ G(t) = b_1 \dfrac{t^{-\nu}}{\Gamma(1-\nu)}; \end{cases} \quad (3.16b)$$

$$fractional\ Voigt\ model : \sigma(t) = m\, \epsilon(t) + b_1 \frac{d^\nu \epsilon}{dt^\nu}, \quad (3.17a)$$

$$\begin{cases} J(t) = \dfrac{1}{m}\left\{1 - \mathrm{E}_\nu\left[-(t/\tau_\epsilon)^\nu\right]\right\}, \\ G(t) = m + b_1 \dfrac{t^{-\nu}}{\Gamma(1-\nu)}, \end{cases} \quad (3.17b)$$

where $(\tau_\epsilon)^\nu = b_1/m$;

$fractional\ Maxwell\ model\ :\ \sigma(t) + a_1 \dfrac{d^\nu \sigma}{dt^\nu} = b_1 \dfrac{d^\nu \epsilon}{dt^\nu}\,,$ (3.18a)

$$\begin{cases} J(t) = \dfrac{a}{b_1} + \dfrac{1}{b}\dfrac{t^\nu}{\Gamma(1+\nu)}\,, \\ G(t) = \dfrac{b_1}{a_1}\,E_\nu\left[-(t/\tau_\sigma)^\nu\right]\,, \end{cases}$$ (3.18b)

where $(\tau_\sigma)^\nu = a_1$;

$fractional\ Zener\ model\ :$

$$\left[1 + a_1 \dfrac{d^\nu}{dt^\nu}\right]\sigma(t) = \left[m + b_1 \dfrac{d^\nu}{dt^\nu}\right]\epsilon(t)\,,$$ (3.19a)

$$\begin{cases} J(t) = J_g + J_1\left[1 - E_\nu\left[-(t/\tau_\epsilon)^\nu\right]\right]\,, \\ G(t) = G_e + G_1\,E_\nu\left[-(t/\tau_\sigma)^\nu\right]\,, \end{cases}$$ (3.19b)

where

$$\begin{cases} J_g = \dfrac{a_1}{b_1}\,,\ J_1 = \dfrac{1}{m} - \dfrac{a_1}{b_1}\,,\ \tau_\epsilon = \dfrac{b_1}{m}\,, \\ G_e = m\,,\ G_1 = \dfrac{b_1}{a_1} - m\,,\ \tau_\sigma = a_1\,; \end{cases}$$

$fractional\ anti-Zener\ model\ :$

$$\left[1 + a_1\dfrac{d^\nu}{dt^\nu}\right]\sigma(t) = \left[b_1\dfrac{d\nu}{dt^\nu} + b_2\dfrac{d^{2\nu}}{dt^{2\nu}}\right]\epsilon(t)\,,$$ (3.20a)

$$\begin{cases} J(t) = J_+ \dfrac{t^\nu}{\Gamma(1+\nu)} + J_1\left[1 - E_\nu\left[-(t/\tau_\epsilon^\nu)\right]\right]\,, \\ G(t) = G_- \dfrac{t^{-\nu}}{\Gamma(1-\nu)} + G_1\,E_\nu\left[-(t/\tau_\sigma)^\nu)\right]\,, \end{cases}$$ (3.20b)

where

$$\begin{cases} J_+ = \dfrac{1}{b_1}\,,\ J_1 = \dfrac{a_1}{b_1} - \dfrac{b_2}{b_1^2}\,,\ \tau_\epsilon = \dfrac{b_2}{b_1}\,, \\ G_- = \dfrac{b_2}{a_1}\,,\ G_1 = \dfrac{b_1}{a_1} - \dfrac{b_2}{a_1^2}\,,\ \tau_\sigma = a_1\,. \end{cases}$$

Extending the procedures of the classical mechanical models, we get the *fractional operator equation* in the form that properly generalizes Eq. (2.25):

$$\left[1 + \sum_{k=1}^{p} a_k \frac{d^{\nu_k}}{dt^{\nu_k}}\right] \sigma(t) = \left[m + \sum_{k=1}^{q} b_k \frac{d^{\nu_k}}{dt^{\nu_k}}\right] \epsilon(t), \quad (3.21)$$

with $\nu_k = k + \nu - 1$, so, as a generalization of Eq. (2.21):

$$\begin{cases} J(t) = J_g + \sum_n J_n \left\{1 - \mathrm{E}_\nu\left[-(t/\tau_{\epsilon,n})^\nu\right]\right\} + J_+ \dfrac{t^\nu}{\Gamma(1+\nu)}, \\ G(t) = G_e + \sum_n G_n \, \mathrm{E}_\nu\left[-(t/\tau_{\sigma,n})^\nu\right] + G_- \dfrac{t^{-\nu}}{\Gamma(1-\nu)}, \end{cases} \quad (3.22)$$

where all the coefficients are non-negative. Of course, also for the fractional operator equation (3.21), we distinguish the same four cases of the classical operator equation (2.25), summarized in Table 2.2.

3.2 Analysis of the fractional Zener model

We now focus on the fractional Zener model. From the results for this model we can easily obtain not only those for the most simple fractional models (Scott-Blair, Voigt, Maxwell) as particular cases, but, by extrapolation, also those referring to more general models that are governed by the fractional operator equation (3.21).

3.2.1 *The material and the spectral functions*

We now consider for the fractional Zener model its creep compliance and relaxation modulus with the corresponding time-spectral functions. Following the notation of Section 2.5 we have $J(t) = J_g + J_\tau(t)$ and $G(t) = G_e + G_\tau(t)$ where

$$\begin{cases} J_\tau(t) = J_1 \left\{1 - \mathrm{E}_\nu\left[-(t/\tau_\epsilon)^\nu\right]\right\} = J_1 \displaystyle\int_0^\infty R_\epsilon(\tau)(1 - e^{-t/\tau})d\tau, \\ G_\tau(t) = G_1 \mathrm{E}_\nu\left[-(t/\tau_\sigma)^\nu\right] = G_1 \displaystyle\int_0^\infty R_\sigma(\tau)\, e^{-t/\tau} d\tau, \end{cases} \quad (3.23)$$

with $J_1 = J_e - J_g$, $G_1 = G_g - G_e$. The creep compliance $J(t)$ and the relaxation modulus $G(t)$ are depicted in Fig. 3.2 for some rational values of ν.

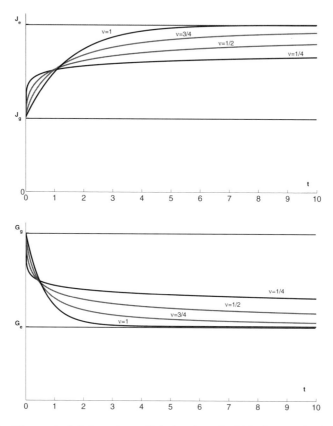

Fig. 3.2 The material functions $J(t)$ (top) and $G(t)$ (bottom) of the fractional Zener model versus t ($0 \leq t \leq 10$) for some rational values of ν, i.e. $\nu = 0.25, 0.50, 0.75, 1$.

Using the method of Laplace transforms illustrated in Section 2.5, we can obtain the time–spectral functions of the fractional Zener model. Denoting the suffixes ϵ, σ by a star, we obtain

$$R_*(\tau) = \frac{1}{\pi \tau} \frac{\sin \nu \pi}{(\tau/\tau_*)^\nu + (\tau/\tau_*)^{-\nu} + 2 \cos \nu \pi}, \qquad (3.24)$$

$$\hat{R}_*(u) = \frac{1}{2\pi} \frac{\sin \nu\pi}{\cosh \nu u + \cos \nu\pi}, \quad u = \log(\tau/\tau_*). \quad (3.25)$$

Plots of the spectral function $R_*(\tau)$ are shown in Fig. 3.3 for some rational values of $\nu \in (0,1)$ taking $\tau_* = 1$.

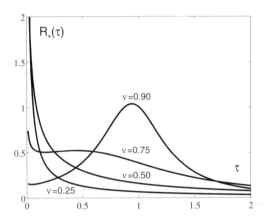

Fig. 3.3 The time–spectral function $\hat{R}_*(\tau)$ of the fractional Zener model versus τ ($0 \leq \tau \leq 2$) for some rational values of ν, i.e. $\nu = 0.25$, 0.50, 0.75, 0.90.

From the plots of the spectra we can easily recognize the effect of a variation of ν on their character; for $\nu \to 1$ the spectra become sharper and sharper until for $\nu = 1$ they reduce to be discrete with a single retardation/relaxation time. In fact we get

$$\lim_{\nu \to 1} R_*(\tau) = \delta(\tau - 1), \quad \lim_{\nu \to 1} \hat{R}_*(u) = \delta(u). \quad (3.26)$$

We recognize from (3.24) that the spectrum $R_*(\tau)$ is a decreasing function of τ for $0 < \nu < \nu_0$ where $\nu_0 \approx 0.736$ is the non-zero solution of equation $\nu = \sin \nu\pi$. Subsequently, with increasing ν, it first exhibits a minimum and then a maximum before tending to the impulsive function $\delta(\tau - 1)$ as $\nu \to 1$. The spectra (3.24) and (3.25) have already been calculated in [Gross (1947a)], where, in the attempt to eliminate the faults which a power law shows for the creep function, B. Gross proposed the Mittag-Leffler function as a general empirical law for both the creep and relaxation functions. Here we have newly derived this result by introducing a memory mechanism into the stress-strain relationships by means of the fractional derivative, following [Caputo and Mainardi (1971a)].

3.2.2 Dissipation: theoretical considerations

Let us now compute the loss tangent for the fractional Zener model starting from its complex modulus $G^*(\omega)$. For this purpose it is sufficient to properly generalize, with the fractional derivative of order ν, the corresponding formulas valid for the standard Zener model, presented in Section 2.8. Following the approach expressed by Eqs. (2.71)-(2.79), we then introduce the parameters

$$\begin{cases} \alpha := 1/\tau_\epsilon^\nu = m/b_1\,, \\ \beta := 1/\tau_\sigma^\nu = 1/a_1\,, \end{cases} \quad \text{with} \quad 0 < \alpha < \beta < \infty\,. \tag{3.27}$$

As a consequence, the constitutive equation (3.19a)-(3.19b) for the fractional Zener model reads

$$\left[1 + \frac{1}{\beta}\frac{d^\nu}{dt^\nu}\right]\sigma(t) = m\left[1 + \frac{1}{\alpha}\frac{d^\nu}{dt^\nu}\right]\epsilon(t)\,, \quad m = G_e = G_g\frac{\alpha}{\beta}\,. \tag{3.28}$$

Then, the complex modulus is

$$G^*(\omega) = G_e \frac{1 + (i\omega)^\nu/\alpha}{1 + (i\omega)^\nu/\beta} = G_g \frac{\alpha + (i\omega)^\nu}{\beta + (i\omega)^\nu}\,, \tag{3.29}$$

henceforth,

$$G^*(\omega) = G'(\omega) + G''(\omega)\,, \quad \text{with} \quad \begin{cases} G'(\omega) = G_g \dfrac{\omega^2 + \alpha\beta}{\omega^2 + \beta^2}\,, \\ G''(\omega) = G_g \dfrac{\omega(\beta - \alpha)}{\omega^2 + \beta^2}\,. \end{cases} \tag{3.30}$$

Finally, the loss tangent is obtained from the known relationship (2.49)

$$\tan\delta(\omega) = \frac{G''(\omega)}{G'(\omega)}\,.$$

Then we get:

fractional Zener model :

$$\tan\delta(\omega) = (\beta - \alpha)\frac{\omega^\nu \sin(\nu\pi/2)}{\omega^{2\nu} + \alpha\beta + (\alpha + \beta)\omega^\nu \cos(\nu\pi/2)}\,. \tag{3.31}$$

For consistency of notations such expression would be compared with (2.75) rather than with (2.70), both valid for the Zener model.

As expected, from Eq. (3.31) we easily recover the expressions of the loss tangent for the limiting cases of the fractional Zener model, that is the loss tangent for the Scott-Blair model (intermediate between the Hooke and Newton models), and for the fractional Voigt and Maxwell models. We obtain:

fractional Newton Scott−Blair model $(0 = \alpha < \beta = \infty)$:
$$\tan \delta(\omega) = \tan(\nu\pi/2) ; \tag{3.32}$$

fractional Voigt model $(0 < \alpha < \beta = \infty)$:
$$\tan \delta(\omega) = \frac{\omega^\nu \sin(\nu\pi/2)}{\alpha + \omega^\nu \cos(\nu\pi/2)}, \tag{3.33}$$

fractional Maxwell model $(0 = \alpha < \beta < \infty)$:
$$\tan \delta(\omega) = \frac{\beta\,\omega^\nu \sin(\nu\pi/2)}{\omega^{2\nu} + \beta\,\omega^\nu \cos(\nu\pi/2)}. \tag{3.34}$$

We note that the Scott-Blair model exhibits a constant loss tangent, that is, quite independent of frequency, a noteworthy property that can be used in experimental checks when ν is sufficiently close to zero. As far as the fractional Voigt and Maxwell models $(0 < \nu < 1)$ are concerned, note that the dependence of loss tangent of frequency is similar but more moderate than those for the standard Voigt and Maxwell models $(\nu = 1)$ described in Eqs. (2.78), (2.79) respectively. The same holds for the fractional Zener model in comparison with the corresponding standard model described in Eq. (2.75).

Consider again the fractional Zener model. Indeed, in view of experimental checks for viscoelastic solids exhibiting a low value for the loss tangent, say less than 10^{-2}, we find it reasonable to approximate the exact expression (3.31) of the loss tangent for the fractional Zener model as follows:
$$\tan \delta(\omega) \simeq (\beta - \alpha) \frac{\omega^\nu \sin(\nu\pi/2)}{\omega^{2\nu} + \alpha^2 + 2\alpha\,\omega^\nu \sin(\nu\pi/2)}. \tag{3.35}$$

This approximation is well justified as soon as the condition
$$\Delta := \frac{\beta - \alpha}{\alpha} \ll 1 \tag{3.36}$$

is satisfied, corresponding to the so-called *nearly elastic* case of our model, in analogy with the standard Zener model (S.L.S.). In such

approximation we set
$$\begin{cases} \omega_0^\nu = \alpha \\ \Delta = \dfrac{\beta - \alpha}{\alpha} \simeq \dfrac{\beta - \alpha}{\sqrt{\alpha \beta}}, \end{cases} \quad (3.37)$$

so that
$$\tan \delta(\omega) \simeq \Delta \frac{(\omega/\omega_0)^\nu \sin(\nu\pi/2)}{1 + (\omega/\omega_0)^{2\nu} + 2(\omega/\omega_0)^\nu \cos(\nu\pi/2)}. \quad (3.38)$$

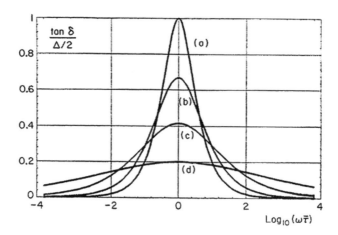

Fig. 3.4 Plots of the loss tangent $\tan \delta(\omega)$ scaled with $\Delta/2$ against the logarithm of $\omega\bar{\tau}$, for some rational values of ν: a) $\nu = 1$, b) $\nu = 0.75$, c) $\nu = 0.50$, d) $\nu = 0.25$.

It is easy to recognize that ω_0 is the frequency at which the loss tangent (3.34) assumes its maximum given by
$$\tan \delta(\omega)|_{max} = \frac{\Delta}{2} \frac{\sin(\nu\pi/2)}{1 + \cos(\nu\pi/2)}. \quad (3.39)$$

It may be convenient to replace in (3.38) the peak frequency ω_0 with $1/\bar{\tau}$ where $\bar{\tau}$ is a characteristic time intermediate between τ_ϵ and τ_σ. In fact, in the approximation $\alpha \simeq \beta$ we get from (3.27)
$$\omega_0 := 1/\tau_\epsilon \simeq 1/\tau_\sigma \simeq 1/\sqrt{\tau_\epsilon \tau_\sigma}. \quad (3.40)$$

Then, in terms of $\bar{\tau}$, the loss tangent in the nearly elastic approximation reads
$$\tan \delta(\omega) \simeq \Delta \frac{(\omega\bar{\tau})^\nu \sin(\nu\pi/2)}{1 + (\omega\bar{\tau})^{2\nu} + 2(\omega\bar{\tau})^\nu \cos(\nu\pi/2)}. \quad (3.38')$$

When the loss tangent is plotted against the logarithm of $\omega/\omega_0 = \omega\bar{\tau}$, it is seen to be a symmetrical function around its maximum value attained at $\omega/\omega_0 = \omega\bar{\tau} = 1$, as shown in Fig. 3.4 for some rational values of ν and for fixed Δ. We note that the peak decreases in amplitude and broadens with a rate depending on ν; for $\nu = 1$ we recover the classical Debye peak of the classical Zener solid.

For the sake of convenience, in view of applications to experimental data, in Fig. 3.5 we report the normalized loss tangent obtained when the maximum amplitude is kept constant, for the previous rational values of ν.

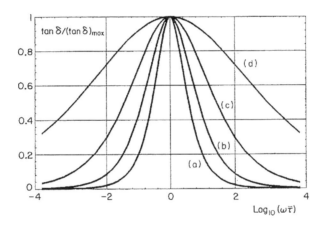

Fig. 3.5 Plots of the loss tangent $\tan\delta(\omega)$ scaled with it maximum against the logarithm of $\omega\bar{\tau}$, for some rational values of ν: a) $\nu = 1$, b) $\nu = 0.75$, c) $\nu = 0.50$, d) $\nu = 0.25$.

3.2.3 Dissipation: experimental checks

Experimental data on the loss tangent are available for various viscoelastic solids; however, measurements are always affected by considerable errors and, over a large frequency range, are scarce because of considerable experimental difficulties. In experiments one prefers to adopt the term *specific dissipation function* Q^{-1} rather than loss tangent, assuming they are equivalent as discussed in Section 2.7, see Eqs. (2.62)-(2.63). We also note that indirect methods of measuring the specific dissipation are used as those based on free oscillations and resonance phenomena, see e.g. [Kolsky (1953); Zener (1948)]. By these methods [Bennewitz-Rotger (1936), (1938)] measured the Q for transverse vibrations in reeds of several metals in the frequency range of three decades. Their data were fitted in [Caputo and Mainardi (1971b)] by using the expression (3.38) in view of the low values of dissipation. Precisely, in their attempt, Caputo and Mainardi computed a fit of (3.38) to the experimental curves by using the parameters Δ, α, ν as follows. From each datum they found ω_0, Q_{max}^{-1} then, (3.39) is a relationship between Δ and ν. The theoretical curve, forced to pass through the maximum of the experimental curve, was then fitted to this by using the other free parameter.

Herewith we report only the fits obtained for brass and steel, as shown in Figs. 3.6 and 3.7, respectively, where a dashed line is used for the experimental curves and a continuous line for the theoretical ones. The values of the parameter ν are listed in Table 3.1.

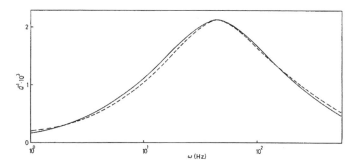

Fig. 3.6 Q^{-1} in brass: comparison between theoretical (continuous line) and experimental (dashed line) curves.

Table 3.1 Parameters for the data fit after Bennewitz and Rotger.

Metal	$\Delta\,(s^{-\nu})$	$\alpha\,(s^{-\nu})$	ν	$f_{max}\,(Hz)$	Q_{max}^{-1}
brass	0.77	153.2	0.90	42.7	$2.14 \cdot 10^{-3}$
steel	0.19	54.3	0.80	23.4	$1.35 \cdot 10^{-3}$

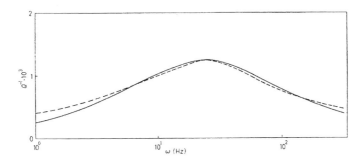

Fig. 3.7 Q^{-1} in steel: comparison between theoretical (continuous line) and experimental (dashed line) curves.

3.3 The physical interpretation of the fractional Zener model via fractional diffusion

According to [Zener (1948)] the physical interpretation of anelasticity in metals is linked to a spectrum of relaxation phenomena. In particular, the thermal relaxation due to diffusion in the thermoelastic coupling is essential to derive the standard constitutive equation (stress-strain relationship) in linear viscoelasticity. This equation corresponds to a simple rheological model (with three independent parameters) known also as Standard Linear Solid (S.L.S.), discussed in Section 2.4, see Eqs. (2.19a)-(2.19b), and in Section 2.8. We now re-write its constitutive equation in the form

$$\sigma + \tau_\epsilon \frac{d\sigma}{dt} = M_r \left(\epsilon + \tau_\sigma \frac{d\epsilon}{dt} \right), \qquad (3.41)$$

where $\sigma = \sigma(t)$ and $\epsilon = \epsilon(t)$ denote the uni-axial stress and strain respectively. The three parameters are M_r, which represents the relaxed modulus, and τ_σ, τ_ϵ, which denote the relaxation times under constant stress and strain respectively; an additional parameter is the unrelaxed modulus M_u given by $\tau_\sigma/\tau_\epsilon = M_u/M_r > 1$.

Following Zener, the model equation (3.41) can be derived from the basic equations of the thermoelastic coupling, provided that τ_σ and τ_ϵ also represent the relaxation times for temperature relaxation at constant stress and strain, respectively, and M_r and M_u represent the isothermal and adiabatic moduli, respectively.

Denoting by ΔT the deviation of the temperature from its standard value, the two basic equations of thermoelasticity are

$$\epsilon = \frac{1}{M_r}\sigma + \lambda \Delta T, \tag{3.42}$$

$$\frac{d}{dt}\Delta T = -\frac{1}{\tau_\epsilon}\Delta T - \gamma \frac{d\epsilon}{dt}, \tag{3.43}$$

where λ is the linear thermal expansion coefficient and $\gamma = (\partial T/\partial \epsilon)_{adiab}$. Equation (3.43) results from the combination of the two basic phenomena which induce temperature changes, (a) relaxation due to diffusion

$$\left(\frac{d}{dt}\Delta T\right)_{diff} = -\frac{1}{\tau_\epsilon}\Delta T, \tag{3.44}$$

and (b) adiabatic strain change

$$\left(\frac{d}{dt}\Delta T\right)_{adiab} = -\gamma \frac{d\epsilon}{dt}. \tag{3.45}$$

Putting $1 + \lambda\gamma = \tau_\sigma/\tau_\epsilon = M_u/M_r$ and eliminating ΔT between (3.42) and (3.43), the relationship (3.41) is readily obtained. In this way the temperature plays the role of a hidden variable.

If now we assume, following [Mainardi (1994b)], that the relaxation due to diffusion is of long memory type and just governed by the fractional differential equation

$$\left(\frac{d^\nu}{dt^\nu}\Delta T\right)_{diff} = -\frac{1}{\overline{\tau}_\epsilon^\nu}\Delta T, \quad 0 < \nu \leq 1, \tag{3.46}$$

where $\overline{\tau}_\epsilon$ is a suitable relaxation time, we allow for a natural generalization of the simple process of relaxation, which now depends on the parameter ν, see e.g. [Mainardi (1996b); Mainardi (1997)]. As a consequence, Eq. (3.43) turns out to be modified into

$$\frac{d^\nu}{dt^\nu}\Delta T = -\frac{1}{\overline{\tau}_\epsilon^\nu}\Delta T - \gamma\frac{d^\nu \epsilon}{dt^\nu}, \tag{3.47}$$

and, *mutatis mutandis*, the stress-strain relationship turns out to be

$$\sigma + \overline{\tau}_\epsilon^\nu \frac{d^\nu \sigma}{dt^\nu} = M_r\left(\epsilon + \overline{\tau}_\sigma^\nu \frac{d^\nu \epsilon}{dt^\nu}\right), \tag{3.48}$$

where we have used $1 + \lambda\gamma = (\overline{\tau}_\sigma/\overline{\tau}_\epsilon)^\nu = M_u/M_r$. So doing, we have obtained the so-called fractional Zener model, analysed in Section 3.2.

3.4 Which type of fractional derivative? Caputo or Riemann-Liouville?

In the previous sections we have investigated some physical and mathematical aspects of the use of fractional calculus in linear viscoelasticity. We have assumed that our systems are at rest for time $t < 0$. As a consequence, there is no need for including the treatment of *pre-history* as it is required in the so-called *initialised fractional calculus*, recently introduced by [Lorenzo and Hartley (2000)] and [Fukunaga (2002)].

We note that the initial conditions at $t = 0^+$ for the stress and strain do not explicitly enter into the fractional operator equation (3.21) if they are taken in the same way as for the classical mechanical models reviewed in the previous chapter (see the remark at the end of Section 2.4). This means that the approach with the Caputo derivative, which requires in the Laplace domain the same initial conditions as the classical models, is quite correct.

On the other hand, assuming the same initial conditions, the approach with the Riemann-Liouville derivative is expected to provide the same results. In fact, in view of the corresponding Laplace transform rule (1.29) for the R-L derivative, the initial conditions do not appear in the Laplace domain. Under such conditions the two approaches appear equivalent.

The equivalence of the two approaches has been noted for the fractional Zener model in a recent note by [Bagley (2007)]. However, for us the adoption of the Caputo derivative appears to be the most suitable choice, since it is fully compatible with the classical approach. We shall return to this matter in Chapter 6, when we consider wave propagation in the Scott-Blair model.

The reader is referred to [Heymans and Podlubny (2006)] for the physical interpretation of initial conditions for fractional differential equations with Riemann-Liouville derivatives, especially in viscoelasticity.

3.5 Notes

During the twentieth-century a number of authors have (implicitly or explicitly) used the fractional calculus as an empirical method of describing the properties of viscoelastic materials. In the first half of that century the early contributors were: Gemant in USA, see [Gemant (1936); (1938)], Scott-Blair in England, see [Scott-Blair (1944); (1947); (1949)], Gerasimov and Rabotnov in the former Soviet Union, see [Gerasimov (1948)], [Rabotnov (1948)].

Gemant published a series of 16 articles entitled *Frictional Phenomena* in Journal of Applied Physics since 1941 to 1943, which were collected in a book of the same title [Gemant (1950)]. In his eighth chapter-paper [Gemant (1942)], p. 220, he referred to his previous articles [Gemant (1936); (1938)] for justifying the necessity of fractional differential operators to compute the shape of relaxation curves for some elasto-viscous fluids. Thus, the words fractional and frictional were coupled, presumably for the first time, by Gemant.

Scott-Blair used the fractional calculus approach to model the observations made by [Nutting (1921); (1943); (1946)] that the stress relaxation phenomenon could be described by fractional powers of time. He noted that time derivatives of fractional order would simultaneously model the observations of Nutting on stress relaxation and those of Gemant on frequency dependence. It is quite instructive to cite some words by Scott-Blair quoted in [Stiassnie (1979)]:

I was working on the assessing of firmness of various materials (e.g. cheese and clay by experts handling them) these systems are of course both elastic and viscous but I felt sure that judgments were made not on an addition of elastic and viscous parts but on something in between the two so I introduced fractional differentials of strain with respect to time. Later, in the same letter Scott-Blair added: *I gave up the work eventually, mainly because I could not find a definition of a fractional differential that would satisfy the mathematicians.*

The 1948 the papers by Gerasimov and Rabotnov were published in Russian, so their contents remained unknown to the majority of western scientists up to the translation into English of the treatises by Rabotnov, see [Rabotnov (1969); (1980)]. Whereas Gerasimov explicitly used a fractional derivative to define his model of viscoelasticity (akin to the Scott-Blair model), Rabotnov preferred to use the Volterra integral operators with weakly singular kernels that could be interpreted in terms of fractional integrals and derivatives. After the appearance of the books by Rabotnov it has became common to speak about *Rabotnov's theory of hereditary solid mechanics*. The relation between Rabotnov's theory and the models of fractional viscoelasticity has been briefly recalled in the recent paper [Rossikhin and Shitikova (2007)]. According to these Russian authors, Rabotnov could express his models in terms of the operators of the fractional calculus, but he considered these operators only as some mathematical abstraction.

In the late sixties, formerly Caputo, see [Caputo (1966); (1967); (1969)], then Caputo and Mainardi, see [Caputo and Mainardi (1971a); (1971b)], explicitly suggested that derivatives of fractional order (of Caputo type) could be successfully used to model the dissipation in seismology and in metallurgy. In this respect the present author likes to recall a correspondence carried out between himself (as a young post-doc student) and the Russian Academician Rabotnov, related to two courses on Rheology held at CISM (International Centre for Mechanical Sciences, Udine, Italy) in 1973 and 1974, where Rabotnov was an invited speaker but without participating, see [Rabotnov (1973); (1974)]. Rabotnov recognized the relevance of the review paper [Caputo and Mainardi (1971b)], writing in his unpublished 1974 CISM Lecture Notes:

That's why it was of great interest for me to know the paper of Caputo and Mainardi from the University of Bologna published in 1971. These authors have obtained similar results independently without knowing the corresponding Russian publications..... Then he added: *The paper of Caputo and Mainardi contains a lot of experimental data of different authors in support of their theory. On the other hand a great number of experimental curves obtained by Postnikov and his coworkers as well as by foreign authors can be found in numerous papers of Shermergor and Meshkov.*

Unfortunately, the eminent Russian scientist did not cite the 1971 paper by Caputo and Mainardi (presumably for reasons independently from his willing) in the Russian and English editions of his later book [Rabotnov (1980)].

Nowadays, several articles (originally in Russian) by Shermergor, Meshkov and their associated researchers have been re-printed in English in Journal of Applied Mechanics and Technical Physics (English translation of Zhurnal Prikladnoi Mekhaniki i Tekhnicheskoi Fiziki), see e.g. [Shermergor (1966)], [Meshkov et al. (1966)], [Meshkov (1967)], [Meshkov and Rossikhin (1968)], [Meshkov (1970)], [Zelenev et al. (1970)], [Gonsovskii and Rossikhin (1973)], available at the URL: http://www.springerlink.com/. On this respect we cite the recent review papers [Rossikhin (2010)], [Rossikhin and Shitikova (2010)] where the works of the Russian scientists on fractional viscoelasticity are examined.

The beginning of the modern applications of fractional calculus in linear viscoelasticity is generally attributed to the 1979 PhD thesis by Bagley (under supervision of Prof. Torvik), see [Bagley (1979)], followed by a number of relevant papers, e.g. [Bagley and Torvik (1979); (1983a); (1983b)] and [Torvik and Bagley (1984)]. However, for the sake of completeness, one would recall also the 1970 PhD thesis of Rossikhin under the supervision of Prof. Meshkov, see [Rossikhin (1970)], and the 1971 PhD thesis of the author under the supervision of Prof. Caputo, summarized in [Caputo and Mainardi (1971b)].

To date, applications of fractional calculus in linear and nonlinear viscoelasticity have been considered by a great and increasing number of authors to whom we have tried to refer in our huge (but not exhaustive) bibliography at the end of the book.

Chapter 4

Waves in Linear Viscoelastic Media: Dispersion and Dissipation

In this chapter we review the main aspects of wave propagation in homogeneous, semi-infinite, linear viscoelastic media. In particular, we consider the so-called *impact waves*, so named since they are generated by impact on an initially quiescent medium. The use of the techniques of integral transforms (of Laplace and Fourier type) allows us to obtain integral representations of these waves and leads in a natural way to the concepts of *wave–front velocity* and *complex index of refraction*. We will discuss the phenomena of *dispersion and dissipation* that accompany the evolution of these waves. We will extend the concepts of *phase* and *group velocity* related to dispersion to take into account the presence of dissipation characterized by the *attenuation coefficient* and the *specific dissipation function*. We also discuss the peculiar notion of *signal velocity* introduced by Brillouin.

4.1 Introduction

Impact waves in linear viscoelastic media are a noteworthy example of linear dispersive waves in the presence of dissipation. They are obtained from a one-dimensional initial-boundary value problem that we are going to deal with the techniques of Laplace and Fourier transforms. We already know from Chapter 2 that these techniques are suited to deal with the various constitutive equations for linear viscoelastic bodies.

In Section 4.2 we consider the structure of the wave equations in the original space-time domain after inversion from the Laplace domain; in particular, we provide the explicit wave equations for the most used viscoelastic models.

In Section 4.3 we introduce the Fourier integral representation of the solution which leads to the notion of the *complex refraction index*. The dispersive and dissipative properties of the viscoelastic waves are then investigated by considering *phase velocity, group velocity* and *attenuation coefficient* of the wave-mode solutions for some relevant models of viscoelasticity. We also discuss these properties related to the Klein-Gordon equation with dissipation: such equation is relevant to provide instructive examples both of normal dispersion (usually met in the absence of dissipation) and anomalous dispersion (always present on viscoelastic waves).

In Section 4.4 we deal with the problem of finding a suitable definition of the *signal velocity* for viscoelastic waves. In fact, because of the presence of anomalous dispersion and dissipation, the identification of the group velocity with the signal velocity is lost and the subject matter must be revisited. This argument is dealt with, following the original idea of Brillouin that is based on the use of the steepest–descent path to compute the solution generated by a sinusoidal impact.

4.2 Impact waves in linear viscoelasticity

4.2.1 *Statement of the problem by Laplace transforms*

Problems of impact waves essentially concern the response of a long viscoelastic rod of uniform small cross-section to dynamical (uniaxial) loading conditions. According to the elementary theory, the rod is taken to be homogeneous (of density ρ), semi-infinite in extent ($x \geq 0$), and undisturbed for $t < 0$. For $t \geq 0$ the end of the rod (at $x = 0$) is subjected to a disturbance (the input) denoted by $r_0(t)$. The response variable (the output) denoted by $r(x,t)$ may be either the displacement $u(x,t)$, the particle velocity $v(x,t) = \frac{\partial}{\partial t} u(x,t)$, the stress $\sigma(x,t)$, or the strain $\epsilon(x,t)$.

The mathematical problem consists in finding a solution for $r(x,t)$ in the region $x > 0$ and $t > 0$, which satisfies the following field equations:
- *the equation of motion*

$$\frac{\partial}{\partial x}\sigma(x,t) = \rho \frac{\partial^2}{\partial t^2} u(x,t), \qquad (4.1)$$

- *the kinematic equation*

$$\epsilon(x,t) = \frac{\partial}{\partial x} u(x,t), \qquad (4.2)$$

- *the stress–strain relationship* $F[\sigma(x,t), \epsilon(x,t)] = 0$,
which, in virtue of Eqs. (2.5a)-(2.5b), reads either in creep representation or in relaxation representation as

$$\epsilon(x,t) = \sigma(x,0^+)\,J(t) + \int_0^t J(t-\tau)\frac{\partial}{\partial \tau}\sigma(x,\tau)\,d\tau, \qquad (4.3a)$$

$$\sigma(x,t) = \epsilon(x,0^+)\,G(t) + \int_0^t G(t-\tau)\frac{\partial}{\partial \tau}\epsilon(x,\tau)\,d\tau, \qquad (4.3b)$$

with *boundary conditions*

$$r(0,t) = r_0(t), \quad \lim_{x\to\infty} r(x,t) = 0, \quad t > 0, \qquad (4.4)$$

and homogeneous *initial conditions*

$$r(x,0^+) = \frac{\partial}{\partial t} r(x,t)\big|_{t=0^+} = 0, \quad x > 0. \qquad (4.5)$$

The stress–strain relationship is known to describe the mechanical properties of the rod and, therefore, it is the constitutive equation for the assumed viscoelastic model, uniquely characterized by the *creep compliance* $J(t)$ or by the *relaxation modulus* $G(t)$. The relationship is most conveniently treated using the Laplace transform as shown in Chapter 2, see Eqs. (2.7a) and (2.7b). Then, the creep and relaxation representations read

$$\widetilde{\epsilon}(x,s) = s\widetilde{J}(s)\,\widetilde{\sigma}(x,s), \qquad (4.6a)$$

$$\widetilde{\sigma}(x,s) = s\widetilde{G}(s)\,\widetilde{\epsilon}(x,s). \qquad (4.6b)$$

Applying the Laplace transform to the other field equations and using Eqs. (4.4) and (4.5), we obtain
$$\frac{\partial}{\partial x^2}\tilde{r}(x,s) - [\mu(s)]^2\,\tilde{r}(x,s) = 0 \implies \tilde{r}(x,s) = \tilde{r}_0(s)\,\mathrm{e}^{-\mu(s)x}, \quad (4.7)$$
where $\mu(s)$ can be written in creep or relaxation representation as
$$\mu(s) = \sqrt{\rho}\,s\left[s\,\tilde{J}(s)\right]^{1/2} = \sqrt{\rho}\,s\left[s\,\tilde{G}(s)\right]^{-1/2}, \quad (4.8)$$
with
$$\mu(s) \geq 0 \quad \text{for} \quad s \geq 0, \qquad \overline{\mu(s)} = \mu(\bar{s}) \quad \text{for} \quad s \in \mathbb{C}, \quad (4.9)$$
where the over-bar denotes the complex conjugate.

The solution $r(x,t)$ is therefore given by the Bromwich representation,
$$r(x,t) = \frac{1}{2\pi i}\int_{Br}\tilde{r}_0(s)\,\mathrm{e}^{st - \mu(s)x}\,ds, \quad (4.10)$$
in which Br denotes the Bromwich path, i.e. a vertical line lying to the right of all singularities of $\tilde{r}_0(s)$ and of $\mu(s)$. Because of (4.8), the singularities of $\mu(s)$ result from the explicit expressions of the functions $s\,\tilde{J}(s)$, $s\,\tilde{G}(s)$ and, thus, of the material functions $J(t)$ and $G(t)$, respectively.

From the analysis carried out in Chapter 2, we obtain the expressions valid for viscoelastic models with *discrete* or *continuous* distributions of retardation and relaxation times, respectively. We recognize that, in the *discrete* case $s\,\tilde{J}(s)$ and $s\,\tilde{G}(s)$ are rational functions with zeros and poles interlacing along the negative real axis, while in the *continuous* case they are analytic functions with a cut in the negative real axis. Correspondingly, $\mu(s)$ exhibits on the negative real axis of the s-plane a series of cuts (connecting the zeros and poles of $s\,\tilde{J}(s)$ and $s\,\tilde{G}(s)$) or the entire semi-infinite cut.

It may be convenient in (4.10) to introduce the so-called *impulse response* (or *Green function*)
$$\mathcal{G}(x,t) \div \tilde{\mathcal{G}}(x,s) := \mathrm{e}^{-\mu(s)x}, \quad (4.11)$$
that is the solution corresponding to $r_0(t) = \delta(t)$. Consequently, we can write
$$r(x,t) = \int_0^t \mathcal{G}(x, t-\tau)\,r_0(\tau)\,d\tau = \mathcal{G}(x,t) * r_0(t), \quad (4.12)$$
where $*$ denotes as usual the (Laplace) time convolution.

A general result, which can be easily obtained from the Laplace representation (4.10), concerns the velocity of propagation of the head (or *wave-front velocity*) of the disturbance. This velocity, denoted by c, is readily obtained by considering the following limit in the complex s-plane

$$\frac{\mu(s)}{s} \to \frac{1}{c} > 0 \quad \text{as} \quad \mathcal{R}e\,[s] \to +\infty. \qquad (4.13)$$

According to the analysis carried out in Chapter 2, the limit in (4.13) holds true for viscoelastic models of types I and II, which exhibit an instantaneous elasticity, namely $0 < G(0^+) = 1/J(0^+) < \infty$, for which $s\,\widetilde{G}(s) \to G(0^+)$ and $s\,\widetilde{J}(s) \to J(0^+)$ as $\mathcal{R}e\,[s] \to +\infty$. Then, setting $G_g = G(0^+)$ and $J_g = J(0^+)$, we easily recognize

$$c = 1/\sqrt{\rho J_g} = \sqrt{G_g/\rho}, \qquad (4.14)$$

and, from the application of Cauchy's theorem in (4.10), $r(x,t) \equiv 0$ for $t < x/c$. Therefore, c represents the wave-front velocity, i.e. the maximum velocity exhibited by the wave precursors. On the other hand, the viscoelastic bodies of types III and IV, for which $J_g = 0$, exhibit an infinite wave–front velocity. Because it is difficult to conceive a body admitting an infinite propagation velocity in this respect, c is assumed finite throughout the remainder of the present analysis, with the exception of certain isolated cases.

Using Eqs. (4.8), (4.13) and (4.14), we find it convenient to set

$$\mu(s) = \frac{s}{c}\,n(s), \qquad (4.15)$$

where

$$n(s) := \left[s\widetilde{J}(s)/J_g\right]^{1/2} = \left[s\widetilde{G}(s)/G_g\right]^{-1/2} \to 1 \quad \text{as} \quad \mathcal{R}e\,[s] \to +\infty, \quad (4.16)$$

so that the solution (4.10) assumes the instructive representation:

$$r(x,t) = \frac{1}{2\pi i} \int_{Br} \widetilde{r}_0(s)\,e^{s[t - (x/c)\,n(s)]}\,ds. \qquad (4.17)$$

Of course, $n(s)$ takes on itself the multivalued nature of $\mu(s)$, exhibiting the same branch cut on the negative real axis and the same positivity and crossing-symmetry properties, i.e.

$$n(s) \geq 0 \quad \text{for} \quad s \geq 0, \quad \overline{n(s)} = n(\overline{s}), \quad s \in \mathbb{C}. \qquad (4.18)$$

We will see in Section 4.3 that $n(s)$ is related to the so-called *complex refraction index*. In the limiting case of a perfectly linear elastic medium we get $n(s) \equiv 1$.

4.2.2 The structure of wave equations in the space-time domain

After having reported on the integral representation of impact waves in the Laplace domain, here we derive the evolution equations in the original space-time domain and discuss their mathematical structure.

For this purpose, we introduce the following non-dimensional functions, related to the material functions $J(t)$ and $G(t)$:

a) Rate of Creep

$$\Psi(t) := \frac{1}{J_g} \frac{dJ}{dt} \geq 0, \qquad (4.19a)$$

b) Rate of Relaxation

$$\Phi(t) := \frac{1}{G_g} \frac{dG}{dt} \leq 0. \qquad (4.19b)$$

The Laplace transforms of these functions turn out to be related to $n(s)$ through Eq. (4.16); we obtain

$$[n(s)]^2 := \frac{s\,\widetilde{J}(s)}{J_g} = 1 + \widetilde{\Psi}(s), \qquad (4.20a)$$

and

$$[n(s)]^{-2} := \frac{s\,\widetilde{G}(s)}{G_g} = 1 + \widetilde{\Phi}(s). \qquad (4.20b)$$

As a consequence of (4.7), (4.16), (4.20a) and (4.20b), we obtain the creep and relaxation representation of the wave equations in the Laplace domain as follows

$$\frac{\partial^2}{\partial x^2} \widetilde{r}(x,s) - \frac{s^2}{c^2}\left[1 + \widetilde{\Psi}(s)\right] \widetilde{r}(x,s) = 0, \qquad (4.21a)$$

$$\left[1 + \widetilde{\Phi}(s)\right] \frac{\partial^2}{\partial x^2} \widetilde{r}(x,s) - \frac{s^2}{c^2} \widetilde{r}(x,s) = 0. \qquad (4.21b)$$

Thus, the required wave equations in the space-time domain can be obtained by inverting (4.21a) and (4.21b), respectively.

We get the following integro-differential equations of convolution type:

a) Creep representation

$$[1 + \Psi(t) *] \frac{\partial^2 r}{\partial t^2} = c^2 \frac{\partial^2 r}{\partial x^2}, \quad r = r(x,t), \qquad (4.22a)$$

b) *Relaxation representation*:
$$\frac{\partial^2 r}{\partial t^2} = c^2 \left[1 + \Phi(t)*\right] \frac{\partial^2 r}{\partial x^2}, \quad r = r(x,t). \quad (4.22b)$$

We recall that, in view of their meaning, the kernel functions $\Psi(t)$ and $\Phi(t)$ are usually referred to as the *memory functions* of the creep and relaxation representations, respectively.

4.2.3 Evolution equations for the mechanical models

Let us now consider the case of mechanical models treated in Chapter 2, Section 2.4. We recall that $s\widetilde{J}(s)$ and $s\widetilde{G}(s)$ turn out to be *rational* functions in $\mathbb{C}$ with simple poles and zeros interlacing along the negative real axis and, possibly, with a simple pole or a simple zero at $s = 0$, respectively. In particular, we write

$$s\widetilde{J}(s) = \frac{1}{s\widetilde{G}(s)} = \frac{P(s)}{Q(s)}, \text{ where } \begin{cases} P(s) = 1 + \sum_{k=1}^{p} a_k s^k, \\ Q(s) = m + \sum_{k=1}^{q} b_k s^k, \end{cases} \quad (4.23)$$

with $q = p$ for models exhibiting a glass compliance, i.e. instantaneous elasticity ($J_g = a_p/b_p > 0$) and $q = p+1$ for the others.

For all these models the general evolution equation (4.7) with (4.8) and (4.9) in the Laplace domain can be easily inverted into the space-time domain and reads

$$\left[m + \sum_{k=1}^{q} b_k \frac{\partial^k}{\partial t^k}\right] \frac{\partial^2 r}{\partial x^2} = \rho \left[1 + \sum_{k=1}^{p} a_k \frac{\partial^k}{\partial t^k}\right] \frac{\partial^2 r}{\partial t^2}, \quad r = r(x,t). \quad (4.24)$$

This is to say that for the mechanical models the integral convolutions entering the evolution equations (4.22a)-(4.22b) can be eliminated to yield the time derivatives present inside the square parenthesis in Eq. (4.24). We note that this is due to the fact that the creep and relaxation functions reduce to linear combinations of exponentials.

We now turn to the most elementary mechanical models, reporting for each of them the corresponding evolution equation for the response variable $r = r(x,t)$.

$Newton$: $\quad \sigma(t) = b \dfrac{d\epsilon}{dt}$,

$$\dfrac{\partial r}{\partial t} = D \dfrac{\partial^2 r}{\partial x^2}, \qquad D = \dfrac{b}{\rho}. \qquad (4.25)$$

In this case we obtain the classical *diffusion equation*, which is of parabolic type.

$Voigt$: $\quad \sigma(t) = m\,\epsilon(t) + b \dfrac{d\epsilon}{dt}$,

$$\dfrac{\partial^2 r}{\partial t^2} = c_0^2 \left(1 + \tau_\epsilon \dfrac{\partial}{\partial t}\right) \dfrac{\partial^2 r}{\partial x^2}, \qquad c_0^2 = \dfrac{m}{\rho}, \qquad \tau_\epsilon = \dfrac{b}{m}. \qquad (4.26)$$

Also in this case we obtain a parabolic equation, but of the third order.

$Maxwell$: $\quad \sigma(t) + a \dfrac{d\sigma}{dt} = b \dfrac{d\epsilon}{dt}$,

$$\dfrac{\partial^2 r}{\partial t^2} + \dfrac{1}{\tau_\sigma}\dfrac{\partial r}{\partial t} = c^2 \dfrac{\partial^2 r}{\partial x^2}, \qquad c^2 = \dfrac{b}{a\rho}, \qquad \tau_\sigma = a. \qquad (4.27)$$

This is the so-called *telegraph equation*, which is of hyperbolic type.

$Zener$: $\quad \left[1 + a\dfrac{d}{dt}\right]\sigma(t) = \left[m + b\dfrac{d}{dt}\right]\epsilon(t), \quad 0 < m < \dfrac{b}{a}$,

$$\dfrac{\partial}{\partial t}\left(\dfrac{\partial^2 r}{\partial t^2} - c^2 \dfrac{\partial^2 r}{\partial x^2}\right) + \dfrac{1}{a}\left(\dfrac{\partial^2 r}{\partial t^2} - c_0^2 \dfrac{\partial^2 r}{\partial x^2}\right) = 0, \quad \begin{cases} c^2 = \dfrac{b}{a\rho}, \\ c_0^2 = \dfrac{m}{\rho}. \end{cases} \qquad (4.28)$$

We recall that for the Zener model $\tau_\epsilon = b/m$, $\tau_\sigma = a$, so that $c_0/c = \tau_\sigma/\tau_\epsilon = am/b$. This common ratio will be denoted by χ, where $0 < \chi < 1$. The Maxwell model is recovered as a limit case when $\chi \to 0$. Here we have an hyperbolic equation of the third order with characteristics (related to c) and sub-characteristics (related to c_0) as pointed out in [Chin (1980)].

We note that the Maxwell and Zener models are the simplest viscoelastic models that exhibit a proper wave character with a finite wave-front velocity.

4.3 Dispersion relation and complex refraction index

4.3.1 *Generalities*

For linear problems, dispersive waves are usually recognized by the existence of elementary solutions in the form of sinusoidal wave trains whose frequency ω and wave number κ are related between them through an equation, referred to as the *dispersion relation*,

$$\mathcal{D}(\omega, \kappa) = 0. \tag{4.29}$$

The function $\mathcal{D}$ is determined by the particular equations of the problem. We write these elementary solutions in one of the following ways

$$\begin{cases} \mathcal{R}e\left\{e^{i\,[\kappa x - \omega(\kappa)t]}\right\} = \cos\left\{\kappa\,[x - V(\kappa)t]\right\},\ V(\kappa) = \dfrac{\omega(\kappa)}{\kappa}, \\ \mathcal{R}e\left\{e^{i\,[\omega t - \kappa(\omega)x]}\right\} = \cos\left\{\omega\left[t - \dfrac{x}{V(\omega)}\right]\right\},\ \dfrac{1}{V(\omega)} = \dfrac{\kappa(\omega)}{\omega}, \end{cases} \tag{4.30}$$

where the constant amplitude has been set to 1 and $V = \omega/\kappa$ denotes the *phase velocity*. The two ways correspond to the fact the dispersion relation may be solved in the form of *real roots* $\omega = \omega(\kappa)$ or $\kappa = \kappa(\omega)$, correspondingly. There will be a number of such solutions, in general, with different functions $\omega(\kappa)$ or $\kappa(\omega)$. We refer to these as *modes* and the corresponding wave trains as *wave-mode solutions*. These solutions are thus monochromatic waves propagating with phase velocity. The concept of *group velocity* $U = d\omega/d\kappa$ is usually associated to the concept of *phase velocity* $V = \omega/\kappa$ to analyse the dispersion properties of the wave-packet constructed as a superposition of monochromatic waves over a range of κ or ω. Both concepts are supposed to be familiar to the reader, who, for further details, is referred to any good treatise or survey on wave propagation, e.g. [Lighthill (1965)], [Whitham (1974)], [Thau (1974)], [Baldock and Bridgeman (1981)].

In the presence of dissipation, however, ω and κ cannot be both real and we need to distinguish which one is to be chosen as the independent real variable and which one is the dependent complex variable. This choice is related to the type of boundary value problem under consideration. For the problem of impact waves, as stated in

Section 4.2, we must assume ω real so $\kappa = \kappa(\omega)$ is to be obtained as a specific complex branch of the dispersion relation. In fact, in alternative to Eq. (4.10), there is another (perhaps more common) integral representation of the solution, which is based on the Fourier transform of causal functions. This Fourier integral representation can be formally derived from our Laplace representation setting $s = i\omega$ in (4.10). Recalling our notation for the Fourier transform pair of a generic function $f(t)$, absolutely integrable in $\mathbb{R}$:

$$f(t) \div \widehat{f}(\omega), \quad \begin{cases} \widehat{f}(\omega) = \int_{-\infty}^{+\infty} e^{-i\omega t} f(t)\, dt, \\ f(t) = \dfrac{1}{2\pi} \int_{-\infty}^{+\infty} e^{+i\omega t} \widehat{f}(\omega)\, d\omega, \end{cases} \quad \omega \in \mathbb{R},$$

Eq. (4.10) can be re-written as

$$r(x,t) = \frac{1}{2\pi} \int_{-\infty}^{+\infty} \widehat{r}_0(\omega)\, e^{i\,[\omega t - \kappa(\omega) x]}\, d\omega, \quad \kappa(\omega) = \mu(i\omega), \quad (4.31)$$

where $\widehat{r}_0(\omega)$ denotes the Fourier transform of the (causal) input disturbance $r_0(t)$ and $\kappa(\omega)$ is the *complex wave number* corresponding to the real frequency ω.

To ensure that $r(x,t)$ is a real function of x and t, the *crossing-symmetry relationship* holds

$$\overline{\kappa(\omega)} = -\kappa(-\omega), \quad \omega \in \mathbb{R}, \quad (4.32)$$

which shows that the real (imaginary) part of $\kappa(\omega) = \kappa_r(\omega) + i\,\kappa_i(\omega)$ is odd (even), respectively. We recognize that the *complex wave number* turns out to be related to the *complex modulus* $G^\star(\omega)$ and to the *complex compliance* $J^\star(\omega)$, defined in Chapter 2, see (2.69), by the following relations

$$\kappa(\omega) = \sqrt{\rho}\,\omega\, [G^\star(\omega)]^{-1/2} = \sqrt{\rho}\,\omega\, [J^\star(\omega)]^{1/2}. \quad (4.33)$$

The integral representation based on Laplace transform (4.10) and that based on Fourier transform (4.31) are to be used according to their major convenience. The Fourier representation is mostly used to show the dispersive nature of the wave motion, starting from the exponential integrand in (4.31) that in real form provides the *wave-mode solution*:

$$\begin{aligned} \mathcal{R}e\left\{ e^{i\,[\omega t - \kappa(\omega) x]} \right\} &= \mathcal{R}e\left\{ e^{\kappa_i(\omega) x}\, e^{i[\omega t - \kappa_r(\omega) x]} \right\} \\ &= e^{-\delta(\omega) x}\, \cos\{\omega\,[t - x/V(\omega)]\}, \end{aligned} \quad (4.34)$$

where

$$\begin{cases} V(\omega) := \dfrac{\omega}{\kappa_r(\omega)} \geq 0, \\ \delta(\omega) := -\kappa_i(\omega) \geq 0. \end{cases} \quad (4.35)$$

As a matter of fact the solution (4.34) represents a *pseudo-monochromatic wave* (of frequency ω), which propagates with a *phase velocity* $V(\omega)$ and with an amplitude exponentially decreasing by an *attenuation coefficient* $\delta(\omega)$.

Remark 1: We use for (4.34) the term *pseudo-monochromatic wave*, since it defines a wave strictly periodic only in time (with frequency ω), being in space exponentially attenuated.

Remark 2: Since attenuation and absorption are here considered as synonyms, we can speak about *attenuation coefficient* or *absorption coefficient*, indifferently.

Of course, the function $\mu(s)$ can also be found from the *dispersion relation*. Setting in Eq. (4.29) $\omega = -is$ and solving for $\kappa = \kappa(-is)$, we have $\mu(s) = \pm\kappa(-is)$, where the choice of sign is dictated by the condition that $\mathcal{R}e\,(\mu) > 0$ when $s > 0$, [Thau (1974)].

The Fourier representation allows us to point out the importance of the analytic function $n(s)$ for problems of wave propagation. For this purpose let us consider $n(s)$ on the imaginary axis (the frequency axis) and put it in relation with the complex wave number. Then, using (4.34) with (4.16) and (4.17), we write

$$\kappa(\omega) = \frac{\omega}{c} n^\star(\omega), \text{ where} \\ n^\star(\omega) := n(i\omega) = [G_g/G^\star(\omega)]^{1/2} = [J^\star(\omega)/J_g]^{1/2}. \quad (4.36)$$

The quantity $n^\star(\omega)$ is referred to as the *complex refraction index* of the viscoelastic medium with respect to mechanical waves, in analogy with the optical case. Of course, $n^\star(\omega) \to 1$ as $\omega \to \infty$ and only in the limiting case of a perfectly linear elastic medium $n^\star(\omega) \equiv 1$.

From Eqs. (4.32) and (4.36) we obtain the *crossing-symmetry relationship* for $n^\star(\omega)$, namely

$$\overline{n^\star(\omega)} = n^\star(-\omega) \quad \omega \in \mathbb{R}, \quad (4.37)$$

which shows that the real/imaginary part of the complex refraction index is even/odd, respectively. We note the property

$$\begin{cases} n_r(\omega) := \operatorname{Re}\left[n^\star(\omega)\right] \geq 0 & \text{for } \omega \geq 0, \\ n_i(\omega) := \operatorname{Im}\left[n^\star(\omega)\right] \leq 0 & \text{for } \omega \geq 0, \end{cases} \qquad (4.38)$$

which derives from (4.36) recalling that the same property holds for $J^\star(\omega)$.

We also point out the so-called *Krönig-Kramers* or *K–K relations*, which hold between the real and imaginary parts of the complex refraction index $n^\star(\omega)$ as a consequence of *causality*,

$$\begin{cases} n_r(\omega) &= 1 - \dfrac{2}{\pi} \displaystyle\int_0^\infty \dfrac{\omega' \, n_i(\omega') - \omega \, n_i(\omega)}{\omega'^2 - \omega^2} \, d\omega', \\ n_i(\omega) &= \dfrac{2\,\omega}{\pi} \displaystyle\int_0^\infty \dfrac{n_r(\omega') - n_r(\omega)}{\omega'^2 - \omega^2} \, d\omega', \end{cases} \qquad (4.39)$$

where the integrals are intended as Cauchy principal values. Similar relations are also expected to hold between the real and imaginary parts of the complex wave number $\kappa(\omega)$ and the dynamic functions $J^\star(\omega)$, $G^\star(\omega)$, provided that they refer to *causal* models of viscoelasticity, i.e. to viscoelastic models of type I and II. Furthermore, Eqs. (4.39) imply that the phase velocity $V(\omega)$ and the attenuation coefficient $\delta(\omega)$ are related to each other.

For applications of the *K–K relations* in propagation problems of viscoelastic waves, we refer the interested reader e.g. to [Futterman (1962)], [Strick (1970)], [Chin (1980)], [Aki and Richards (1980)], and [Ben-Menahem and Singh (1981)].

4.3.2 *Dispersion: phase velocity and group velocity*

Extending the definition of the *phase velocity* $V(\omega)$ and the *group velocity* $U(\omega)$ in terms of the real part of the complex wave number and consequently of the complex refraction index we have

$$\begin{cases} V(\omega) := \dfrac{\omega}{\kappa_r(\omega)} = \dfrac{c}{n_r(\omega)}, \\ U(\omega) := \left[\dfrac{d}{d\omega} \kappa_r(\omega)\right]^{-1} = \dfrac{c}{n_r(\omega) + \omega \, \dfrac{dn_r}{d\omega}}. \end{cases} \qquad (4.40)$$

Ch. 4: Waves in Linear Viscoelastic Media: Dispersion and Dissipation

In the presence of dissipation we note that, while the concept of phase velocity retains its kinematic meaning of the *phase speed* in the above definition of group velocity is expected to lose its usual kinematic meaning, the *wave packet speed*. In fact, for purely dispersive waves, according to the classical argument of Lord Rayleigh, the concept of group velocity arises from the consideration of a superposition of two monochromatic waves of equal amplitude and nearly equal frequency and wavelength. When dispersion is accompanied by dissipation, two such waves cannot exist at all times because they are attenuated by different amounts due to the imaginary part of the wave number κ (or of the frequency ω), see e.g. [Bland (1960)].

Concerning the dependence of the phase velocity with frequency, we obtain from the first in Eqs. (4.40):

$$\frac{1}{V}\frac{dV}{d\omega} = -\frac{1}{n_r}\frac{dn_r}{d\omega}. \tag{4.41}$$

Concerning the relation between the phase velocity and the group velocity, we obtain using both Eqs. (4.40):

$$\frac{V}{U} = 1 + \frac{\omega}{n_r}\frac{dn_r}{d\omega} = 1 - \frac{\omega}{V}\frac{dV}{d\omega}. \tag{4.42}$$

As customary, we refer to the case $0 < U < V$ as *normal dispersion*, while the other cases ($U > V > 0$ and $U < 0 < V$) are referred as *anomalous dispersion*. It is easy to recognize from Eqs. (4.42) that the dispersion is normal or anomalous when $dn_r/d\omega > 0$ or $dn_r/d\omega < 0$, respectively or, in other words, when the phase velocity is a decreasing or increasing function of ω.

Because of the mathematical structure of the linear viscoelasticity we expect that

$$i) \ n_r > 1, \ ii) \ \frac{dn_r}{d\omega} \leq 0, \ \text{for} \ 0 \leq \omega < \infty, \tag{4.43}$$

with $n_r = 1$ only in the limit $\omega \to \infty$, where $V(\omega) \to c$. So, including for convenience the value $\omega = \infty$, for any linear viscoelastic solid we can state the following fundamental results for $0 \leq \omega \leq \infty$:

$$n_r(\omega) \geq 1 \iff 0 \leq V(\omega) \leq c, \tag{4.44}$$

and

$$\frac{dn_r}{d\omega} \leq 0 \iff \begin{cases} a) \ V \ \text{increasing function of} \ \omega, \\ b) \ U \geq V, \ \text{anomalous dispersion}. \end{cases} \tag{4.45}$$

For proof we refer to [Mainardi (1983a)]. In the following subsections we will consider some instructive examples of anomalous dispersion present in the classical Zener model and its fractional generalization where the dispersion plots show these peculiar properties of viscoelastic waves.

We thus summarize the dispersion properties of *viscoelastic waves* by stating that their *dispersion is completely anomalous*, i.e. anomalous throughout the full frequency range, and writing

$$0 \leq V(\omega) \leq U(\omega). \tag{4.46}$$

We note that for $\omega \to 0$ and $\omega \to \infty$ we have $U(\omega) \to V(\omega)$, so $\omega = 0$ and $\omega = \infty$ can be considered the non-dispersive limits.

While for $0 < \omega < \infty$ the phase velocity $V(\omega)$ turns out to be an increasing function of ω, never exceeding the wave–front velocity $c = V(\infty)$, the group velocity $U(\omega)$ is expected to increase up to reach a maximum value greater than c and then decreases to get c from above at infinity. As a consequence, there exists a certain value ω_0 of the frequency such that for $\omega > \omega_0$ it turns out that $U(\omega) \geq c$. Since the group velocity of viscoelastic waves may attain non-physical values (i.e. greater than the wave-front velocity), we need to revisit the concept of *signal velocity* usually identified with the group velocity when the dispersion is normal and the dissipation is absent or negligible. We will deal with this interesting topic in the following Section.

4.3.3 *Dissipation: the attenuation coefficient and the specific dissipation function*

The dissipation is characterized by the attenuation coefficient that now we express in terms of the imaginary part of the complex refraction index:

$$\delta(\omega) := -\kappa_i(\omega) = -\frac{\omega\, n_i(\omega)}{c}. \tag{4.47}$$

Another interesting quantity related to attenuation is the *specific dissipation function* $Q^{-1}(\omega)$ that we have seen as equivalent to the *loss tangent* discussed in Chapters 2 and 3 for time oscillations in linear viscoelastic bodies. Now, we have time and space oscillations

Ch. 4: Waves in Linear Viscoelastic Media: Dispersion and Dissipation   91

coupled together in in the steady state of the wave motion, which are attenuated only in space. Because of the interrelations among the loss tangent, the dynamic functions and the complex refraction index, see Eqs. (2.49), (2.63) and (4.36), we write

$$Q^{-1}(\omega) = \frac{\text{Im}\{n^2(-i\omega)\}}{\text{Re}\{n^2(-i\omega)\}} = \frac{2\,n_i(\omega)/n_r(\omega)}{1-[n_i(\omega)/n_r(\omega)]^2}\,. \tag{4.48}$$

So we recognize that now dissipation is related with dispersion through the complex refraction index.

For the *low-loss media* ($Q^{-1}(\omega) \ll 1$) a useful approximation is

$$Q^{-1}(\omega) \simeq 2\,\frac{n_i(\omega)}{n_r(\omega)} = 2\,V(\omega)\,\frac{\delta(\omega)}{\omega}\,. \tag{4.49}$$

If in addition we can neglect dispersion when negligible,

$$n_r(\omega) \simeq 1 \iff V(\omega) \simeq c\,,$$

we obtain the so-called *reduced specific dissipation function* mainly used in seismology, see [Futterman (1962)],

$$Q_0^{-1}(\omega) = 2\,n_i(\omega) = 2\,c\,\frac{\delta(\omega)}{\omega}\,. \tag{4.50}$$

4.3.4 *Dispersion and attenuation for the Zener and the Maxwell models*

In order to show the dispersion and attenuation for viscoelastic waves, the plots reporting frequency versus the phase velocity, the group velocity and the attenuation coefficient are necessary. In Fig. 4.1 we show these quantities for the two simplest viscoelastic models of type I and II, i.e. the *Zener* and the *Maxwell* model, respectively. For this purpose we refer to Eqs. (4.35) and (4.47) based on the complex function $n(s)$ with $s = i\omega$ corresponding to the Zener model and to the Maxwell model. The function $n(s)$ is readily derived from the constitutive equations of the two models, see Eq. (2.19) [reproduced above Eq. (4.28)] and Eq. (2.18) [reproduced above Eq. (4.27)], respectively. As a matter of fact we get

$$Zener\ model \quad n(s) = \left[\frac{s+1/\tau_\sigma}{s+1/\tau_\epsilon}\right]^{1/2},\quad \tau_\epsilon > \tau_\sigma > 0\,, \tag{4.51}$$

$$Maxwell\ model \quad n(s) = \left[1+\frac{1}{\tau_\sigma s}\right]^{1/2},\quad \tau_\sigma > 0\,. \tag{4.52}$$

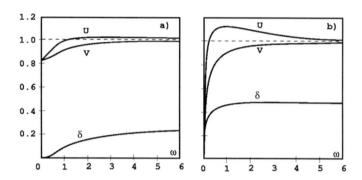

Fig. 4.1 Phase velocity V, group velocity U and attenuation coefficient δ versus frequency ω for a) Zener model, b) Maxwell model.

We note that the expression for the Maxwell model can be derived from that of the Zener model in the limit $\tau_\epsilon \to \infty$. We also point out for this model that the attenuation coefficient turns out to be proportional to the phase velocity. In fact, from Eq. (4.52) with $s = i\omega$ we get:
$$n_r^2(\omega) - n_i^2(\omega) + 2i\, n_r(\omega)\, n_i(\omega) = 1 - i/(\omega \tau_\sigma)\,, \qquad (4.53)$$
so
$$-\omega\, n_i(\omega)\, n_r(\omega) := \delta(\omega)\, c/V(\omega) = 1/(2\tau_\sigma)\,. \qquad (4.54)$$

In plotting we have considered non-dimensional variables, by scaling the frequency with $\omega_\star = 1/\tau_\sigma$, the velocities V and U with c, and the coefficient δ with $\delta_\star = 1/(c\tau_\sigma)$. In practice, we have assumed $c = 1$, $\tau_\sigma = 1$ and $\tau_\epsilon = 1.5$, so for the Maxwell model we recognize that, using this normalization, $\delta(\omega) = V(\omega)/2$.

4.3.5 Dispersion and attenuation for the fractional Zener model

For the fractional Zener model, see Section 3.2.2, we have
$$n(s) = \left[\frac{s^\nu + 1/\tau_\sigma^\nu}{s^\nu + 1/\tau_\epsilon^\nu}\right]^{1/2}\,, \quad 0 < \nu \le 1\,. \qquad (4.55)$$
For convenience, let us define the non-dimensional parameter
$$\gamma := (\tau_\epsilon/\tau_\sigma)^\nu > 1\,, \qquad (4.56)$$
that turns out to be related to the amount of dissipation.

Ch. 4: Waves in Linear Viscoelastic Media: Dispersion and Dissipation 93

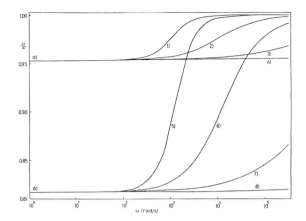

Fig. 4.2 Phase velocity over a wide frequency range for some values of ν with $\tau_\epsilon = 10^{-3}\,s$ and a) $\gamma = 1.1$: 1) $\nu = 1$, 2) $\nu = 0.75$, 3) $\nu = 0.50$, 4) $\nu = 0.25$. b) $\gamma = 1.5$: 5) $\nu = 1$, 6) $\nu = 0.75$, 7) $\nu = 0.50$, 8) $\nu = 0.25$.

In Figs. 4.2 and 4.3 we show the dispersion and the attenuation plots for the fractional Zener model taking: $\nu = 0.25, 0.50, 0.75, 1$. As an example we fix $\tau_\epsilon = 10^{-3}\,s$ and take $\gamma = 1.1, 1.5$ for the dispersion plots and $\gamma = 1.5$ for the attenuation plots.

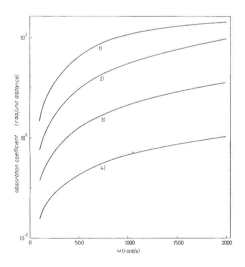

Fig. 4.3 Attenuation coefficient over a wide frequency range for some values of ν with $\tau_\epsilon = 10^{-3}\,s$, $\gamma = 1.5$: 1) $\nu = 1$, 2) $\nu = 0.75$, 3) $\nu = 0.50$, 4) $\nu = 0.25$.

4.3.6 The Klein-Gordon equation with dissipation

In a series of papers the author [Mainardi (1983a); (1983b); (1984)] has considered the wave equation
$$\frac{\partial^2 r}{\partial t^2} + 2\alpha \frac{\partial r}{\partial t} + \beta^2 r = c^2 \frac{\partial^2 r}{\partial x^2}, \; r = r(x,t), \; \alpha, \beta \geq 0, \quad (4.57)$$
that reduces to the linear Klein-Gordon equation for $\alpha = 0$ and to the telegraph equation for $\beta = 0$. Since the parameter α is related to dissipation, we refer to it as to the *Klein-Gordon equation with dissipation*.

This equation is quite interesting since it is simple enough to admit, as we will see here after, the *closed-form solution* suitable to any problem of impact waves as stated in Subsection 4.2.1 via a convolution integral with the input disturbance $r_0(t)$, and the *algebraic dispersion equation* easy to be discussed and interpreted. We are going to show the occurrence of *normal dispersion* if $0 \leq \alpha < \beta$, and *anomalous dispersion* for $0 \leq \beta < \alpha$. We note that the case $\alpha = \beta$ corresponds to the *distortion-less* wave propagation where there is attenuation without any dispersion.

As a matter of fact the case $0 \leq \beta < \alpha$ turns out to be akin to that of viscoelastic waves in view of the common anomalous dispersion in the full range of frequencies; furthermore the special case $\beta = 0$ reproduces the Maxwell model of viscoelasticity.

The function $n(s)$. Before starting our analysis we must consider the fundamental quantity $n(s)$ corresponding to Eq. (4.57), which embodies the dispersion and attenuation properties. Applying the Laplace transform to (4.57) we obtain
$$n(s) = \left[1 + 2\frac{\alpha}{s} + \frac{\beta^2}{s^2}\right]^{1/2} = \frac{\left[(s+\alpha)^2 \pm \chi^2\right]^{1/2}}{s}, \quad (4.58)$$
where
$$\chi^2 = \beta^2 - \alpha^2, \text{ hence } \chi = \sqrt{|\beta^2 - \alpha^2|}. \quad (4.59)$$

In Eq. (4.58) the opposite signs in front of χ^2 correspond to physically distinct cases that we refer as cases $(+)$ and $(-)$, respectively. As a matter of fact we will see how the parameter χ turns to be fundamental to characterize the solutions and the dispersion properties of the Klein-Gordon equation with dissipation.

The solution. Denoting as usual by $r_0(t)$ the input disturbance at $x = 0$, the solution of the corresponding signalling problem is found according to (4.17) by inverting the Laplace transform,

$$\tilde{r}(x,s) = \tilde{r}_0(s) e^{s[t - (x/c)n(s)]}, \tag{4.60}$$

where $n(s)$ is given by Eq. (4.58). Then, by introducing the new space and time variables

$$\xi := x/c \quad \tau = t - x/c, \tag{4.61}$$

and recalling the Laplace transforms pairs of the Bessel functions, see Appendix B, after simple manipulation we get for $\tau > 0$:

$$r(\xi, \tau) = e^{-\alpha \xi} \left[r_0(\tau) \mp \chi \xi \int_0^\tau e^{-\alpha \tau'} F^\pm(\xi, \tau') r_0(\tau - \tau') d\tau' \right], \tag{4.62}$$

with

$$\begin{cases} F^+(\xi, \tau) = \dfrac{J_1\left\{[\tau(\tau+\xi)]^{1/2}\right\}}{[\tau(\tau+\xi)]^{1/2}}, & \beta^2 - \alpha^2 > 0, \\[1em] F^-(\xi, \tau) = \dfrac{I_1\left\{[\tau(\tau+\xi)]^{1/2}\right\}}{[\tau(\tau+\xi)]^{1/2}}, & \beta^2 - \alpha^2 < 0, \end{cases} \tag{4.63}$$

where J_1 and I_1 denote the ordinary and the modified Bessel functions of order 1, respectively. In addition to χ it may be convenient to introduce the parameter

$$m := \frac{\alpha}{\beta} \tag{4.64}$$

so that the case $0 < m < 1$ corresponds to $(+)$, whereas $1 < m < \infty$ to $(-)$.

The solution thus consists of two terms: the first represents the input signal, propagating at velocity c and exponentially attenuated in space; the second is responsible for the distortion of the signal, that depends on the position and on the time elapsed from the wave front. The amount of distortion can be measured by the parameter χ that indeed vanishes for $\alpha = \beta$, the *distortion-less* case.

Dispersion and attenuation properties. As it is known, the function $n(s)$, as given in Eq. (4.58), provides the complex refraction index for $s = \pm i\omega$, from which, as described through Eqs. (4.40)-(4.47), we derive the dispersion and attenuation characteristics, namely the phase velocity $V(\omega)$, the group velocity $U(\omega)$ and the attenuation coefficient $\delta(\omega)$. In fact, from

$$n(\pm i\omega) = c \left[\frac{1}{V(\omega)} \mp i \frac{\delta(\omega)}{\omega} \right] = \left[1 \mp 2i \frac{\alpha}{\omega} - \frac{\beta^2}{\omega^2} \right]^{1/2}, \qquad (4.65)$$

we obtain $V(\omega)$ and $\delta(\omega)$ by solving the system

$$\begin{cases} c^2 \left[\dfrac{\omega^2}{V^2(\omega)} - \delta^2(\omega) \right] = \omega^2 - \beta^2, \\[2mm] c^2 \dfrac{\delta(\omega)}{V(\omega)} = \alpha. \end{cases} \qquad (4.66)$$

Then, the group velocity $U(\omega)$ can be computed from $V(\omega)$ using e.g. Eq. (4.42).
- For $\alpha = 0$, i.e. in the absence of dissipation, a cut-off occurs at $\omega = \beta$, namely

$$\begin{cases} 0 \le \omega < \beta, & \delta(\omega) = \sqrt{\beta^2 - \omega^2}/c, \quad \text{no propagaton}; \\ \beta < \omega \le \infty, & V(\omega) = c\omega/\sqrt{\omega^2 - \beta^2}, \quad \text{no attenuation}. \end{cases} \qquad (4.67)$$

- For $\alpha \ne 0$, i.e. in the presence of dissipation, we obtain for any ω:

$$\begin{cases} \delta(\omega) = \alpha V(\omega)/c^2, \\ \omega^2 \left[c^2/V^2(\omega) - 1 \right] = \alpha^2 V^2(\omega)/c^2 - \beta^2. \end{cases} \qquad (4.68)$$

We note that the attenuation coefficient is proportional to the phase velocity. In the distortion-less case $\alpha = \beta$ ($m = 1$) we easily derive from (4.68) constant attenuation ($\delta = \alpha/c^2$) and no dispersion ($V = U = c$).

If $\alpha \ne \beta$ ($m \ne 1$) some manipulations are necessary to derive from (4.68) the explicit expression of $V(\omega)$ (and consequently of $\delta(\omega)$), that we leave as an exercise together with the following interesting relationship between V and U (shown in [Mainardi (1984)]):

$$U = V \left[1 + \frac{\alpha^2(1 - V^2/c^2) + \beta^2(1 - c^2/V^2)}{\alpha^2 - \beta^2} \right]. \qquad (4.69)$$

When $\beta = 0$ (Telegraph equation), Eq. (4.69) reduces to the expression derived in [Carrier et al. (1966)]:

$$U = 2V - V^3/c^2, \qquad (4.70)$$

whereas when $\alpha = 0$ (Klein-Gordon equation), it reduces to the well-known expression

$$UV = c^2. \qquad (4.71)$$

As a matter of fact, we obtain the following inequalities

$$\begin{cases} 0 \leq \alpha < \beta, \quad 1 \leq \dfrac{V}{c} \leq \dfrac{\beta}{\alpha}, \ U \leq V \ : \text{normal dispersion;} \\[2mm] 0 \leq \beta < \alpha, \quad \dfrac{\beta}{\alpha} \leq \dfrac{V}{c} \leq 1, \ U \geq V \ : \text{anomalous dispersion.} \end{cases} \qquad (4.72)$$

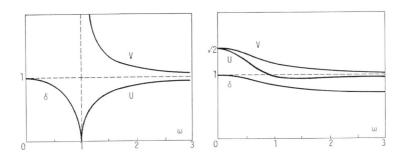

Fig. 4.4 Dispersion and attenuation plots: $m = 0$ (left), $m = 1/\sqrt{2}$ (right).

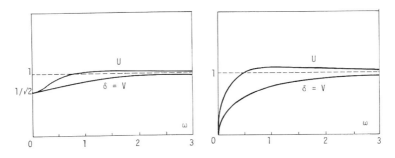

Fig. 4.5 Dispersion and attenuation plots: $m = \sqrt{2}$ (left), $m = \infty$ (right).

In Figs. 4.4 and 4.5 we show the dispersion and attenuation plots for normal and anomalous dispersion, corresponding to cases (+) and (-), respectively. For case (+) we chose $\beta = 1$ and $m = 0, 1/\sqrt{2}$, whereas for case (-) $\alpha = 1$ and $m = \sqrt{2}$. In the last case we note consequently $\delta(\omega) = V(\omega)$, that for $\beta = 0$ provides the Maxwell model of viscoelasticity with $\tau_\sigma = 1/2$, compare with Eq. (4.54). The dispersion plots confirm that in regime of normal/anomalous dispersion the phase velocity is indeed a decreasing/increasing function of the frequency according to the fundamental relationships (4.42).

4.4 The Brillouin signal velocity

4.4.1 *Generalities*

We have already pointed out that for a given linear viscoelastic medium the complex refraction index $n^\star(\omega)$ characterizes the dispersion and attenuation of pseudo-monochromatic waves propagating in it. Based on the pioneering analysis of A. Sommerfeld and L. Brillouin, see [Brillouin (1960)], carried out in 1914 for electromagnetic waves propagating in a dielectric, we recognize that these pseudo-monochromatic waves are obtained as a steady–state response to a sinusoidal excitation of a given frequency Ω provided at $x = 0$ for $t \geq 0$ of a viscoelastic medium. In fact, adopting the Laplace representation[1] stated in Eq. (4.17), where we assume

$$r_0(t) = \cos(\Omega t)\,\Theta(t) \div \tilde{r}_0(s) = \frac{s}{s^2 + \Omega^2}, \qquad (4.73)$$

the solution admits the complex integral representation

$$r(x,t) = \frac{1}{2\pi i}\int_{Br}\frac{s}{s^2+\Omega^2}\,e^{s[t-(x/c)\,n(s)]}\,ds\,. \qquad (4.74)$$

We refer to this signalling problem as the *Brillouin problem*.

---

[1] The original analysis by Brillouin was made by using the Fourier integral and deforming the path of integration into the upper complex half plane. It may be shown by a simple change of variable that the complex Fourier integral employed by Brillouin is exactly the Bromwich integral (4.74) with the path deformed into the left complex half plane. Our approach with the Laplace transform is consistent with that sketched in [Stratton (1941)], see Chapter 5, Section 18, where the arguments by Brillouin are summarized.

The application of the Cauchy theorem requires the analysis of the singularities of the integrand in (4.74) that are laying to the left of the Bromwich path, namely for $\mathcal{R}e\,[s] \leq 0$. We note that they are represented by the branch cut of the complex function $n(s)$ and the two simple poles on the imaginary axis, $s_\pm = \pm i\Omega$, exhibited by the Laplace transform of the sinusoidal input disturbance. Then, by applying the Cauchy theorem, one gets the following picture of the course of the mechanical disturbance $r(x,t)$ at a distance $x > 0$. Up to time $t = x/c$ no motion occurs; then for $t \geq x/c$ the wave motion starts from a certain amplitude (that could also be very small or even vanishing) and consists of two parts: the contribution from the branch cut, and that from the two poles. These parts of are usually referred to as the *transient state* $r_T(x,t)$ and the *steady state* $r_S(x,t)$, respectively, since the former is expected to vanish at any x as $t \to \infty$, while the latter is oscillating in space and time and exponentially attenuated only in space. As a consequence the steady state gives the limiting value of the total response variable at any x as $t \to \infty$.

Writing
$$n(\pm i\Omega) = n_r(\Omega) \pm i\, n_i(\Omega) = \frac{c}{\Omega}\left[\kappa_r(\Omega) \pm i\,\kappa_i(\Omega)\right], \qquad (4.75)$$

we obtain, as a simple exercise of complex analysis, that the sum contribution of the two poles $s = \pm i\Omega$ is given by
$$r_S(x,t) = e^{-\delta(\Omega)\,x} \cos\Omega\left[t - x/V(\Omega)\right], \qquad (4.76)$$

where
$$\begin{cases} V(\Omega) = \dfrac{c}{n_r(\Omega)} = \dfrac{\Omega}{\kappa_r(\Omega)} \geq 0, \\ \delta(\Omega) = -\Omega\,\dfrac{n_i(\Omega)}{c} = -\kappa_i(\Omega) \geq 0. \end{cases} \qquad (4.77)$$

Thus we recognize that *the steady–state response to a sinusoidal impact of frequency Ω is a pseudo-monochromatic wave, travelling with phase velocity $V(\Omega)$ with an amplitude exponentially attenuated in space with attenuation coefficient $\delta(\Omega)$*. This is to say that the steady-state response is a particular wave-mode solution, whose complex wave number and real frequency satisfy the dispersion equation (4.29) with $\omega = \Omega$.

4.4.2 Signal velocity via steepest–descent path

It is known that the group velocity for electromagnetic waves propagating in a dielectric may attain non-physical values (being in some frequency ranges negative or greater than the wave–front velocity) and thus cannot be identified with any signal velocity. This fact led Sommerfeld and Brillouin, after the advent of the (restricted) theory of relativity, to investigate the subject matter more carefully. From 1914 these scientists analysed the propagation of an electro-magnetic signal of type (4.73) in a dielectric obeying the *Lorentz-Lorenz dispersion equation*. The contributions were translated into English and collected in the sixties by Brillouin in a relevant booklet [Brillouin (1960)]. In particular, Brillouin introduced a suitable definition of signal velocity in order to meet the physical requirement to be positive; lessen the wave-front velocity (the velocity of light in the vacuum) in the frequency range where the dispersion is anomalous and the absorption is high. The equivalence of the signal velocity with the group velocity was effectively proved only in the presence of negligible dissipation. In regions of high absorption some estimations of the appropriate signal velocity were provided in a 1914 article, reported in [Brillouin (1960)], and then improved by [Baerwald (1930)], see also [Elices and García-Moliner (1968)]. For a more recent and exhaustive discussion, see the interesting book [Oughstun and Sherman (1994)].

For viscoelastic waves that exhibit only anomalous dispersion and relatively high absorption, the problem of the identification of a suitable signal velocity turns out to be of some relevance. Hereafter we summarize the main results applicable to the class of viscoelastic waves obtained by the present author, see [Mainardi (1983a); (1983b); (1984)], who extended the classical arguments of Brillouin valid for electromagnetic waves. For this purpose, let us consider the Brillouin signalling problem, represented in the Laplace domain through Eqs. (4.73) and (4.74).

Following the idea of Brillouin, we intend to use the path of steepest descent in order to evaluate the complex Bromwich integral in

(4.74). For this purpose, let us change the original variables x, t into

$$\xi = x/c, \quad \theta = ct/x > 1, \qquad (4.78)$$

so that the solution reads

$$r(\xi, \theta) = \frac{1}{2\pi i} \int_{Br} \frac{s}{s^2 + \Omega^2} e^{\xi F(s;\theta)} ds, \qquad (4.79)$$

where

$$F(s; \theta) = \Phi(s, \theta) + i \Psi(s, \theta) := s\left[\theta - n(s)\right]. \qquad (4.80)$$

After this, the original fixed path Br is to be deformed into the new, moving path $L(\theta)$, that is of steepest descent for the real part of $F(s; \theta)$ in the complex s-plane. Restricting our attention to the simplest case, for any given $\theta > 1$ this path L turns out to be defined by the following properties:

- (i) L passes through the saddle point s_0 of $F(s; \theta)$, i.e.

$$\left.\frac{dF}{ds}\right|_{s=s_0} = 0 \iff n(s) + s\frac{dn}{ds} = \theta \quad \text{at} \quad s = s_0(\theta); \quad (4.81)$$

- (ii) the imaginary part of $F(s; \theta)$ is constant on L, i.e.

$$\Psi(s; \theta) = \Psi(s_0; \theta), \quad s \in L; \qquad (4.82)$$

- (iii) the real part of $F(s; \theta)$ attains its maximum at s_0 along L, i.e.

$$\Phi(s; \theta) < \Phi(s_0; \theta), \quad s \in \{L - s_0\}; \qquad (4.83)$$

- (iv) the integral on L is equivalent either to the original one or differs from it by the contribution due to the residues at the poles $s = \pm i\Omega$.

In our case the integral on $L(\theta)$ turns out to be equal to the Bromwich integral, or different from it by the steady state solution (4.76), according to $1 \leq \theta < \theta_s(\Omega)$ or $\theta > \theta_s(\Omega)$, respectively, where $\theta_s(\Omega)$ is the value of θ for which $L(\theta)$ intersects the imaginary axis at the frequency $\pm\Omega$.

Therefore, the representation of the wave motion by the integral on $L(\theta)$, which we refer to as the *Brillouin representation*, allows one to recognize the arrival of the steady state and, following Brillouin, to define the *signal velocity* as

$$S(\Omega) := \frac{c}{\theta_s(\Omega)}. \qquad (4.84)$$

The condition $\theta \geq 1$ ensures that, in any dispersive motion where the representation (4.74), namely (4.79), holds, the signal velocity $S(\Omega)$ is always less than c, the wave–front velocity; this property is independent of the fact that the group velocity $U(\Omega)$ can a priori be greater than c or be negative.

We note that the analytical determination of the path L and its evolution is in general a difficult task even for simple models of viscoelasticity, so special numerical techniques can be envisaged. For the Zener model (S.L.S.) with $n(s)$ provided by (4.51) [Mainardi (1972)] was able to show in several plates the evolution of the corresponding path L of steepest descent, in a way similar to that by Brillouin for the Lorentz-Lorenz dispersion equation. More recently, in [Mainardi and Vitali (1990)] the authors found it convenient to use their specific routine to determine the path L in the complex s−plane by solving Eq. (4.82) numerically, based on rational approximations (Padé Approximants of type II), according to an algorithm that ensures an exponential convergence.

The exact evaluation of the signal velocity S may appear only as a numerical achievement. In other words, for a general linear dispersive motion, the definition of signal velocity by Brillouin appears as a computational prescription. Its evaluation, in general depending on the explicit knowledge of $n(s)$ in the complex s-plane, appears more difficult than the evaluation of the phase velocity V and group velocity U, both of which depend on the values of $n(s)$ only in the imaginary axis.

Hereafter we find it instructive to illustrate the Brillouin method in the cases of *full normal dispersion* and *full anomalous dispersion* by considering again the *Klein-Gordon equation with dissipation* (4.57). Such an equation can be assumed as a simple prototype of the two regimes of dispersion that depend on the relative weight of their two independent parameters α and β, better summarized in a unique parameter: χ given by (4.59) or m given by (4.64).

The evolution of the steepest–descent path. In order to depict the evolution of the steepest descent path $L(\theta)$ for our equation (4.65) according to the idea of Brillouin, we must keep the cases (+) and (-) distinct. To this aim we recall Eq. (4.58), i.e.,

$$n(s) = \frac{\left[(s+\alpha)^2 \pm \chi^2\right]^{1/2}}{s}, \quad \chi^2 = |\beta^2 - \alpha^2|,$$

where the opposite signs in front of χ^2 correspond to the cases (+) $[0 < m < 1]$ and (−) $[m > 1]$, respectively. As a consequence, the branch points of $n(s)$ turn out to be

$$\text{case }(+) : s^\pm = -\alpha \pm i\chi, \quad \text{case }(-) : s^\pm = -\alpha \pm \chi, \quad (4.85)$$

so that they are complex conjugate in the case (+), and real negative in the case (-). Following the instructions previously illustrated, see Eqs. (4.80)-(4.83), we write for our cases

$$F(s,\theta) := s[\theta - n(s)] = -\alpha\theta + (s+\alpha)\theta - \left[(s+\alpha)^2 \pm \chi^2\right]^{1/2}. \quad (4.86)$$

Setting for convenience

$$s + \alpha = p = u + iv, \quad R(\theta) = (\theta^2 - 1)^{1/2}, \quad (4.87)$$

the saddle points and the steepest–descent paths through them in the complex p-plane are:

$$\text{case }(+) : p_0^\pm = \pm i\chi\theta/R(\theta), \quad (4.88)$$

$$\begin{cases} \mathcal{I}m\{p\theta - (p^2 + \chi^2)^{1/2}\} = \pm\chi R(\theta), \\ \mathcal{R}e\{p\theta - (p^2 + \chi^2)^{1/2}\} \leq 0; \end{cases} \quad (4.89)$$

and

$$\text{case }(-) : p_0^\pm = \pm\chi\theta/R(\theta), \quad (4.90)$$

$$\begin{cases} \mathcal{I}m\{p\theta - (p^2 - \chi^2)^{1/2}\} = 0, \\ \mathcal{R}e\{p\theta - (p^2 = \chi^2)^{1/2}\} \leq \pm\chi R(\theta). \end{cases} \quad (4.91)$$

For case (+) the path $L(\theta)$ consists of two branches, the upper one L^+ passing through p_0^+ with the direction $\gamma = 3\pi/4$, and the lower one L^-, passing through p_0^- with $\gamma = \pi/4$, which is the mirror image of L^+. In particular L^+ intersects line $u = 0$ at two points with $v = \chi\theta/R(\theta)$ (the saddle point) and $v = \chi R(\theta)/\theta$.

For the case (–) the path $L(\theta)$ consists of an ellipse of equation
$$(\theta^2 - 1)[u^2 + \theta^2 v^2] = (\chi\theta)^2, \qquad (4.92)$$
passing through $p_0^{\pm}$ with $\gamma = \pm\pi/2$. We notice that on L, the real part in (4.86) has in p_0^+ its maximum while in p_0^- its minimum. The ellipse intersects line $u = 0$ at point $v = \pm\chi/R(\theta)$.

Figures 4.6 and 4.7 show the steepest–descent paths in the upper complex s-plane for the cases (+) and (–), respectively, with arrows showing the direction of ascent through the saddle points.

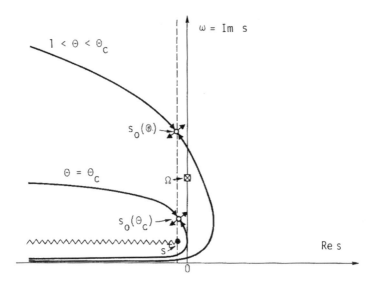

Fig. 4.6 The evolution of the steepest–descent path $L(\theta)$: case (+).

From the above analysis we can infer that the saddle points leave the infinity point at $\theta = 1$ and, with θ increasing, they move towards the branch points of $n(s)$ on the line connecting these points. The path $L(\theta)$ intersects the imaginary axis (the frequency axis) for any θ such that $1 < \theta \le \theta_c$, where θ_c denotes the critical value of θ for which the corresponding path is tangent to the axis. This implies for the signal velocity $S(\Omega)$ (for a sinusoidal signal of frequency Ω) the following range
$$c/\theta_c \le S(\Omega) < c. \qquad (4.93)$$

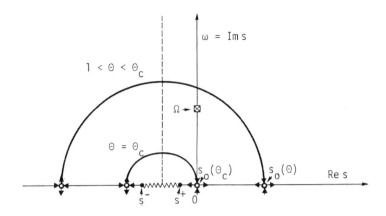

Fig. 4.7 The evolution of the steepest–descent path $L(\theta)$: case $(-)$.

When $\alpha = 0$ in the case $(+)$, the saddle points move along the imaginary axis $s = \pm i\omega$, with $\omega > \chi = \beta$. This fact will be shown to imply that for $\Omega > \chi$ *the signal velocity coincides with the group velocity*. In fact, because the lines of steepest descent for $\Phi = \mathcal{R}e\{F(s,,\theta)\}$ cross the axis at the saddle points with slope angles $\pm \pi/4$, the axis is a line of steepest descent for $\Psi = \mathcal{I}m\{F(s,\theta)\}$. This means that along the imaginary axis Ψ is stationary at the saddle points. When the saddle points meet the poles $s = \pm i\Omega$, from (4.8) we obtain:

$$\left.\frac{d\Psi}{d\omega}\right|_{\omega=\Omega} = \left.\frac{d}{d\omega}\{\omega[\theta - n_r(\omega)]\}\right|_{\omega=\Omega} = 0, \qquad (4.94)$$

with $\theta = \theta_s(\Omega)$. Accounting for (4.40) and (4.84) we finally get

$$S(\Omega) := \frac{c}{\theta_s(\Omega)} = \frac{c}{[n_r(\omega) + \omega\, dn_r/d\omega]_{\omega=\Omega}} := U(\Omega). \qquad (4.95)$$

This result is in agreement with that given by the method of stationary phase [Lighthill (1965)]. However, as pointed out in [Brillouin (1960)], only in the absence of dissipation (i.e. $n(i\Omega) = n_r(\Omega)$) are the points of stationary phase saddle points, and the two methods give the same result for the signal velocity. If $\Omega < \chi = \beta$ we have $n(i\Omega) = in_i(\Omega)$, that is dissipation is present without dispersion. In this case the signal is a standing wave decaying exponentially with distance, that appears for $\theta \geq \theta_s(\Omega) = c/U(\chi^2/\Omega)$ [Thau (1974)].

In the more general case (+) with $0 < m < 1$, we notice that the identification of the group velocity with the signal velocity cannot be valid for all frequencies. In fact, as it is visible in the right plate of Fig. 4.4 ($m = 1/\sqrt{2}$), we note that $U > c$ for $0 \leq \omega < \omega_*$, where ω_* is a certain frequency depending on m. Here we do not consider the evaluation of the Brillouin signal velocity for $0 < m < 1$, limiting ourselves to state that the group velocity may represent the signal velocity only if the dissipation effects are sufficiently small and the frequency is sufficiently high.

Here, however, we deal with the case (−) where $1 < m \leq \infty$, because it is just akin to that of viscoelastic waves where the phase velocity is an increasing function over the full range of the frequency. In particular, for $m = \infty$ ($\beta = 0$) we recover the telegraph equation that governs the viscoelastic waves in the Maxwell model. For the entire range $m > 1$, viewing the evolution of the steepest–descent path in Fig. 4.7, we recognize that the relevant saddle point s_0 moves from $+\infty$ (at the wave front, $\theta = 1$) up to the largest branch point of $n(s)$ (for $\theta = \infty$), and that the path $L(\theta)$ of steepest descent through s_0 is a curve that encloses the branch cut of $n(s)$. This situation is similar to that we have verified for the Zener model [Mainardi (1972)], which also includes the Maxwell model. In other words, we may assume that this evolution of the steepest–descent path is common to viscoelastic models that exhibit branch cuts in the *finite* part of the negative real axis.

Thus, when the path $L(\theta)$ intersects the imaginary axis at the poles $s = \pm i\Omega$, we obtain

$$Im\,[F(s_0, \theta_S)] = Im\,[F(\pm i\Omega, \theta_S)] = \pm \Omega\,[\theta_S - n_r(\Omega)] = 0. \quad (4.96)$$

This means that

$$S(\Omega) := \frac{c}{\theta_s(\Omega)} = \frac{c}{n_r(\Omega)} := V(\Omega). \quad (4.97)$$

In the case of absence of dissipation the identification of signal velocity with group velocity in quite clear: at the wave front it appears an oscillating forerunner (or percursor) that matches with the steady state solution after a time $\tau_U = x/U(\Omega) - x/c$. Being the percursor of small and highly decaying amplitude, the steady state solution is well recognized as the main signal.

In the case of full anomalous dispersion, the identification of signal velocity with phase velocity is to be interpreted differently. We observe here at the wave front the appearance of a non-oscillating forerunner that matches with the steady state solution after a time $\tau_V = x/V(\Omega) - x/c$. Being the precursor of substantial (weakly damped) amplitude, the steady solution is not well recognized. As a matter of fact, this precursor is known to be the source of great disturbance for transmissions. In other words, the appearance of a strong forerunner can lead to a wrong identification of the input signal. In this respect the interested reader is referred to [Mainardi (1984)] and to the more recent paper [Hanyga (2002c)].

Note again that the relevant example of anomalous dispersion is provided by the Maxwell model of viscoelasticity, that is known to be governed by the telegraph equation, see Eq. (4.57) with $\beta = 0$. Some authors, [Carrier et al. (1966)] and [Thau (1974)], have considered this equation to evaluate the Brillouin signal velocity but have overlooked the identification with the phase velocity. This fact has been pointed out by [Mainardi (1983b)].

4.5 Notes

An interesting problem related to the identification of the signal velocity is that of the *energy velocity*. This topic is much more subtle since the energy is not well defined in the presence of dissipation. The author has also considered the problem of characterizing the energy propagation by means suitable definitions for energy velocity, in the presence of anomalous dispersion and dissipation, see [Mainardi (1987); Mainardi (1993)], [Mainardi and Van Groesen (1989); Van Groesen and Mainardi (1989); Van Groesen and Mainardi (1990)] [Mainardi et al. (1991); Mainardi et al. (1992); Mainardi and Tocci (1993)]. Formerly, [Brillouin (1960)] devoted great attention to this topic, providing interesting results for electromagnetic waves in dielectrics. For pseudo-monochramatic viscoelastic waves [Bland (1960)] has shown that energy velocity is identified with phase velocity, a result that was later confirmed through an independent approach by [Mainardi (1973)].

Chapter 5

Waves in Linear Viscoelastic Media: Asymptotic Representations

In this chapter we consider some mathematical methods that allow us to derive asymptotic expansions in space-time domains for impact viscoelastic waves, starting from their Laplace transform representations. We first deal with recursive series methods which, after inversion of the Laplace transform, yield asymptotic or convergent expansions, suitable in a space-time domain close to the wave front. Then, we apply the saddle–point method to the Bromwich representation of the inverse Laplace transform, which, for the evolution of impact waves, provides approximations suitable far from the wave front. Because the numerical convergence of wave–front expansions usually does not allow for matching with saddle-point approximation in any space-time domain, we suggest the acceleration technique of rational Padè approximants as a good candidate to achieve this goal.

5.1 The regular wave–front expansion

The full analytical models of viscoelasticity We now consider viscoelastic models for which the *material functions* $J(t)$ and $G(t)$ with $J_g := J(0^+) > 0$ and $G_g := G(0^+) > 0$ are *entire functions of exponential type*. We recall from [Widder (1971)] that an entire function $f(z)$ with $z \in \mathbb{C}$ is said to be of exponential type and we write $f(z) \in \{1, \delta\}$ with $\delta > 0$, if its Taylor series is such that

$$f(z) = \sum_{k=0}^{\infty} a_k \, z^k/k!\,, \text{ with } \varlimsup_{k\to\infty} |a_k|^{1/k} \leq \delta\,.$$

Since the material functions are defined as causal functions, their analytic continuation from $t \in \mathbb{R}^+$ to $t \in \mathbb{C}$ is understood. Henceforth, we shall refer to the corresponding models as *full analytical models of viscoelasticity*.

It is now convenient to consider non-dimensional memory functions defined in Eqs. (4.19) as

$$\Psi(t) := \frac{1}{J_g}\frac{dJ}{dt}, \quad \Phi(t) := \frac{1}{G_g}\frac{dG}{dt} \leq 0.$$

We will refer to them simply as *rate of creep* and *rate of relaxation*, respectively. They are entire functions of exponential type as well, so that their Laplace transforms turn out to be analytic functions at infinity according to a known theorem, see e.g. [Widder (1971)]. More precisely we write

$$\Psi(t) \in \{1,\delta_\psi\} \iff \widetilde{\Psi}(s) \in \mathcal{A} \quad \text{for} \quad |s| > \delta_\psi, \quad \widetilde{\Psi}(\infty) = 0, \quad (5.1a)$$

and

$$\Phi(t) \in \{1,\delta_\phi\} \iff \widetilde{\Phi}(s) \in \mathcal{A} \quad \text{for} \quad |s| > \delta_\phi, \quad \widetilde{\Phi}(\infty) = 0, \quad (5.1b)$$

where $\mathcal{A}$ denotes the class of analytic functions in the complex s-plane and δ_ψ and δ_ϕ are suitable positive numbers. Assuming the *creep representation* we can write

$$J(t) = J_g\left[1 + \sum_{k=1}^{\infty}\psi_k\frac{t^k}{k!}\right], \quad \psi_k := \frac{1}{J_g}\left[\frac{d^k J}{dt^k}\right]_{t=0^+}. \quad (5.2a)$$

Thus, recalling Eq. (4.20a), we get in the Laplace domain

$$[n(s)]^2 := \frac{s\widetilde{J}(s)}{J_g} = 1 + \widetilde{\Psi}(s) = 1 + \sum_{k=1}^{\infty}\frac{\psi_k}{s^k}, \quad |s| > \delta_\psi. \quad (5.3a)$$

Similarly, assuming the *relaxation representation*, we write

$$G(t) = G_g\left[1 + \sum_{k=1}^{\infty}\phi_k\frac{t^k}{k!}\right], \quad \phi_k := \frac{1}{G_g}\left[\frac{d^k G}{dt^k}\right]_{t=0^+}. \quad (5.2b)$$

Thus, in view of Eq. (4.20b), we have

$$[n(s)]^{-2} := \frac{s\widetilde{G}(s)}{G_g} = 1 + \sum_{k=1}^{\infty}\frac{\phi_k}{s^k}, \quad |s| > \delta_\phi. \quad (5.3b)$$

We refer to coefficients ψ_k and ϕ_k of the memory functions as *creep* and *relaxation* coefficients, respectively. In view of Eqs. (5.3) also the functions $[n(s)]^{\pm 2}$ and consequently $n(s)$ turn out to be analytic in the s-plane, regular at infinity where they assume the value 1.

As a matter of fact, to ensure that $\widetilde{\Psi}(s)$ and $\widetilde{\Phi}(s)$ are analytic and vanishing at infinity and, hence, $\Psi(t)$ and $\Phi(t)$ are entire functions of exponential type, we have to assume that for the corresponding viscoelastic models the retardation/relaxation spectra, either *discrete* or *continuous*, are such that $n(s)$ provided by Eq. (4.16) exhibit a *finite* branch cut on the negative real axis. This means that a positive number δ can be found so that $n(s)$ is represented by a power series, absolutely convergent for $|s| > \delta$,

$$n(s) = 1 + \frac{n_1}{s} + \frac{n_2}{s^2} + \dots, \quad |s| > \delta. \tag{5.4}$$

Of course, the series coefficients n_k ($k = 1, 2, \dots$) can be obtained either from the creep coefficients or from the relaxation coefficients. Setting $c = 1$ for convenience, and recalling from Eqs. (5.3)

$$n(s) = \left[1 + \sum_{k=1}^{\infty} \frac{\psi_k}{s^k}\right]^{1/2} = \left[1 + \sum_{k=1}^{\infty} \frac{\phi_k}{s^k}\right]^{-1/2}, \tag{5.5}$$

we easily obtain the first few coefficients of $n(s)$, e.g.

$$n_1 = \frac{1}{2}\psi_1, \; n_2 = \frac{1}{2}\psi_2 - \frac{1}{8}\psi_1^2, \; \dots \tag{5.6a}$$

and

$$n_1 = -\frac{1}{2}\phi_1, \; n_2 = -\frac{1}{2}\phi_2 + \frac{3}{8}\phi_1^2, \; \dots \tag{5.6b}$$

Recalling from Section 2.5 that the creep compliance $J(t)$ is a Bernstein function and the relaxation modulus $G(t)$ is a CM function, from Eqs. (2.31) we get $(-1)^n \psi_n \leq 0$ and $(-1)^n \phi_n \geq 0$, $n \geq 1$, so that we deduce for consistency that in Eqs. (5.6): $n_1 \geq 0$ and $n_2 \leq 0$.

The recursive series method. We now present an efficient recursive series method, which allows us to obtain a *convergent wave-front expansion* for the solution $r(x, t)$, starting from the creep coefficients, as proposed by [Mainardi and Turchetti (1975)].

For this purpose, we start from the solution of the general evolution equation (4.7) in the Laplace domain, that is, after (4.15)-(4.16) and (5.4)-(5.6),

$$\widetilde{r}(x,s) = \widetilde{r}_0(s)\,e^{-\mu(s)x} = \widetilde{r}_0(s)\,e^{-sn(s)x} \text{ with}$$

$$\mu(s) = s + \alpha - \frac{\beta}{s} + \ldots \quad n(s) = 1 + \frac{\alpha}{s} - \frac{\beta}{s^2} + \ldots, \qquad (5.7)$$

where, for convenience, we have assumed $c = 1$ and set

$$\alpha = n_1 = \frac{1}{2}\psi_1 \geq 0, \quad \beta = -n_2 = -\frac{1}{2}\psi_2 + \frac{1}{8}\psi_1^2 \geq 0. \qquad (5.8)$$

Let us re-write the transform solution as follows

$$\widetilde{r}(x,s) := \widetilde{r}_0(s)\,e^{-\mu(s)x} = e^{-\alpha x}\left[e^{-sx}\,\widetilde{R}(x,s)\right], \qquad (5.9)$$

which yields in the time domain

$$r(x,t) = e^{-\alpha x}\,R(x,t-x). \qquad (5.10)$$

The purpose of the exponentials in (5.9) is to isolate the wave front propagating with velocity $c = 1$ and with amplitude that exponential decays in space by an attenuation coefficient α.

The equation satisfied by $\widetilde{R}(x,s)$ is, from Eqs. (4.7) and (5.9),

$$\left\{\frac{d^2}{dx^2} - 2(s+\alpha)\frac{d}{dx} - \left[\mu^2(s) - (s+\alpha)^2\right]\right\}\widetilde{R}(x,s) = 0, \qquad (5.11)$$

subjected to the initial condition $\widetilde{R}(0,s) = \widetilde{r}_0(s)$. After simple manipulations we obtain

$$\left[\mu^2(s) - (s+\alpha)^2\right] = -2\beta + \sum_{j=1}^{\infty}\frac{\psi_{j+2}}{s^j}. \qquad (5.12)$$

This shows how the differential operator acting on $\widetilde{R}(x,s)$ in (5.11) depends on the creep coefficients ψ_j that are easily obtained from the creep law.

For $\widetilde{R}(x,s)$ we now seek a series expansion in integer powers of $1/s$. Based on the theory of Laplace transforms, see e.g. [Doetsch (1974)], one is led to think that the term-by-term inversion of such series provides in the time domain an expansion for $R(x,t)$ *asymptotic* as $t \to (x/c)^+$. Consequently, a formal wave-front expansion for $r(x,t)$ is expected to be of the kind considered in [Achenbach and Reddy (1967)] and [Sun (1970)] by using the theory of propagating surfaces of discontinuity, and in [Buchen (1974)] by using the ray-series method. Here, however, we show that a term-by-term inversion yields an expansion which is *convergent* in any space-time domain. For this purpose we use the following theorem, stated in [Mainardi and Turchetti (1975)].

Theorem. *Let $\widetilde{R}(x,s)$ be analytic for $|s| > \delta$, uniformly continuous in x for $0 \le x \le X$, with $\widetilde{R}(x,\infty) = 0$, whose expansion reads*

$$\widetilde{R}(x,s) = \sum_{k=0}^{\infty} \frac{w_k(x)}{s^{k+1}}, \quad |s| > \delta, \tag{5.13}$$

then the inverse Laplace transform $R(x,t)$ is an entire function of t of exponential type, whose expansion reads

$$R(x,t) = \sum_{k=0}^{\infty} w_k(x) \frac{t^k}{k!}, \quad uniformly\ in\ x, \tag{5.14}$$

and vice versa.

We easily recognize that the conditions of the above theorem are fulfilled if:
(i) the creep compliance is an entire function of exponential type, i.e.

$$J(t) \in \{1, \delta_\psi\} \iff s\widetilde{J}(s) = J_g \left[1 + \sum_{k=1}^{\infty} \frac{\psi_k}{s^k} \right], \quad |s| > \delta_\psi;$$

(ii) the input $r_0(t)$ is an entire function of exponential type, i.e.

$$r_0(t) \in \{1, \delta_0\} \iff \widetilde{r}_0(s) = \sum_{k=0}^{\infty} \frac{\rho_k}{s^{k+1}}, \quad |s| > \delta_0. \tag{5.15}$$

While assumption (i) has been requested since the beginning, assumption (ii) can be released by considering the *impulse response* (or *Green function*) $\mathcal{G}(x,t)$ as shown in Section 4.2, Eq. (4.12), and performing a suitable convolution with $r_0(t)$.

We now illustrate the procedure to find the coefficients $w_k(x)$ by recurrence from the creep coefficients. The required expansion can be determined by substituting the expansions (5.12)-(5.13) into Eq. (5.11) and collecting like powers of s. For this purpose let us expand the operator in brackets at L.H.S of Eq. (5.11) in power series of s. Dividing by the highest power of s, i.e. $-2s$, and defining this operator as $\hat{L}(x,s)$, we have

$$\hat{L}(x,s) = \sum_{i=0}^{\infty} \hat{L}_i(x)\, s^{-i}, \qquad (5.16)$$

where, using (5.8) and (5.12),

$$\hat{L}_0 = \frac{d}{dx}, \; \hat{L}_1 = -\frac{1}{2}\frac{d^2}{dx^2} + \alpha\frac{d}{dx} - \beta, \; \hat{L}_i = \frac{1}{2}\psi_{i+1}, \; i \geq 2. \quad (5.17)$$

With a minimum effort the coefficients $w_k(x)$ in Eq. (5.13) prove to be solutions of a recursive system of linear first-order differential equations with initial conditions $w_k(0) = \rho_k$ ($k = 0, 1, 2, \ldots$). In fact, since the expansion of $\hat{L}\{R(x,s)\}$ is given by termwise application of the L_i to the $w_k(x)$, the coefficients $w_k(x)$ satisfy the recursive system of equations

$$\begin{cases} \hat{L}_0\, w_0 = 0, \\ \hat{L}_0\, w_k = -\sum_{i,j} \hat{L}_i\, w_j, \quad k \geq 1, \end{cases} \qquad (5.18)$$

where the summation is taken over values of i, j for which $i+j = k+1$. The solutions of this system are easily seen to be polynomials in x of degree k, which we write in the form

$$w_k(x) = \sum_{h=0}^{k} A_{k,h}\, \frac{x^h}{h!}, \quad w_k(0) = \rho_k, \qquad (5.19)$$

where the $A_{k,h}$ are obtained from the initial data $A_{k,0} = \rho_k$ by the following recurrence relations, with $1 \leq h \leq k$,

$$\begin{aligned} A_{k,h} &= \frac{1}{2} A_{k-1,h+1} - \alpha\, A_{k-1,h} + \beta\, A_{k-1,h-1} \\ &\quad -\frac{1}{2} \sum_{j=2}^{k-h+1} \psi_{j+1}\, A_{k-j,h-1}, \; 1 \leq h \leq k. \end{aligned} \qquad (5.20)$$

Finally, we obtain the following series representation

$$R(x,t) = \sum_{k=0}^{\infty} \sum_{h=0}^{k} A_{k,h}\, \frac{x^h}{h!}\, \frac{t^k}{k!}, \qquad (5.21)$$

that with (5.10) provides the requested solution. This solution is easy to handle for numerical computations since it is obtained in a recursive way. However, the numerical convergence of the series is expected to fall down far from the wave front, as we will see later.

Examples: the Zener and the Maxwell models. By way of example, let us consider the Zener model (S.L.S.) for which $n(s)$ is provided in Section 4.3 by Eq. (4.51). Taking $\tau_\sigma = 1$ and $\tau_\epsilon = 1/a > 1$, we easily know the corresponding rate of creep in time and Laplace domains and consequently, we derive the creep coefficients. We have

$$\Psi(t) = (1-a)e^{-at} \implies \psi_k = (1-a)(-a)^{k-1}, \quad k = 1, 2, \dots \quad (5.22)$$

with $\delta_\Psi = 1$ as requested by Eq. (5.3a) In particular, we obtain the coefficients $\alpha = (1-a)/2$ and $\beta = (1-a)(1+3a)/8$ as requested by Eqs. (5.7)-(5.8).

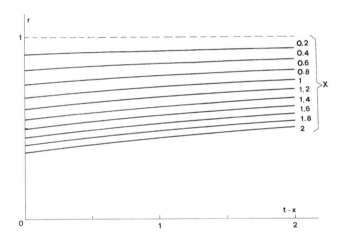

Fig. 5.1 The pulse response for the Maxwell model depicted versus $t - x$ for some fixed values of x.

For $a = 0$ the Zener model is known to reduce to the Maxwell model, see Eq. (4.52), so we simply get

$$\Psi(t) = 1, \implies \psi_k = \delta_{k\,1}, \quad k = 1, 2, \dots \quad (5.23)$$

Consequently, for the Maxwell model the recurrence relation (5.20) simplifies to

$$A_{k,h} = \frac{1}{2} A_{k-1,h+1} - \frac{1}{2} A_{k-1,h} + \frac{1}{8} A_{k-1,h-1}. \quad (5.24)$$

The computations for solutions corresponding to an initial Heaviside step function $r_0(t) = \Theta(t)$ provide the plots in Fig. 5.1 depicted versus time elapsed from the wave front for some values of x.

5.2 The singular wave–front expansion

We now consider viscoelastic models for which the *material functions* $J(t)$, $G(t)$ and consequently the respective *memory functions* $\Psi(t)$, $\Phi(t)$ are no longer entire functions of exponential type.

In this framework we discuss the general method developed by [Buchen and Mainardi (1975)], which provides an asymptotic series solution when the creep compliance $J(t)$ of the viscoelastic medium exhibits at the time origin a behaviour of the form

$$J(t) = J_g + O(t^\gamma) \quad \text{as} \quad t \to 0^+, \tag{5.25}$$

where $J_g \geq 0$ is the instantaneous compliance and $0 < \gamma \leq 1$.

Creep compliances which conform to this representation cover a wide class of viscoelastic materials including the models for which $J(t)$ has a well-defined expansion of about $t = 0^+$ in fractional non-negative powers of t with $J_g \geq 0$.

The starting point of the Buchen–Mainardi method is the asymptotic behaviour of $\mu(s)$ provided by Eq. (4.8) as $s \to \infty$. Note that in our present case we cannot use the function $n(s)$ and the consequent normalization $c = 1$, since when $J_g = 0$ the wave–front velocity is infinite. In general, for $\mu(s)$ we have an expansion in decreasing powers of s of the form

$$\mu(s) := \sqrt{\rho}\, s\, [s\widetilde{J}(s)]^{1/2} \sim \sum_{k=0}^{\infty} b_k s^{1-\beta_k}, \; 0 = \beta_0 < \beta_1 < \ldots. \tag{5.26}$$

Denoting by $\mu_+(s)$ the sum of the first $(m+1)$ terms with $\beta_k \leq 1$ ($k = 0, 1, \ldots, m$) and by $\mu_-(s)$ the remainder of the series, we can write the transform solution as

$$\widetilde{r}(x,s) = \widetilde{r}_0(s)\, e^{-x\mu(s)} = \widetilde{r}_0(s)\, e^{-x\mu_+(s)}\, e^{-x\mu_-(s)}. \tag{5.27}$$

Let us now set

$$\widetilde{R}(x,s) := e^{-x\mu_-(s)}, \tag{5.28}$$

and, without loosing generality, we agree to take as input the unit step Heaviside function, $r_0(t) = \Theta(t)$ so that $\widetilde{r}_0(s) := 1/s$. The function $\mu_+(s)$ is referred to as the principal part of the expansion of $\mu(s)$. It can be obtained with a minimum effort from the first few terms of the expansion of $J(t)$ as $t \to 0^+$. From the asymptotic limits (4.8), (4.15) and (5.25) we easily infer

$$\begin{cases} J_g = 0 \implies b_0 = 0, \quad \beta_1 = \gamma/2, \\ J_g > 0 \implies b_0 = \sqrt{\rho J_g} = 1/c, \quad \beta_1 = \gamma. \end{cases} \quad (5.29)$$

For $\widetilde{R}(x,s)$ we seek an asymptotic expansion in negative powers of s as $s \to \infty$ of the kind

$$\widetilde{R}(x,s) \sim \sum_{k=0}^{\infty} w_k(x) s^{-\lambda_k}, \quad 0 \le \lambda_0 < \lambda_1 < \dots. \quad (5.30)$$

Furthermore, let us set

$$\widetilde{\Phi}_k(x,s) := s^{-(\lambda_k + 1)} e^{-x[\mu_+(s) - s/c]}, \quad (5.31)$$

with the convention $1/c = 0$ when $J_g = 0$. Then, from Eqs. (5.27)-(5.28) and (5.30)-(5.31) *the Laplace transform of the complete solution admits the following asymptotic expansion as $s \to \infty$*:

$$\widetilde{r}(x,s) \sim e^{-xs/c} \sum_{k=0}^{\infty} w_k(x) \widetilde{\Phi}_k(x,s). \quad (5.32)$$

The purpose of the exponential function in (5.32) is to isolate, upon inversion, the wave front propagating with velocity c ($0 < c \le \infty$).

From Eqs. (5.26), (5.31) two things are evident for $k = 0, 1, \dots$, as $s \to \infty$

$$\begin{aligned} &(i) \ \widetilde{\Phi}_{k+1}(x,s) = o\left(\widetilde{\Phi}_k(x,s)\right), \\ &(ii) \ e^{\delta s} \widetilde{\Phi}_k(x,s) \to \infty, \ \forall \delta > 0. \end{aligned} \quad (5.33)$$

Because of a lemma from [Erdélyi (1956)], we expect that the conditions (5.33) will allow a term-by-term inversion of (5.32), which provides the required asymptotic solution in the time domain, as $t \to (x/c)^+$. Therefore, finally, our asymptotic solution reads in time-space domain

$$r(x,t) \sim \sum_{k=0}^{\infty} w_k(x) \Phi_k(x, t - x/c), \text{ as } t \to (x/c)^+. \quad (5.34)$$

We now discuss the recursive methods to determine
a) the functions $w_k(x)$ and the exponents λ_k;
b) the functions $\Phi_k(x,t)$, the Laplace inverse of (5.31).

As far as the first goal is concerned, we note that $\widetilde{R}(x,s)$ formally satisfies the following differential equation obtained from the general equation (4.7) with the positions (5.27) and (5.28),

$$\left\{ \frac{d^2}{dx^2} - 2\left[\mu_+(s)\right]\frac{d}{dx} - \left[\mu^2(s) - \mu_+^2(s)\right] \right\} \widetilde{R}(x,s) = 0, \qquad (5.35)$$

subjected to the initial condition $\widetilde{R}(0,s) = 1$.

In order to obtain the coefficients $w_k(x)$ and exponents λ_k in the asymptotic expansion (5.30), we will use an argument which generalizes the one followed in the previous Section in order to allow non-integer powers of s. This argument is based on a general theorem stated by [Friedlander and Keller (1955)], which we report for convenience.

Theorem by Friedlander and Keller. *Let $\hat{L}$ be a linear operator that admits an asymptotic expansion with respect to a parameter ϵ as $\epsilon \to 0$ of the form $\hat{L} \sim \sum_{i=0}^{\infty} \epsilon^{\nu_i} \hat{L}_i$ with $0 = \nu_0 < \nu_1 < \dots$, and v a solution of $\hat{L}v = 0$, with asymptotic expansion as $\epsilon \to 0$,*

$$v \sim \sum_{k=0}^{\infty} \epsilon^{\lambda_k} v_k, \quad \lambda_0 < \lambda_1 < \dots$$

If the asymptotic expansion of $\hat{L}v$ is given by termwise application of the $\hat{L}_i$ to the v_j and if $\hat{L}_0 v_k \neq 0$ for $k > 0$, then the coefficients v_k satisfy the recursive system of equations

$$\begin{cases} \hat{L}_0 v_0 = 0, \\ \hat{L}_k v_k = -\sum_{i,j} \hat{L}_i v_j, \quad k = 1, 2, \dots, \end{cases}$$

where the summation is taken over values of i, j for which $\nu_i + \lambda_j = \lambda_k$. The constant λ_0 is arbitrary but λ_k for $k > 0$ is the $(k+1)$-st number in the increasing sequence formed from the set of numbers $\lambda_0 + \sum_{i=1}^{\infty} m_i \nu_i$ where the m_i are any non-negative integers.

In the application of this theorem to our problem, we recognize that the $\hat{L}_i$ are differential operators with respect to x, and

$$\epsilon = s^{-1}, \quad v = \widetilde{R}(x,s), \quad v_k = w_k(x).$$

Ch. 5: Waves in Linear Viscoelastic Media: Asymptotic Representations 119

If in the differential equation (5.35) governing $\widetilde{R}(x,s)$ we expand the coefficients and divide the L.H.S. by the term containing the highest power of s (i.e. $-2b_0 s$ if $J_g > 0$ or $-2b_1 s^{1-\beta_1}$ if $J_g = 0$), we obtain quite generally:
$$\begin{cases} \hat{L}_0 = \dfrac{d}{dx}, \\ \hat{L}_i = p_i \dfrac{d^2}{dx} + q_i \dfrac{d}{dx} + r_i, \quad i = 1, 2, \ldots, \end{cases} \tag{5.36}$$
where p_i, q_i, r_i, like the exponents ν_i, can be determined from the behaviour of the creep compliance $J(t)$ as $t \to 0^+$, in view of the previous considerations. This fact will be clarified by some examples.

Now, taking into account the asymptotic expansion (5.30) for $\widetilde{R}(x,s)$ in negative powers of s with the initial condition $\widetilde{R}(0,s) = 1$, the application of the Friedlander-Keller theorem leads to the determination of the exponents λ_k and of the coefficients $w_k(x)$ with $k = 0, 1, 2, \ldots$. In fact, we have for the constants λ_k the recursive algebraic system
$$\begin{cases} \lambda_0 = 0, \\ \lambda_k = \displaystyle\sum_{i=1}^{\infty} m_i \nu_i, \quad k = 1, 2, \ldots, \end{cases} \tag{5.37}$$
and for the functions $w_k(x)$ the recursive system of linear differential equations
$$\begin{cases} \dfrac{dw_0}{dx} = 0, \\ \dfrac{dw_k}{dx} = -\displaystyle\sum_{i,j} \left(p_i \dfrac{d^2}{dx} + q_i \dfrac{d}{dx} + r_i \right) w_j(x), \quad k = 1, 2, \ldots. \end{cases} \tag{5.38}$$
subjected to the initial conditions
$$w_k(0) = \delta_{k0}. \tag{5.39}$$
The solutions of the differential system are easily seen to be polynomials in x of order k, which we write in the form
$$w_k(x) = \sum_{h=0}^{k} A_{k,h} \frac{x^k}{k!}, \quad k = 0, 1, \ldots, \tag{5.40}$$
where the coefficients $A_{k,h}$ are obtained from the initial data $A_{k,0} = \delta_{k0}$ by the following recurrence relations
$$A_{k,h} = -\sum_{i,j} (p_i A_{j,h+1} + q_i A_{j,k} + r_i A_{j,h-1}), \quad 1 \le h \le k. \tag{5.41}$$

Let us now consider the functions $\Phi_k(x,t)$ whose Laplace transforms are provided by Eqs. (5.31) with (5.26). We now derive a recursive method for their determination. We have

$$\widetilde{\Phi}_k(x,s) = s^{-(\lambda_k+1)} e^{-xb_1 s^{1-\beta_1}} \ldots e^{-xb_m s^{1-\beta_m}}. \qquad (5.42)$$

We first set, for sake of convenience, with $i = 1, 2, \ldots, m$,

$$y_i = x\, b_i\,, \quad \gamma_i = 1 - \beta_i \quad 1 > \gamma_1 > \ldots \gamma_m \geq 0\,,$$

and

$$\widetilde{G}_{\gamma_i}(y_i, s) = \exp\left(-y_i s^{\gamma_i}\right). \qquad (5.43)$$

Then, inverting the Laplace transforms, the required functions $\Phi_k(x,t)$ read as

$$\Phi_k(x,t) = \frac{t^{\lambda_k}}{\Gamma(\lambda_k+1)} * G_{\gamma_1}(y_1,t) * \ldots * G_{\gamma_m}(y_m,t), \qquad (5.44)$$

where $*$ denotes the convolution from 0 to t. For any fixed $\gamma_i = \gamma$ and $y_i = y$, we recognize that the generic function $G_\gamma(y,t)$ turns out to be related to the auxiliary F_γ-Wright function, see in Appendix F Eqs. (F.12) and (F.28), so

$$G_\gamma(y,t) = \frac{1}{t} F_\gamma\left(\frac{y}{t^\gamma}\right) = \sum_{n=1}^{\infty} \frac{(-y)^n\, t^{-(\gamma n+1)}}{n!\, \Gamma(-\gamma n)}, \quad t > 0\,. \qquad (5.45)$$

We note that [Buchen and Mainardi (1975)], albeit unaware of the Wright functions, have provided analytical representations of $G_\gamma(y,t)$ in the following special cases

$$\begin{cases} G_{1/3}(y,t) = \dfrac{y}{3^{1/3}\, t^{4/3}}\, \text{Ai}\left(\dfrac{y}{3^{1/3}\, t^{1/3}}\right), \\ \\ G_{1/2}(y,t) = \dfrac{y}{2\sqrt{\pi}\, t^{3/2}}\, \exp\left(-\dfrac{y^2}{4t}\right), \end{cases} \qquad (5.46)$$

where Ai is the Airy function. More importantly, they have sought efficient methods of obtaining the inversion formulas. For this purpose they have introduced the functions

$$H_\gamma(z, \lambda_k) = t^{-\lambda_k}\left[\frac{t^{\lambda_k}}{\Gamma(\lambda_k+1)} * G_\gamma(y,t)\right], \quad z = \frac{y}{t^\gamma}, \qquad (5.47)$$

and have provided the following recurrence relation

$$\lambda_k\, H_\gamma(z, \lambda_k) = -\gamma\, z\, H_\gamma(z, \lambda_k - \gamma) + H_\gamma(z, \lambda_k - 1), \qquad (5.48)$$

which can simplify the determination of the functions $\Phi_k(x,t)$.

The expansions for analytical models. Particular examples are the models for which $J(t)$ admits a Taylor expansion about $t = 0^+$ with $J_g > 0$, without necessarily being an entire function of exponential type; we refer to these models as *simply analytical models*, to be distinguished from the *full analytical models* considered in the previous section. It is clear that for the *full* or *simply analytical models* the method must provide the solution (5.10) with a wave–front expansion of the type (5.21). Of course, this expansion is expected to be convergent or only asymptotic, correspondingly. In fact, in these cases, we obtain

$$\begin{cases} \nu_i = i, & \lambda_k = k, \\ p_1 = -1/2, & q_1 = \alpha, & r_1 = -\beta, \\ p_i = q_i = 0, & r_i = \psi_{i+1}/2, & i \geq 2, \end{cases} \quad (5.49)$$

and

$$\Phi_k(x,t) = e^{-\alpha x} \frac{t^k}{k!}. \quad (5.50)$$

Thus, the wave–front expansion is originated by the recurrence relation (5.20), but with $A_{k,0} = \delta_{k0}$.

The expansions for non-analytical models. To better illustrate the importance of the Buchen–Mainardi method we need to consider *non-analytical models*, for which we may have, for example,

$$\mu^+(s) = b_0 + b_1 s^{\beta_1}, \quad 0 < \beta_1 < 1. \quad (5.51)$$

Instructive examples that conform to (5.51) are the simple *Voigt model*, see in Section 2.4 Eqs. (2.17), and the *fractional Maxwell model of order* $1/2$, see in Section 3.1 Eqs. (3.18) with $\nu = 1/2$. In fact, after a suitable normalization, these models are described as follows.
(a) *Voigt model*:

$$\sigma(t) = \epsilon(t) + \frac{d\epsilon}{dt} \implies J(t) = 1 - e^{-t}, \quad (5.52a)$$

so that

$$\mu(s) = s\,(s+1)^{-1/2} \implies b_0 = 0, \, b_1 = 1, \, \beta_1 = 1/2; \quad (5.53a)$$

(b) *fractional Maxwell model of order* $1/2$:
$$\sigma(t) + \frac{d^{1/2}\sigma}{dt^{1/2}} = \frac{d^{1/2}\epsilon}{dt^{1/2}} \implies J(t) = 1 + \frac{t^{1/2}}{\Gamma(3/2)}, \quad (5.52b)$$
so that
$$\mu(s) = s\,(1+s^{-1/2})^{1/2} \implies b_0 = -1/8\,,\; b_1 = 1/2\,,\; \beta_1 = 1/2\,. \quad (5.53b)$$

As a consequence, for the above models, we respectively obtain
$$\begin{cases} \nu_1 = 1/2,\; \nu_2 = 1,\; \nu_3 = 3/2,\; \nu_i = 0 \text{ for } i \geq 4, \\ \lambda_k = k/2,\; j = k-1,\, k-2,\, k-3; \end{cases} \quad (5.54a)$$
and
$$\begin{cases} \nu_1 = 1/2,\; \nu_2 = 1,\; \nu_i = 0 \text{ for } i \geq 3, \\ \lambda_k = k/2,\; j = k-1,\, k-2. \end{cases} \quad (5.54b)$$

Then the coefficients $A_{k,h}$ of the polynomials $w_k(x)$ in (5.38)-(5.40) turn out to be obtained from the initial data $A_{k,0} = \delta_{k0}$ by the following recurrence relations
$$A_{k,h} = \frac{1}{2} A_{k-1,h+1} - \frac{1}{2} A_{k-1,h-1} - A_{k-2,h} + \frac{1}{2} A_{k-3,h+1}, \quad (5.55a)$$
$$1 \leq h \leq k\,,$$
and
$$A_{k,h} = -\frac{1}{2} A_{k-1,h} + \frac{1}{16} A_{k-1,h-1} - \frac{1}{2} A_{k-2,h+1}$$
$$-\frac{1}{8} A_{k-2,h} - \frac{1}{128} A_{k-2,h-1},\; 1 \leq h \leq k\,. \quad (5.55b)$$

Furthermore, for both models we obtain
$$\Phi_k(x,t) = e^{-b_0 x}\, t^{k/2}\, H_{1/2}(z, k/2)\,, \quad (5.56)$$
with
$$z = \frac{b_1\, x}{t^{1/2}}\,,\; H_{1/2}(z, k/2) = 2^k\, I^k \text{erfc}\left(\frac{z}{2}\right). \quad (5.57)$$

These repeated integrals of the error function are easily computed from the following recurrence relations found in Appendix C, see also [Abramowitz and Stegun (1965)], which are a particular case of the most general relations (5.47). Setting for convenience
$$K_k(z) := H_{1/2}(z, k/2) \quad (5.58)$$

we have
$$\begin{cases} K_{-1}(z) = \frac{1}{\sqrt{\pi}} \exp\left(-\frac{z^2}{4}\right), & K_0(z = \operatorname{erfc}\left(\frac{z}{2}\right), \\ K_k(z) = -\frac{z}{k} K_{k-1}(z) + \frac{2}{k} K_{k-2}(z), & k \geq 1. \end{cases} \quad (5.59)$$

In conclusion, we report below the whole asymptotic expansion for the two models.

(a) *Voigt model* as $t \to x^+$:

$$r(x,t) \sim \sum_{k=0}^{\infty} \sum_{h=0}^{k} A_{k,h} \frac{x^h}{h!} t^{k/2} K_k\left(\frac{x}{\sqrt{t}}\right), \quad (5.60a)$$

where the functions $K_k(z)$ are defined by (5.57)-(5.58) and the coefficients $A_{k,h}$ are obtained from the recurrence relation (5.56a). The Voigt model exhibits a response $\forall t > 0$ and has a non-analytical expansion, which we call *diffusion-like* response. Figure 5.2 displays the essential character for times up to twice the retardation time.

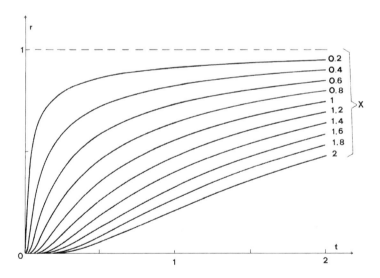

Fig. 5.2 The pulse response for the Voigt model depicted versus t.

(b) *Fractional Maxwell model of order* $1/2$ as $t \to x^+$:

$$r(x,t) \sim e^{x/8} \sum_{k=0}^{\infty} \sum_{h=0}^{k} A_{k,h} \frac{x^h}{h!} (t-x)^{k/2} K_k\left(\frac{x}{2\sqrt{t-x}}\right), \quad (5.60b)$$

where the functions $K_k(z)$ are still defined by (5.57)-(5.58) and the coefficients $A_{k,h}$ are obtained from the recurrence relation (5.55b). The fractional Maxwell model displays features in common to both the simple Maxwell and simple Voigt models. There is no motion for $t < x$, but the response at the front $t = x^+$ is zero for $\forall x > 0$ and its expansion is non-analytic, which we call *wave-diffusion like* response. Figure 5.3 shows the essential characteristics of the pulse in the neighbourhood of the onset.

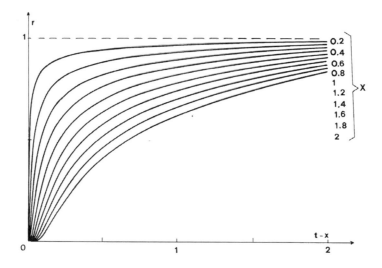

Fig. 5.3   The pulse response for the Maxwell 1/2 model depicted versus $t - x$.

Discussion on wave–front expansions. We have presented a general asymptotic theory, in the neighbourhood of the onset, for the propagation of an initial step pulse $r_0(t) = \Theta(t)$ in a semi-infinite viscoelastic rod. We have considered fairly general classes of models which are characterized by a creep compliance having power law behaviour for small times. These include both analytic and non-analytic behaviour at the origin and it is the nature of this behaviour which determines the character of the solution at the onset.

As instructive examples we have provided details for three viscoelastic models; one analytic: the simple Maxwell model, and two singular: the simple Voigt model and the fractional Maxwell model of order 1/2. Hereafter, for brevity, we will list the particular results for $r(x,t)$ when we take only the first term of our asymptotic solutions. We obtain (in dimensionless form):

(i) simple Maxwell model (analytic) at $t \simeq x^+$:

$$r(x,t) \simeq \exp\left(-\frac{x}{2}\right) \Theta(t-x); \qquad (5.61)$$

(ii) simple Voigt model (non-analytic) at $t \simeq 0^+$:

$$r(x,t) \simeq \operatorname{erfc}\left(\frac{x}{2\sqrt{t}}\right) \Theta(t) \simeq \frac{2}{\sqrt{\pi}} \frac{\sqrt{t}}{x} \exp\left(-\frac{x^2}{4t}\right) \Theta(t); \qquad (5.62)$$

(iii) fractional Maxwell model of order 1/2 (non-analytic) at $t \simeq x^+$:

$$\begin{aligned} r(x,t) &\simeq \exp\left(\frac{x}{2}\right) \operatorname{erfc}\left(\frac{x}{4\sqrt{t-x}}\right) \Theta(x-t) \\ &\simeq \frac{4}{\sqrt{\pi}} \frac{\sqrt{t-x}}{x} \exp\left[\frac{x}{8} - \frac{x^2}{16(t-x)}\right] \Theta(x-t). \end{aligned} \qquad (5.63)$$

Our asymptotic representation, though strictly valid only near the pulse onset, is particularly suitable for the study of pulse transmission in viscoelastic media of long relaxation or retardation times and provides a useful method for determining how the pulse develops from the onset. A great bonus of the method is its particularly suitable form for numerical computation. What is otherwise a complicated problem, has been reduced to a series in which everything is determined by simple recurrence relations. The series was found to display remarkable convergence, particularly for the smaller times and distances involved.

The three models chosen for detailed study represent the different types of solution which relate to the given creep compliance. These are the wave type, the diffusion type and the wave–diffusion type. The wave-type solution, corresponding to the analytic creep models, reduces to the ray–series solution obtained by [Buchen (1974)]. The other types represent an extension of the ray–series method for non-analytic creep models. Finally, we remark that the solution for an arbitrary source function $r_0(t)$ can be obtained by convolution with our asymptotic series.

5.3 The saddle–point approximation

5.3.1 *Generalities*

The wave-front expansions, even when mathematically convergent in any space-time domain, cannot be used to represent the wave evolution sufficiently far from the onset of the wave front. In fact, their numerical convergence is expected to slow down by increasing the time elapsed from the wave front, with a rate depending on the spatial coordinate. Thus, it is customary to use the saddle-point method to invert the Laplace representation of the response variable, which provides a suitable approximation of the wave evolution sufficiently far from the wave front. Precisely, using the notation of Chapter 4, see Eq. (4.17) we apply the saddle point method to the integral in (4.17), that we re-write as

$$r(x,t) = \frac{1}{2\pi i} \int_{Br} \widetilde{r}_0(s) \, e^{s[t - (x/c)\, n(s)]} \, ds$$
$$= \frac{1}{2\pi i} \int_{Br} \widetilde{r}_0(s) \, e^{(x/c)\, s[\theta - n(s)]} \, ds \, , \ \theta = ct/x > 1 \, . \quad (5.64)$$

For the sake of convenience we have considered a finite wave–front velocity c so that all considerations for the signal velocity illustrated in Subsection 4.4.3, including the introduction of the non-dimensional parameter θ and the steepest–descent path $L(\theta)$, will apply. If $s_0(\theta)$ denotes the relevant saddle point of

$$F(s;\theta) := s\,[\theta - n(s)] \, , \quad (5.65)$$

the method just consists in taking the dominant contribution of s_0 to the integral along the steepest–descent path. In the cases when this path is equivalent to the Bromwich path and the relevant saddle point is not close to any singularities of $\widetilde{r}_0(s)$, the saddle-point method allows us to write, see e.g. [Bleistein and Handelsman (1986)],

$$r(x,\theta) \sim \frac{\widetilde{r}(s_0)}{2\pi i} e^{(x/c)F(s_0)} \int_{L(\theta)} e^{(x/c)F''(s_0)(s - s_0)^2} \, ds \, ,$$
$$F'(s_0) = \left.\frac{dF}{ds}\right|_{s_0} = 0 \, , \ F''(s_0) = \left.\frac{d^2 F}{ds^2}\right|_{s_0} \neq 0 \, , \quad (5.66)$$

where on $L(\theta)$ the $\mathcal{R}e\,[F]$ attains its maximum value at s_0 and $\mathcal{I}m\,[F]$ is constant.

Then, recalling the Gauss integral $\int_{-\infty}^{\infty} \exp{-|a|u^2}\,du = \sqrt{\pi/|a|}$, we obtain the required asymptotic representation as

$$r(x,\theta) \sim \frac{\widetilde{r}(s_0)}{\sqrt{2\pi |F''(s_0)|}}\exp[(x/c)F(s_0)]\,,\quad s_0 = s_0(\theta)\,. \tag{5.67}$$

We note that for this method a detailed knowledge of the path L is not usually required, while the location of the relevant saddle point with respect to the singularities of $\widetilde{r}_0(s)$ is relevant to get a uniform approximation.

As instructive exercises, let us now consider the saddle–point method for two particular impact problems in a Maxwell and in a Zener viscoelastic solid, in order to evaluate its applicability. These problems have been denoted respectively as Lee-Kanter and Jeffreys problems from the names of the scientists who have formerly investigated them.

5.3.2 *The Lee-Kanter problem for the Maxwell model*

We know from the analysis in Chapter 4 that the Maxwell model is the simplest viscoelastic solid that admits a finite wave–front velocity. Recalling its characteristic function $n(s)$ from (4.52) with $\tau_\sigma = 1$,

$$n(s) = (1+1/s)^{1/2}\,, \tag{5.68}$$

the Laplace representation (4.17) for the general impact problem, using non-dimensional variables with $c=1$, reads

$$r(x,t) = \frac{1}{2\pi i}\int_{Br} \widetilde{r}_0(s)\,e^{s[t - x(1+1/s)^{1/2}]}\,ds\,. \tag{5.69}$$

This formula is indeed sufficiently simple to allow us to determine the solution in the space–time domain. In fact, writing

$$sn(s) = s(1+1/s)^{1/2} = (s^2+1/s)^{1/2} = \left[(s+1/2)^2 - 1/4\right]^{1/2}, \tag{5.70}$$

we can apply the Laplace transform pair (B.64) to (5.69)-(5.70). We obtain for $t \geq x$:

$$\begin{aligned}r(x,t) &= e^{-x/2}r_0(t-x)\\ &+ \frac{x}{2}\int_x^t e^{-t'/2}r_0(t-t')\frac{I_1\left[(t'^2-x^2)^{1/2}/2\right]}{(t'^2-x^2)^{1/2}}\,dt'\,,\end{aligned} \tag{5.71}$$

whre I_1 denotes the modified Bessel function of order 1.

For the so-called Lee-Kanter problem, see [Lee and Kanter (1956)], the impact problem (in non-dimensional variables) consists in finding the stress when the semi-infinite bar is subjected at the accessible end $x = 0$ to a unit step of velocity $v(t) = \Theta(t)$. In our notations, recalling Eqs. (4.1)-(4.8), this means to find the Laplace inversion of

$$\widetilde{r}(x,s) = \widetilde{r}_0(s)\exp[-xn(s)] \quad \text{with} \quad \widetilde{r}_0(s) = \frac{1}{s\,n(s)}, \qquad (5.72)$$

where $sn(s)$ is given by (5.70). Then, by recalling the Laplace transform pair (B.62), we have the solution in explicit form through a modified Bessel function of order 0,

$$r(x,t) = e^{-t/2} I_0\left[\frac{\sqrt{t^2 - x^2}}{2}\right], \; t \geq x. \qquad (5.73)$$

The simplicity of the Lee-Kanter problem provides us with the opportunity to evaluate for it the evolution of the steepest–descent path and to compare the saddle–point approximation of the solution with the closed exact expression (5.73). For this purpose we first determine the saddle points and the steepest–decent path through them.

For the Maxwell model Eq. (4.81) for the saddle points reads

$$n(s) + s\frac{dn}{ds} = \left(1 + \frac{1}{s}\right)^{1/2}\left[1 - \frac{1}{2(1+s)}\right] = 0, \qquad (5.74)$$

that, when rationalized, is a quadratic equation in s with two real solutions. These solutions are the required saddle points, which read

$$s^{\pm} = -\frac{1}{2}\left[1 \mp \frac{\theta}{\sqrt{\theta^2 - 1}}\right]. \qquad (5.75)$$

Therefore, these saddle points move on the real axis from $\pm\infty$ (for $\theta = 1$) to the branch points of $n(s)$: $s_\star^+ = 0$, $s_\star^- = -1$ (for $\theta = \infty$).

The path of steepest descent through $s^{\pm}$ is defined by the condition (4.82), that now reads

$$\mathcal{I}m\,[F(s,\theta)] = \mathcal{I}m\,\left\{s\theta - [s(s+1)]^{1/2}\right\} = 0. \qquad (5.76)$$

Setting $s = \xi + i\eta$, this condition leads to the following algebraic equation

$$\frac{(\xi + 1/2)^2}{a^2} + \frac{\eta^2}{b^2} = 1, \quad a = \frac{\theta}{2\sqrt{\theta^2 - 1}}, \quad b = \frac{1}{2\sqrt{\theta^2 - 1}}. \quad (5.77)$$

Thus, the curve L represented by (5.77) is an ellipse, with axes $2a$, $2b$ and foci in the branch points $s_\star^\pm$, that intersects the real axis in the saddle points $s^\pm$.

In Fig. 5.4 we show the evolution of the path $L(\theta)$ in the s-complex plane for the Maxwell model for the following five values of $1/\theta$: 1) 0.85, 2) 0.75, 3) 0.50, 4) 0.25, 5) 0.15. From the figure we easily understand the evolution of the ellipse from a big circle at infinity (for $\theta = 1$) to the segment of the branch cut (for $\theta = \infty$).

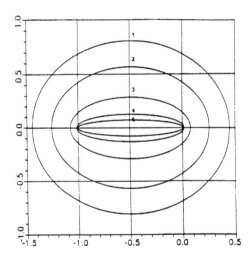

Fig. 5.4 The evolution of the steepest-descent path $L(\theta)$ for the Maxwell model.

When the integration is carried out on the *entire* curve L, the exact result is expected to be found for any x and t, without any problem related to an oscillating integrand since on L the imaginary part of $F(s, \theta)$ is constant.

For large x and t we can presume that the dominant contribution to the integral along L comes from a neighborhood of s_0: this yields the leading term in the asymptotic expansion of the wave form, that can easily be found by the *saddle-point method*. It is easy to prove that s^+ is the relevant saddle point s_0, being a maximum of $Re\{F(s;\theta)\}$ on L, and that the original Bromwich path Br can be deformed into the ellipse L. Thus, the saddle–point method provides the following approximate representation

$$r(\theta,t) \sim \sqrt{\frac{2\theta}{\pi t}}\,[s_0(s_0+1)]^{1/4}\,e^{t\{s_0 - [s_0(s_0+1)]^{1/2}/\theta\}}. \qquad (5.78)$$

We can easily check the range of validity of the saddle-point method. Indeed, in Fig. 5.5, we compare the exact solution [continuous line] with the approximate solution [dashed line] for some fixed values of time $[t = 2, 4, 6, 8, 10]$. We recognize that the saddle-point method provides a satisfactory approximation (i.e. a discrepancy of less than one per cent) for any x, only if $t \geq 8$. For smaller time values the contribution to the integral from a neighborhood of the relevant saddle point s_0 is thus inadequate to represent the solution.

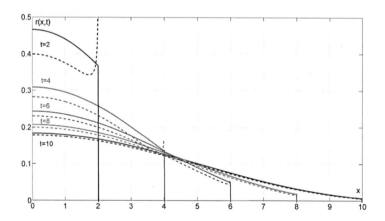

Fig. 5.5   The Lee-Kanter pulse for the Maxwell model depicted versus x.

The saddle-point approximation is of course not uniform on the wave front, since for $\theta \to 1$ the two saddle points coalesce at infinity. However, it can be proved to be regular for $\theta \to \infty$; for $\theta = \infty$ (i.e. at $x = 0$) it yields $r(x=0,t) \sim 1/\sqrt{\pi t} \leq \exp(-t/2)\,I_0(t/2)$, where on the R.H.S. the exact value is quoted from (5.73).

5.3.3 The Jeffreys problem for the Zener model

Early attempts to explain the duration of the seismogram led Sir Harold Jeffreys, the great applied mathematician and precursor of the modern seismology, to investigate the effect of viscoelasticity on the propagation of the seismic pulse, see [Jeffreys (1931); Jeffreys (1932)]. He adopted the saddle–point approximation to evaluate the effect on an initial Heaviside step pulse propagating in the simplest viscoelastic models. In the first paper, Jeffreys considered the Voigt model, whereas in the second, the Maxwell model and the combination of the previous ones (known later as Standard Linear Solid or Zener model). He calls the Voigt effect, "firmoviscosity", and the Maxwell effect, "elastovicosity". The second analysis, revisited later by the present author [Mainardi (1972)], is now presented to illustrate the saddle-point approximation for the Zener model.

As usual, let us start with the Bromwich representation of the response variable for a unit step impact $r_0(t) = \Theta(t) \div r_0(s) = 1/s$. From Eq. (4.17) we have

$$\widetilde{r}(x,s) = \frac{1}{2\pi i} \int_{Br} \frac{1}{s} \exp\left\{s\left[t - \frac{x}{c} n(s)\right]\right\} ds, \qquad (5.79)$$

where, recalling Eq. (4.51),

$$n(s) = \left[\frac{s + 1/\tau_\sigma}{s + 1/\tau_\epsilon}\right]^{1/2}, \quad \tau_\epsilon > \tau_\sigma > 0. \qquad (5.80)$$

Then, for our case, Eq. (4.81) for the saddle points reads

$$\theta = n(s) + s\frac{dn}{ds}$$
$$= \left[\frac{s + 1/\tau_\sigma}{s + 1/\tau_\epsilon}\right]^{1/2}\left[1 - \frac{1}{2(s\tau_\sigma + 1)} + \frac{1}{2(s\tau_\epsilon + 1)}\right]. \qquad (5.81)$$

When rationalized, Eq. (5.81) gives a quartic equation for s, with two real roots which represent the required saddle points. Their exact location can be obtained from θ graphically. Figure 5.6 shows an example for $\tau_\epsilon/\tau_\sigma = 1.5$. For $\theta = 1$ ($t = x/c$) they are at $\pm\infty$, then, with increasing time, come closer to the two branch points of $n(s)$, ($s = -1/\tau - \epsilon$, $s = -1/\tau_\sigma$), tending to them for $\theta \to \infty$ ($t \to \infty$). Setting

$$n_0 = n(0) = \sqrt{\tau_\epsilon/\tau_\sigma} > 1, \quad w := c/n_0 < c, \qquad (5.82)$$

we recognize that for $\theta = n_0$ ($t = x/w$) the saddle point is at $s = 0$, which is the pole of the integrand. For [Jeffreys (1932)] the saddle point was erroneously located between the pole ($s = 0$) and the branch point ($s = -1\tau_\epsilon$); then, there is no distinction between the instants $t = x/c$ ($\theta = 1$) and $t = x/w$ ($\theta = n_0$), and the contribution of the saddle point for $t \simeq x/w$ is missed. Then for $1 < \theta < n_0$

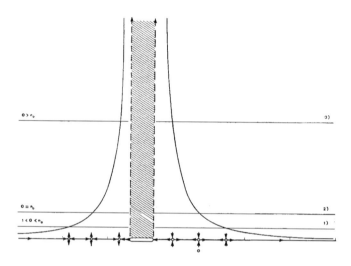

Fig. 5.6 The position of the saddle points as a function of time elapsed from the wave front: 1) $1 < \theta < n_0$; 2) $\theta = n_0$; 3) $\theta > n_0$; where $n_0 = \sqrt{1.5}$.

the path of steepest descent is equivalent to the Bromwich path; for $\theta > n_0$ one must add the contribution from the pole which represents the elastic solution.

For $\theta \simeq n_0$ the saddle-point method is expected to give the main contribution, since we have the coupled contributions from the saddle point and the pole. But, because of the vicinity of the two points, we have to adopt a modified version of the standard saddle-point method in order to have a uniform approximation. Based on the method by [Van der Waerden (1951)], [Mainardi (1972)] got the final result

$$r(x,t) = \frac{1}{2}\left[1 + \mathrm{erf}\left(\frac{t - x/w}{\Delta}\right)\right], \qquad (5.83)$$

where erf denotes the Error function, see Appendix C, and

$$\Delta = \sqrt{2(\tau_\epsilon - \tau_\sigma)\, x/c}\,. \tag{5.84}$$

In view of the behaviour of Error function, the result (5.83)-(5.84) implies that the pulse is not sharp; it begins a little before the instant $t = x/w$ and increases continuously tending to the initial value 1, the greater part of the change being spread over an interval of order 2Δ. This reduction of the abruptness of the pulse is analogous to the effect found by Jeffreys with the Voigt model [Jeffreys (1931)], referred to as the firmoviscous effect by him. As emphasized in [Jeffreys (1931)], the effect of firmoviscosity is of more direct seismological interest than that of elastoviscosity provided by the Maxwell model that practically implies a reduction of the amplitude of the shock. The reduction of the abruptness of the shock is suggestive, because the seismic (S and P) waves in distant earthquakes begin less abruptly than in near ones. The measure of the broadening, found for sufficient great distances, is given by Δ; it is proportional to the square root of the distance, as seen from the expression of Δ in (5.84). For more details we refer the interested reader to [Mainardi (1972)].

5.4 The matching between the wave–front and the saddle–point approximations

Padè Approximants. To obtain the solution to a given impact problem in any desired space–time domain the two approximations investigated in the previous sections are not always sufficient because the matching between them can be lost in some intermediate regions.

In order to meet the requirement of a matching, Mainardi and Turchetti, see [Mainardi and Turchetti (1975); Turchetti and Mainardi (1976)], have proposed to accelerate the numerical convergence of series of the wave-front approximation with the technique of *Padè approximants*, henceforth referred to as PA.

For details on this technique the interested reader is referred to specialized treatises e.g. [Baker (1975)], [Baker and Gammel (1970)], [Bender and Orszag (1987)].

We now limit ourselves to provide the basic ideas of Padé approximation. We begin by pointing out that PA are a noteworthy rational approximation to a function represented by a power (convergent or asymptotic) series. When a power series representation of a function diverges, it indicates the presence of singularities, so that the divergence of the series reflects the inability of a polynomial to approximate a function adequately near a singularity. However, even when the power series representation converges, as it is the case of the Taylor series for an entire function exponential type, which are bounded at infinity, the polynomials are not able to adequately represent the function. We are faced with the same difficulty as in the evaluation of $\exp(-x)$ using its Taylor expansion when x is large.

The basic idea of Padé summation is to replace a power series, say $\sum_{n=0}^{\infty} a_n z^n$ with a sequence of rational functions of the form

$$P_N^M(z) = \frac{\sum_{n=0}^{M} A_n z^n}{\sum_{n=0}^{N} B_n z^n}, \qquad (5.85)$$

where we choose $B_0 = 1$ without loss of generality. We choose the remaining $(M + N + 1)$ coefficients $A_0, A_1, \ldots A_M, B_1, B_2, \ldots B_N$, so that the first $(M + N + 1)$ terms in the Taylor series expansion of $P_N^M(z)$ match the first $(M + N + 1)$ terms of the power series $\sum_{n=0}^{\infty} a_n z^n$. The resulting rational function $P_N^M(z)$ is called *Padè approximant*.

As a matter of fact, the construction of $P_N^M(z)$ becomes very useful. If $\sum_{n=0}^{\infty} a_n z^n$ is a power series representation of the function $f(z)$, then in many instances $P_N^M(z) \to f(z)$ as $M, N \to \infty$ even if the power series is divergent. Usually one considers only the convergence of Padè sequences $P_0^J, P_1^{1+J}, P_2^{2+J}, \ldots$ having $M = N + J$ with J fixed and $N \to \infty$. The special sequence $J = 0$ is called *diagonal sequence*. The full power series representation of a function need not be known to construct a PA, just the first $M + N + 1$ terms. Since PA involve only algebraic operations, they are determined by a simple sequence of matrix operations.

Ch. 5: Waves in Linear Viscoelastic Media: Asymptotic Representations   135

Pulse responses in the Zener viscoelastic model. For our impact problems concerning viscoelastic waves, the technique of diagonal PA allowed Mainardi and Turchetti to use a reasonable number of series terms of the wave-front approximation (no more than 20) in order to have a matching with the long-time asymptotic solution, obtained by the saddle-point method. So, a representation of the wave phenomenon in any space-time domain of interest may be achieved. In particular, these authors obtained noteworthy results (indeed in perfect agreement with closed-form solutions when available) to represent the Jeffreys pulse response and the Lee-Kanter pulse response for the Zener (S.L.S.) model, as illustrated in Figs. 5.7 and 5.8.

There we have considered non dimensional space-time variables by setting $c = 1$ and $\tau_\sigma = 1$, and we have fixed the parameter $a := \tau_\sigma/\tau_\epsilon = 0.5$ in Eq. (5.22).

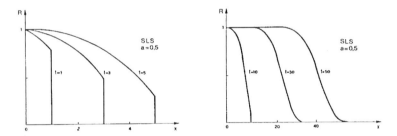

Fig. 5.7 The step-pulse response for the Zener (S.L.S.) model depicted versus x: for small times (left) and for large times (right).

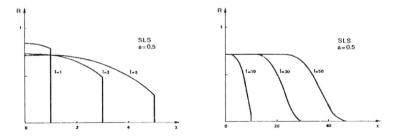

Fig. 5.8 The Lee-Kanter pulse response for the Zener (S.L.S.) model depicted versus x: for small times (left) and for large times (right).

Chapter 6

Diffusion and Wave–Propagation via Fractional Calculus

In this chapter we analyse some boundary value problems for partial differential equations of fractional order in time. These are fundamental for understanding phenomena of anomalous diffusion or intermediate between diffusion and wave propagation. A typical process of the second kind is provided by the one-dimensional propagation of stress pulses in a linear viscoelastic medium with constant Q, which is quite relevant in seismology. This process is indeed governed by an evolution equation of fractional order in time, which interpolates the Fourier heat equation and the D'Alembert wave equation. We show that the fundamental solutions for the corresponding Cauchy and Signalling problems are expressed in terms of functions of the Wright type in the similarity variable and their behaviour turns out to be intermediate between those for the limiting cases of a perfectly viscous fluid and a perfectly elastic solid.

6.1 Introduction

The evolution equations of fractional order. It is known that the standard partial differential equations governing the basic phenomena of diffusion and wave propagation are the Fourier diffusion equation and the D'Alembert wave equation, respectively.

Denoting as usual x, t the space and time variables, and $r = r(x,t)$ the response variable, these equations read:

$$\frac{\partial r}{\partial t} = d\,\frac{\partial^2 r}{\partial x^2}, \qquad (6.1)$$

$$\frac{\partial^2 r}{\partial t^2} = c^2 \frac{\partial^2 r}{\partial x^2}. \qquad (6.2)$$

In Eq. (6.1) the constant d denotes a diffusivity coefficient, whereas in Eq. (6.2) c denotes a characteristic velocity. In this Chapter we consider the family of evolution equations

$$\frac{\partial^\beta r}{\partial t^\beta} = a \frac{\partial^2 r}{\partial x^2}, \quad 0 < \beta \leq 2, \qquad (6.3)$$

where *the time derivative of order β is intended in the Caputo sense*, namely is the operator ${}_0^*D_t^\beta$, introduced in Chapter 1, Eq. (1.17), and a is a positive constant of dimension $L^2 T^{-\beta}$. We must distinguish the cases $0 < \beta \leq 1$ and $1 < \beta \leq 2$. We have

$$\frac{\partial^\beta r}{\partial t^\beta} := \begin{cases} \dfrac{1}{\Gamma(1-\beta)} \displaystyle\int_0^t \left[\dfrac{\partial}{\partial \tau} r(x,\tau)\right] \dfrac{d\tau}{(t-\tau)^\beta}, & 0 < \beta < 1, \\[2mm] \dfrac{\partial r}{\partial t}, & \beta = 1; \end{cases} \qquad (6.4a)$$

$$\frac{\partial^\beta r}{\partial t^\beta} := \begin{cases} \dfrac{1}{\Gamma(2-\beta)} \displaystyle\int_0^t \left[\dfrac{\partial^2}{\partial \tau^2} r(x,\tau)\right] \dfrac{d\tau}{(t-\tau)^{\beta-1}}, & 1 < \beta < 2, \\[2mm] \dfrac{\partial^2 r}{\partial t^2}, & \beta = 2. \end{cases} \qquad (6.4b)$$

It should be noted that in both cases $0 < \beta \leq 1$, $1 < \beta \leq 2$, the time fractional derivative in the L.H.S. of Eq. (6.3) can be removed by a suitable fractional integration[1], leading to alternative forms where the necessary initial conditions at $t = 0^+$ explicitly appear. As a matter fact, we get the integro-differential equations:
if $0 < \beta \leq 1$:

$$r(x,t) = r(x,0^+) + \frac{a}{\Gamma(\beta)} \int_0^t \left(\frac{\partial^2 r}{\partial x^2}\right)(t-\tau)^{\beta-1}\, d\tau; \qquad (6.5a)$$

---

[1] We apply to Eq. (6.3) the fractional integral operator of order β, namely ${}_0I_t^\beta$. For $\beta \in (0,1]$ we have:

$${}_0I_t^\beta \circ {}_0^*D_t^\beta\, r(x,t) = {}_0I_t^\beta \circ {}_0I_t^{1-\beta} D_t^1\, r(x,t) = {}_0I_t^1 D_t^1\, r(x,t) = r(x,t) - r(x,0^+).$$

For $\beta \in (1,2]$ we have:

$${}_0I_t^\beta \circ {}_0^*D_t^\beta\, r(x,t) = {}_0I_t^\beta \circ {}_0I_t^{2-\beta} D_t^2\, r(x,t) = {}_0I_t^2 D_t^2\, r(x,t)$$
$$= r(x,t) - r(x,0^+) - r_t(x,0^+).$$

if $1 < \beta \leq 2$:
$$r(x,t) = r(x,0^+) + t\frac{\partial}{\partial t}r(x,t)|_{t=0^+} + \frac{a}{\Gamma(\beta)}\int_0^t \left(\frac{\partial^2 r}{\partial x^2}\right)(t-\tau)^{\beta-1}\,d\tau.$$
(6.5b)

The plan of the Chapter. In Section 6.2, we analyse the fractional evolution equation (6.3) in the general case $0 < \beta < 2$, essentially based on our previous works, see e.g. [Mainardi (1994a); (1995a); (1995b); (1996a); (1996b); (1997)]. We first consider the two basic boundary-value problems, referred to as the *Cauchy problem* and the *Signalling problem*, by the technique of the Laplace transforms, and we derive the transformed expressions of the respective fundamental solutions, often referred to as the *Green functions* of the corresponding problems. Then we carry out the inversion of the relevant transforms and we outline a *reciprocity relation* between the Green functions in the space-time domain. In view of this relation, the Green functions can be expressed in terms of two interrelated *auxiliary functions* in the similarity variable $\xi = |x|/(\sqrt{a}t^\nu)$, where $\nu = \beta/2$. These auxiliary functions can be considered as restrictions on the real line of entire complex functions of the Wright type, see Appendix F.

In Section 6.3, we outline the scaling properties of the fundamental solutions and we exhibit their evolution for some values of the order ν. We also show how the fundamental solutions can be interpreted as probability density functions related to certain Lévy stable distributions with index of stability depending on the order of the fractional derivative.

Finally, in Section 6.4, we deal with the signalling problem for uniaxial stress waves in a viscelastic solid exhibiting a creep compliance proportional to t^γ with $0 < \gamma < 1$. The evolution equation is shown to be of type (6.3) with $1 < \beta := 2 - \gamma < 2$. Since $1 < \beta < 2$, the behaviour of the Green function turns out to be intermediate between diffusion (found for a viscous fluid) and wave-propagation (found for an elastic solid), so that it is common to speak about *fractional diffusive waves*. We conclude the chapter with a section devoted to historical and bibliographical notes.

6.2 Derivation of the fundamental solutions

Green functions for the Cauchy and Signalling problems.
In order to guarantee the existence and the uniqueness of the solution, we must equip (6.1) with suitable data on the boundary of the space-time domain. The basic boundary-value problems for diffusion are the so-called Cauchy and Signalling problems. In the Cauchy problem, which concerns the space-time domain $-\infty < x < +\infty$, $t \geq 0$, the data are assigned at $t = 0^+$ on the whole space axis (initial data). In the Signalling problem, which concerns the space-time domain $x \geq 0$, $t \geq 0$, the data are assigned both at $t = 0^+$ on the semi-infinite space axis $x > 0$ (initial data) and at $x = 0^+$ on the semi-infinite time axis $t > 0$ (boundary data); here, as mostly usual, the initial data are assumed to vanish.

Denoting by $f(x)$, $x \in \mathbb{R}$ and $h(t)$, $t \in \mathbb{R}^+$ sufficiently well-behaved functions, the basic problems are thus formulated as following, assuming $0 < \beta \leq 1$,

a) Cauchy problem

$$r(x,0^+) = f(x)\,, \; -\infty < x < +\infty\,;\; r(\mp\infty,t) = 0\,,\; t > 0; \quad (6.6a)$$

b) Signalling problem

$$r(x,0^+) = 0\,,\; x > 0;\; r(0^+,t) = h(t)\,,\; r(+\infty,t) = 0\,,\; t > 0. \quad (6.6b)$$

If $1 < \beta < 2$, we must add into (6.6a) and (6.6b) the initial values of the first time derivative of the field variable, $r_t(x,0^+)$, since in this case the fractional derivative is expressed in terms of the second order time derivative. To ensure the continuous dependence of our solution with respect to the parameter β also in the transition from $\beta = 1^-$ to $\beta = 1^+$, we agree to assume

$$\frac{\partial}{\partial t} r(x,t)|_{t=0^+} = 0\,, \text{ for } 1 < \beta \leq 2\,, \quad (6.7)$$

as it turns out from the integral forms (6.5a)-(6.5b).

In view of our subsequent analysis we find it convenient to set

$$\nu := \beta/2\,, \text{ so } \begin{cases} 0 < \nu \leq 1/2\,, \iff 0 < \beta \leq 1\,, \\ 1/2 < \nu \leq 1\,, \iff 1 < \beta \leq 2\,, \end{cases} \quad (6.8)$$

and from now on to add the parameter ν to the independent space-time variables x, t in the solutions, writing $r = r(x,t;\nu)$.

For the Cauchy and Signalling problems we introduce the so-called Green functions $\mathcal{G}_c(x,t;\nu)$ and $\mathcal{G}_s(x,t;\nu)$, which represent the respective fundamental solutions, obtained when $f(x) = \delta(x)$ and $h(t) = \delta(t)$. As a consequence, the solutions of the two basic problems are obtained by a space or time convolution according to

$$r(x,t;\nu) = \int_{-\infty}^{+\infty} \mathcal{G}_c(x-\xi,t;\nu)\, f(\xi)\, d\xi\,, \qquad (6.9a)$$

$$r(x,t;\nu) = \int_{0^-}^{t^+} \mathcal{G}_s(x,t-\tau;\nu)\, h(\tau)\, d\tau\,. \qquad (6.9b)$$

It should be noted that in (6.9a) $\mathcal{G}_c(x,t;\nu) = \mathcal{G}_c(|x|,t;\nu)$ because the Green function of the Cauchy problem turns out to be an even function of x. According to a usual convention, in (6.9b) the limits of integration are extended to take into account for the possibility of impulse functions centred at the extremes.

Reciprocity relation and auxiliary functions for the standard diffusion and wave equations. First, let us consider the particular cases $\nu = 1/2$ and $\nu = 1$, which correspond to the standard diffusion equation (6.1) with $a = d$ and to the standard wave equation (6.3) with $a = c^2$. For these cases the two Green functions for the Cauchy and Signalling problems are usually derived in classical texts of mathematical physics by using the techniques of Fourier transforms in space and Laplace transforms in time, respectively.

Then, using the notation $\mathcal{G}^d_{c,s}(x,t) := \mathcal{G}_{c,s}(x,t;1/2)$, the two Green functions of the *standard diffusion equation* read:

$$\mathcal{G}^d_c(x,t) = \frac{1}{2\sqrt{\pi a}}\, t^{-1/2}\, e^{-x^2/(4\,a\,t)} \stackrel{\mathcal{F}}{\leftrightarrow} e^{-at\kappa^2}\,, \qquad (6.10a)$$

$$\mathcal{G}^d_s(x,t) = \frac{x}{2\sqrt{\pi a}}\, t^{-3/2}\, e^{-x^2/(4\,a\,t)} \stackrel{\mathcal{L}}{\leftrightarrow} e^{-(x/\sqrt{a})s^{1/2}}\,. \qquad (6.10b)$$

From the explicit expressions (6.10a)-(6.10b) we recognize *the reciprocity relation* between the two Green functions, for $x > 0$, $t > 0$:

$$x\,\mathcal{G}^d_c(x,t) = t\,\mathcal{G}^d_s(x,t) = F^d(\xi) = \frac{1}{2}\,\xi\,M^d(\xi)\,, \qquad (6.11)$$

where
$$M^d(\xi) = \frac{1}{\sqrt{\pi}} e^{-\xi^2/4}, \quad \xi = \frac{x}{\sqrt{a}\, t^{1/2}} > 0. \tag{6.12}$$

The variable ξ plays the role of *similarity variable* for the standard diffusion, whereas the two functions $F^d(\xi)$ and $M^d(\xi)$ can be considered the *auxiliary functions* for the diffusion equation itself because each of them provides the fundamental solutions through (6.11). We note that the function $M^d(\xi)$ satisfies the normalization condition $\int_0^\infty M^d(\xi)\, d\xi = 1$.

For the *standard wave equation*, using the notation $\mathcal{G}_{c,s}^w(x,t) := \mathcal{G}_{c,s}(x,t;1)$, the two Green functions read

$$\mathcal{G}_c^w(x,t) = \frac{1}{2}\left[\delta(x - \sqrt{a}\,t) + \delta(x + \sqrt{a}\,t)\right]$$
$$\overset{\mathcal{F}}{\leftrightarrow} \frac{1}{2}\left[e^{+i\sqrt{a}\,t\kappa} + e^{-i\sqrt{a}\,t\kappa}\right], \tag{6.13a}$$

$$\mathcal{G}_s^w(x,t) = \delta(t - x/\sqrt{a}) \overset{\mathcal{L}}{\leftrightarrow} e^{-(x/\sqrt{a})s}. \tag{6.13b}$$

From the explicit expressions (6.13a)-(6.13b) we recognize *the reciprocity relation* between the two Green functions, for $x > 0$, $t > 0$:

$$2x\, \mathcal{G}_c^w(x,t) = t\, \mathcal{G}_s^w(x,t) = F^w(\xi) = \xi\, M^w(\xi), \tag{6.14}$$

where
$$M^w(\xi) = \delta(1-\xi), \quad \xi = \frac{x}{\sqrt{a}\, t} > 0. \tag{6.15}$$

Even if ξ does not appear as a similarity variable in the ordinary sense, we attribute to ξ and to $\{F^w(\xi), M^w(\xi)\}$ the roles of similarity variable and auxiliary functions of the standard wave equation.

Reciprocity relation and auxiliary functions for the fractional diffusion and wave equations. We now properly extend the previous results to the general case $0 < \nu \leq 1$ by determining the two Green functions through the technique of Laplace transforms.

We show how this technique allows us to obtain the transformed functions $\widetilde{\mathcal{G}}_c(x,s;\nu)$, $\widetilde{\mathcal{G}}_s(x,s;\nu)$, by solving ordinary differential equations of the second order in x and then, by inversion, the required Green functions $\mathcal{G}_c(x,t;\nu)$ and $\mathcal{G}_s(x,t;\nu)$ in the space-time domain.

For the Cauchy problem (6.6a) with $f(x) = \delta(x)$, the application of the Laplace transform to Eqs. (6.3)-(6.4) with $r(x,t) = \mathcal{G}_c(x,t;\nu)$ and $\nu = \beta/2$ leads to the non-homogeneous differential equation satisfied by the image of the Green function, $\widetilde{\mathcal{G}}_c(x,s;\nu)$,

$$a \frac{d^2 \widetilde{\mathcal{G}}_c}{dx^2} - s^{2\nu} \widetilde{\mathcal{G}}_c = -\delta(x) s^{2\nu-1}, \quad -\infty < x < +\infty. \quad (6.16)$$

Because of the singular term $\delta(x)$ we have to consider the above equation separately in the two intervals $x < 0$ and $x > 0$, imposing the boundary conditions at $x = \mp\infty$, $\mathcal{G}_c(\mp\infty, t;\nu) = 0$, and the necessary matching conditions at $x = 0^{\pm}$. We obtain

$$\widetilde{\mathcal{G}}_c(x,s;\nu) = \frac{1}{2\sqrt{a}\, s^{1-\nu}} e^{-(|x|/\sqrt{a})s^\nu}, \quad -\infty < x < +\infty. \quad (6.17)$$

A different strategy to derive $\mathcal{G}_c(x,t;\nu)$ is to apply the Fourier transform to Eqs. (6.3)-(6.4) as illustrated in [Mainardi et al. (2001)], [Mainardi and Pagnini (2003)], to which the interested reader is referred for details.

For the Signalling problem (6.6b) with $h(t) = \delta(t)$, the application of the Laplace transform to Eqs. (6.3)-(6.4) with $r(x,t) = \mathcal{G}_s(x,t;\nu)$ and $\nu = \beta/2$, leads to the homogeneous differential equation

$$a \frac{d^2 \widetilde{\mathcal{G}}_s}{dx^2} - s^{2\nu} \widetilde{\mathcal{G}}_s = 0, \quad x \geq 0. \quad (6.18)$$

Imposing the boundary conditions $\mathcal{G}_s(0^+, t;\nu) = h(t) = \delta(t)$ and $\mathcal{G}_s(+\infty, t;\nu) = 0$, we obtain

$$\widetilde{\mathcal{G}}_s(x,s;\nu) = e^{-(x/\sqrt{a})s^\nu}, \quad x \geq 0. \quad (6.19)$$

The transformed solutions provided by Eqs. (6.17) and (6.19) must be inverted to provide the requested Green functions in the space-time domain. For this purpose we recall the Laplace transform pairs related to the transcendental functions $F_\nu(r)$, $M_\nu(r)$ of the Wright type, discussed in Appendix F, see Eqs. (F.28)-(F.29), where r stands for the actual time coordinate t. In fact, the Laplace transform pairs in Appendix F imply

$$\frac{1}{\nu} F_\nu(1/t^\nu) = \frac{1}{t^\nu} M_\nu(1/t^\nu) \div \frac{e^{-s^\nu}}{s^{1-\nu}}, \quad 0 < \nu < 1. \quad (6.20)$$

$$\frac{1}{t} F_\nu(1/t^\nu) = \frac{\nu}{t^{\nu+1}} M_\nu(1/t^\nu) \div e^{-s^\nu}, \quad 0 < \nu < 1, \quad (6.21)$$

so that these formulas can be used to invert the transforms in Eqs. (6.17) and (6.19).

Then, introducing for $x > 0$, $t > 0$, the *similarity variable*
$$\xi := x/(\sqrt{a}\, t^\nu) > 0 \qquad (6.22)$$
and recalling the rules of scale-change in the Laplace transform pairs[2], after some manipulation we obtain the Green functions in the space-time domain in the form
$$\mathcal{G}_c(x,t;\nu) = \frac{1}{2\nu x} F_\nu(\xi) = \frac{1}{2\sqrt{a}\, t^\nu} M_\nu(\xi), \qquad (6.23a)$$
$$\mathcal{G}_s(x,t;\nu) = \frac{1}{t} F_\nu(\xi) = \frac{\nu x}{\sqrt{a}\, t^{1+\nu}} M_\nu(\xi). \qquad (6.23b)$$
We also recognize the following *reciprocity relation* for the original Green functions,
$$2\nu x\, \mathcal{G}_c(x,t;\nu) = t\, \mathcal{G}_s(x,t;\nu) = F_\nu(\xi) = \nu\xi\, M_\nu(\xi). \qquad (6.24)$$
Now $F_\nu(\xi)$, $M_\nu(\xi)$ are the *auxiliary functions* for the general case $0 < \nu \le 1$, which generalize those for the standard diffusion given in Eqs. (6.11)-(6.12) and for the standard wave equation given in Eqs. (6.14)-(6.15). In fact, for $\nu = 1/2$ and for $\nu = 1$ we recover the expressions of $M^d(\xi)$ and $M^w(\xi)$, respectively given by (6.12) and (6.15), as it can be easily verified using the formulas provided in Appendix F.

Hereafter, for the reader's convenience, we provide the series expansions for the auxiliary functions in powers of the similarity variable $\xi > 0$ as deduced from the corresponding series expansions in the complex domain given in Eqs. (F.12)-(F.13). We have:
$$\begin{aligned} F_\nu(\xi) &= \sum_{n=1}^{\infty} \frac{(-r)^n}{n!\, \Gamma(-\nu n)} \\ &= -\frac{1}{\pi} \sum_{n=1}^{\infty} \frac{(-\xi)^n}{n!} \Gamma(\nu n + 1)\, \sin(\pi\nu n), \end{aligned} \qquad (6.25)$$
$$\begin{aligned} M_\nu(\xi) &= \sum_{n=0}^{\infty} \frac{(-\xi)^n}{n!\, \Gamma[-\nu n + (1-\nu)]} \\ &= \frac{1}{\pi} \sum_{n=1}^{\infty} \frac{(-\xi)^{n-1}}{(n-1)!} \Gamma(\nu n)\, \sin(\pi\nu n). \end{aligned} \qquad (6.26)$$

---
[2] $f(t) \div \widetilde{f}(s)$, $f(bt) \div \frac{1}{b}\widetilde{f}(s/b)$, $\frac{1}{b}f(t/b) \div \widetilde{f}(bs)$, with $b > 0$.

Although convergent in all of $\mathbb{R}$, the series representations in (6.25)-(6.26) can be used to provide a numerical evaluation of our auxiliary functions only for relatively small values of ξ, so that asymptotic evaluations as $\xi \to +\infty$ are required. Following the considerations in Appendix F, see Eqs. (F.20)-(F.21), and choosing as a variable ξ/ν rather than ξ, the asymptotic representation for the M_ν function obtained by the standard saddle–point method reads:

$$M_\nu(\xi/\nu) \sim \frac{\xi^{(\nu - 1/2)/(1 - \nu)}}{\sqrt{2\pi\,(1-\nu)}} \exp\left[-\frac{1-\nu}{\nu}\,\xi^{1/(1-\nu)}\right]. \quad (6.27)$$

We note that the standard saddle-point method for $\nu = 1/2$ provides the exact result (6.12), i.e. $M_{1/2}(\xi) = M^d(\xi) = (1/\sqrt{\pi})\exp(-\xi^2/4)$, but breaks down for $\nu \to 1^-$. The case $\nu = 1$, (namely $\beta = 2$) for which Eq. (6.3) reduces to the standard wave equation (6.2), is of course a singular limit since $M_1(\xi) = \delta(1 - \xi)$. We postpone the discussion of this limit to the next Section where it is relevant for pulse propagation in certain viscoelastic solids.

Equation (6.27) along with Eqs. (6.22) and (6.23a), (6.23b) allows us to note the exponential decay of $\mathcal{G}_c(x, t; \nu)$ as $x \to +\infty$ (at fixed t) and the algebraic decay of $\mathcal{G}_s(x, t; \nu)$ as $t \to +\infty$ (at fixed x), for $0 < \nu < 1$. In fact, we get

$$\mathcal{G}_c(x, t; \nu) \sim A(t)\, x^{(\nu-1/2)/(1-\nu)}\, e^{-B(t)x^{1/(1-\nu)}}, \quad x \to \infty, \quad (6.28a)$$

$$\mathcal{G}_s(x, t; \nu) \sim C(x)\, t^{-(1+\nu)}, \quad t \to \infty, \quad (6.28b)$$

where $A(t)$, $B(t)$ and $C(x)$ are positive functions.

6.3 Basic properties and plots of the Green functions

Scaling properties of the Green functions. Looking at Eqs. (6.23a)-(6.23b) we recognize that for the Green function of the Cauchy [Signalling] problem the time [spatial] shape is the same at each position [instant], the only changes being due to space [time] - dependent changes of width and amplitude. The maximum amplitude in time [space] varies precisely as $1/x$ [$1/t$].

The two fundamental solutions exhibit *scaling properties* that make easier their plots versus distance (at fixed instant) and versus time (at fixed position). In fact, using the well-known scaling properties of the Laplace transform in (6.17) and (6.19) (see footnote [2]), we easily prove, for any $p, q > 0$, that

$$\mathcal{G}_c(px, qt; \nu) = \frac{1}{q^\nu} \mathcal{G}_c(px/q^\nu, t; \nu), \qquad (6.29a)$$

$$\mathcal{G}_s(px, qt; \nu) = \frac{1}{q} \mathcal{G}_s(px/q^\nu, t; \nu), \qquad (6.29b)$$

and, consequently, in plotting versus the space or time variable we can choose suitable values for the other variable kept fixed.

Plots of the Green functions. In Fig. 6.1, as an example, we compare versus $|x|$, at fixed t, the fundamental solutions of the Cauchy problem with different ν ($\nu = 1/4, 1/2, 3/4$). We consider the range $0 \le |x| \le 4$ for $t = 1$, assuming $a = 1$.

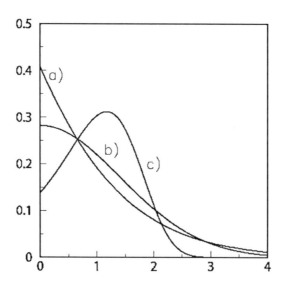

Fig. 6.1 The Cauchy problem for the time-fractional diffusion-wave equation: the fundamental solutions versus $|x|$ with a) $\nu = 1/4$, b) $\nu = 1/2$, c) $\nu = 3/4$.

In Fig. 6.2 as an example, we compare versus t, at fixed x, the fundamental solutions of the Signalling problem with different

ν ($\nu = 1/4$, $1/2$, $3/4$). We consider the range $0 \leq t \leq 3$ for $x = 1$, assuming $a = 1$.

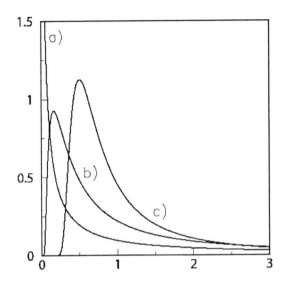

Fig. 6.2 The Signalling problem for the time-fractional diffusion-wave equation: the fundamental solutions versus t with a) $\nu = 1/4$, b) $\nu = 1/2$, c) $\nu = 3/4$.

The Green functions as probability density functions. We intend to show that each Green function that we have previously discussed can be interpreted as a probability density function (pdf): more precisely for the Cauchy problem as a spatial pdf evolving in time whereas that for the Signalling problem as a temporal pdf evolving in space. In virtue of Eqs. (6.23a)-(6.23b) the two Green functions are surely non-negative in the space-time domain being proportional to the auxiliary function M_ν which is positive on real axis as stated in Appendix F. It remains to prove the normalization conditions for both the Green functions. The proof for $\mathcal{G}_c(x, t; \nu)$ is obvious in view of its time evolution self-similarity. In fact, from Eq. (6.22) with (6.23a) and (F.33), we obtain:

$$\int_{-\infty}^{+\infty} \mathcal{G}_c(x, t; \nu) \, dx = \frac{1}{2} \int_{-\infty}^{+\infty} M_\nu(\xi) \, d\xi = 1 \, .$$

For $\mathcal{G}_s(x,t;\nu)$ the proof is less obvious since it requires more manipulation. In fact, from Eqs. (6.22) with (6.23b) and (F.33), the change of variable $t \to \xi$ yields $dt = -\sqrt{a}/(\nu x)\, t^{1+\nu}\, d\xi$, so that we have:

$$\int_0^{+\infty} \mathcal{G}_s(x,t;\nu)\, dt = -\frac{\nu x}{\sqrt{a}} \int_0^{+\infty} t^{-(1+\nu)} M_\nu(\xi)\, dt = \int_0^{+\infty} M_\nu(\xi)\, d\xi = 1.$$

We now show how our Green functions, being expressed in terms of Wright functions, may be related in particular to the class of Lévy stable distributions, following our discussion in Appendix F.

For sake of simplicity let us start with the Green functions of the standard diffusion equation (6.1) provided respectively by Eqs. (6.10a) and (6.10b) for the Cauchy and Signalling problems.

For the Cauchy problem we easily recognize (as indeed it is well known) that the corresponding Green function is the spatial *Gauss* or *normal* pdf, evolving with variance σ^2 proportional to the first power of time. In fact, recalling the general expression of the (symmetric) Gaussian pdf defined for $x \in \mathbb{R}$ with variance σ^2 and of its Fourier transform (the characteristic function), we have:

$$p_G(x;\sigma) := \frac{1}{\sqrt{2\pi}\,\sigma}\, e^{-x^2/(2\sigma^2)} \div e^{-(\sigma^2/2)\kappa^2}. \tag{6.30}$$

Furthermore, by comparing (6.10a) with (6.30) we get the identity

$$\mathcal{G}_c^d(x,t) = \frac{1}{2\sqrt{\pi a t}}\, e^{-x^2/(4at)} = p_G(x;\sigma = \sqrt{2at}). \tag{6.31}$$

We now consider the Green function of the standard diffusion equation for the Signalling problem. It is not so well known that it is related to the temporal *Lévy-Smirnov* pdf, evolving with a median proportional to the second power of space. In fact, recalling the general expression of the (unilateral) Lévy-Smirnov pdf defined for $t \in \mathbb{R}^+$ with parameter μ and of its Laplace transform, see [Feller (1971)],

$$p_{LS}(t;\mu) = \frac{\sqrt{\mu}}{\sqrt{2\pi}\, t^{3/2}}\, e^{-\mu/(2t)} \div e^{-\sqrt{2\mu}\, s^{1/2}}, \tag{6.32}$$

we have

$$\mathcal{G}_s^d(x,t) = \frac{x}{2\sqrt{\pi a}\, t^{3/2}}\, e^{-x^2/(4at)} = p_{LS}(t;\mu = x^2/(2a)). \tag{6.33}$$

Because the Lévy-Smirnov pdf decays at infinity like $t^{-3/2}$, all its moments of positive integer order are infinite. More precisely, we note that the moments of real order δ are finite only if $0 \leq \delta < 1/2$. In particular, for this pdf the mean (i.e. the expectation value) is infinite, but the median is finite, resulting $t_{med} \approx 2\mu$. In fact, recalling the cumulative distribution function

$$\mathcal{P}_{LS}(t;\mu) := \int_0^t p_L(t;\mu)\, dt = \mathrm{erfc}\left(\sqrt{\frac{\mu}{2t}}\right), \qquad (6.34)$$

from $\mathcal{P}_{LS}(t_{med};\mu) = 1/2$, it turns out that $t_{med} \approx 2\mu$, since the complementary error function gets the value $1/2$ where its argument is approximatively $1/2$.

In Probability theory, the *Gauss* and *Lévy–Smirnov* laws are special cases of the important class of *Lévy stable distributions* that we discuss in Appendix F, using the notation L_α^θ for the canonic representation of their densities. Then, the Gauss pdf $p_G(x;\sigma)$ corresponds to $L_2^0(x) \div \exp(-\kappa^2)$ if we set $\sigma^2 = 2$, whereas the Lévy-Smirnov pdf $p_{LS}(t;\mu)$ corresponds to $L_{1/2}^{-1/2}(t) \div \exp(-s^{1/2})$ setting $\mu = 1/2$. This means that, taking into account suitable scale factors, the two Green functions for the standard diffusion equation can be expressed in terms of stable densities as follows,

$$\mathcal{G}_c^d(x,t) = \frac{1}{2\sqrt{\pi a t}}\, e^{-x^2/(4at)} = \frac{1}{\sqrt{at}}\, L_2^0(x/\sqrt{at}), \qquad (6.35)$$

and

$$\mathcal{G}_s^d(x,t) = \frac{x}{2\sqrt{\pi a}\, t^{3/2}}\, e^{-x^2/(4at)} = \frac{a}{x^2}\, L_{1/2}^{-1/2}\left(ta/x^2\right). \qquad (6.36)$$

Now, let us consider the general case $0 < \nu < 1$ for which the two Green functions are given by Eqs. (6.23a)-(6.23b) in terms of our auxiliary Wright functions. In order to relate the Green functions with stable densities. we must recall the relations between our auxiliary functions and the extremal stable densities, proven in Appendix F, see Eqs. (F.48)-(F.49). Consequently we now are in condition to discuss the possibility to interpret our fundamental solutions (6.23a) and (6.23b) in terms of stable pdf's, so generalizing the arguments for the standard diffusion equation.

As far as the Cauchy problem is concerned, we note that the corresponding Green function is a symmetrical pdf in (scaled) distance with two branches, for $x > 0$ and $x < 0$, obtained one from the other by reflection.

For large $|x|$ each branch exhibits an exponential decay according to (6.28a) and, only for $1/2 \le \nu < 1$ it is the corresponding branch of an extremal stable pdf with index of stability $\alpha = 1/\nu$. In fact, from (6.23a) and (F.49) we obtain:

$$\mathcal{G}_c(|x|, t; \nu) = \frac{1}{2\nu} \frac{1}{\sqrt{a}\, t^\nu} L_{1/\nu}^{1/\nu-2}\left(\frac{|x|}{\sqrt{a}\, t^\nu}\right). \qquad (6.37)$$

This property had to the author's knowledge not yet been noted: it properly generalizes the Gaussian property of the pdf found for $\nu = 1/2$ (standard diffusion).

Furthermore, using the expression of the moments of the auxiliary function M_ν in (F.34), the moments (of even order) of $\mathcal{G}_c(x, t; \nu)$ turn out to be

$$\int_{-\infty}^{+\infty} x^{2n} \mathcal{G}_c(x, t; \nu)\, dx = \frac{\Gamma(2n+1)}{\Gamma(2\nu n + 1)} (at^{2\nu})^n, \quad n = 1, 2, \ldots \qquad (6.38)$$

We recognize that the variance is now growing like $t^{2\nu}$, which implies a phenomenon of *slow diffusion* if $0 < \nu < 1/2$ and *fast diffusion* if $1/2 < \nu < 2$.

We now consider the Green function for the Signalling problem. We recognize that it is a unilateral extremal stable pdf in (scaled) time with index of stability $\alpha = \nu$, which decays according to (6.28b) with a power law. In fact, from (6.23b) and (F.48) we obtain:

$$\mathcal{G}_s(x, t; \nu) = \left(\frac{\sqrt{a}}{x}\right)^{1/\nu} L_\nu^{-\nu}\left[t\left(\frac{\sqrt{a}}{x}\right)^{1/\nu}\right]. \qquad (6.39)$$

This property, that will be recalled in the next section, has been noted also by [Kreis and Pipkin (1986)].

6.4 The Signalling problem in a viscoelastic solid with a power-law creep

Let us now consider a Signalling problem of great relevance in seismology, that is the propagation of waves generated by an impact pulse of delta type in a viscoelastic medium exhibiting a loss tangent independent of frequency. Such viscoelastic medium, in view of our analysis in Chapter 3, is characterized by a stress-strain relation of Scott-Blair type introduced in Eqs. (3.16a)-(3.16b), with a creep compliance of power-law type. Taking into account the dependence on the spatial coordinate x, we now write the stress-strain relation as:

$$\sigma(x,t) = b_1 \frac{\partial^\gamma}{\partial t^\gamma} \epsilon(x,t), \ 0 < \gamma \le 1, \tag{6.40}$$

where the fractional derivative is intended in the Caputo sense, and γ stands for its order. Setting for convenience

$$b_1 = \rho a, \tag{6.41}$$

where ρ is the constant density of the medium, the creep compliance of such medium and its Laplace transform read

$$J(t) = \frac{1}{\rho a} \frac{t^\gamma}{\Gamma(\gamma+1)} \div \widetilde{J}(s) = \frac{1}{\rho a} s^{-(\gamma+1)}, \ 0 < \gamma \le 1. \tag{6.42}$$

Recalling Eq. (3.32), such medium exhibits a loss tangent or internal friction, denoted in seismology by Q^{-1} as in Eq. (2.63), which turns quite independent on frequency. We have

$$Q^{-1} = \tan\left(\frac{\pi}{2}\gamma\right). \tag{6.43}$$

Then, using the creep representation in Laplace domain discussed in Chapter 4, Eqs. (4.7)-(4.8), the evolution equation for viscoelastic waves $r(x,t) \div \widetilde{r}(x,s)$ propagating in such medium reads

$$\frac{d}{dx^2} \widetilde{r}(x,s) - \frac{s^{2-\gamma}}{a} \widetilde{r}(x,s) = 0,$$

which implies in the space-time domain

$$\frac{\partial^{2-\gamma}}{\partial t^{2-\gamma}} r(x,t) = a \frac{\partial}{\partial x^2} r(x,t). \tag{6.44}$$

Such an evolution equation is equipped with the conditions typical of the Signalling problem:
$$r(0,t) := r_0(t) = \delta(t) \iff \widetilde{r}(0,s) := \widetilde{r}_0(s) \equiv 1. \tag{6.45}$$
In order to be consistent with previous notations used in Sections 6.2, 6.3, let us set
$$\beta = 2\nu := 2 - \gamma, \quad r(x,t) = \mathcal{G}_s(x,t;\nu). \tag{6.46}$$
In fact, so doing, we recognize the analogy between Eqs. (6.44) with Eqs. (6.3) and (6.18). In view of Eq. (6.23b) we have
$$\mathcal{G}_s(x,t;\nu) = \frac{1}{2\sqrt{a}\,t^\nu} M_\nu \left(\frac{x}{\sqrt{a}\,t^\nu} \right). \tag{6.47}$$
Because the order of the time fractional derivative in (6.44) is included in the interval $(1,2)$, we are in presence of an evolution process that is intermediate between standard diffusion ($\gamma = 1$) and standard wave propagation ($\gamma = 0$). We agree to denote this process as *fractional diffusion-wave phenomenon* and the corresponding solutions as *fractional diffusive waves*.

We point out that such a viscoelastic model is of great interest in seismology and in material sciences. In fact, the independence of the Q from the frequency is experimentally verified in pulse propagation phenomena not only in seismology, see [Kjartansson (1979)], [Strick (1970)] [Strick (1982a)], [Strick and Mainardi (1982)], but also in many materials as formerly shown in [Kolsky (1956)]. From (6.43) we note that Q is also independent on the material constants ρ and a which, however, play a role in the phenomenon of wave dispersion.

The limiting cases of absence of energy dissipation (the elastic energy is fully stored) and of absence of energy storage (the elastic energy is fully dissipated) are recovered from (6.43) for $\gamma = 0$ (perfectly elastic solid) and $\gamma = 1$ (perfectly viscous fluid), respectively.

However, in view of seismological applications, we must take values the parameter γ in the creep law sufficiently close to zero in order to guarantee realistic values for the factor Q^{-1} of the order of 10^{-3}. This means that we deal with *nearly elastic* materials for which we can approximate in Eq. (6.43) the loss tangent with its argument and write
$$Q^{-1} \approx \frac{\pi}{2}\gamma \approx 1.57\,\gamma \iff \gamma \approx \left(\frac{2}{\pi Q} \right). \tag{6.48}$$

As a consequence our parameter ν must be sufficiently close to 1, let us say, in view of (6.46), $\nu = 1 - \epsilon$ with $\epsilon = \gamma/2$. This implies the adoption of the Pipkin method, see [Kreis and Pipkin (1986)], in the evaluation of the Wright M_ν function entering the formula (6.47), as illustrated in appendix F. As instructive examples for a realistic view of the resulting pulse evolution we chose in our plots of the Green function $\epsilon = 0.01$ and $\epsilon = 0.001$, as shown in Fig. 6.3.

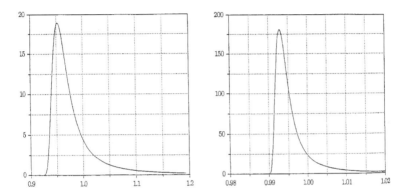

Fig. 6.3 Plots of the fundamental solution $\mathcal{G}_s(x,t;\nu)$ versus t at fixed $x = 1$ with $a = 1$, and $\nu = 1 - \epsilon$ ($\gamma = 2\epsilon$) in the cases: left $\epsilon = 0.01$, right $\epsilon = 0.001$.

We note an asymmetric time evolution of the pulse which, started as a delta function at $x = 0$, moves with a speed theoretically infinite. However, the peak of the pulse moves with a speed close to a, the characteristic velocity of the standard wave equation, being preceded by a non-causal monotonic forerunner

6.5 Notes

There is a huge literature concerning evolution equations of the types discussed above, both with and without explicit reference to the fractional calculus.

Neglecting the papers already cited in the previous sections, we now quote a number of references from the last century that have attracted our attention; [Caputo (1969); Caputo (1996a)], [Meshkov and Rossikhin (1970)], [Gonsovskii and Rossikhin (1973)], [Lokshin and Suvorova (1982)], [Nigmatullin (1986)], [Wyss (1986)], [Schneider and Wyss (1989)], [Fujita (1990a)], [Kochubei (1990)], [Giona and Roman (1992a); Giona and Roman (1992b)], [Prüsse (1983)], [Metzler et al. (1994)], [Gorenflo and Rutman (1994)], [Engler (1997)], [Rossikhin and Shitikova (1997a)], and [Saichev and Zaslavsky (1997)]. Of course, this list is not exhaustive.

The integro-differential equations (6.5a)-(6.5b) were investigated via Mellin transforms and Fox H functions by [Schneider and Wyss (1989)] in their pioneering paper. Starting from late 1992, the author has simplified the approach by Schneider-Wyss by using the more familiar technique of Laplace transform to deal with Eqs. (6.3)-(6.4). He has recognized in an explicit way that the fundamental solutions can be expressed in terms of a special function (of Wright type), a fact not known before. In addition, he has studied the analytical properties of this function and provided for the first time some significant plots. Unfortunately, due to some referees, who in those times were strongly in opposition with fractional calculus[3], his final results have appeared only some years later in refereed international journals, see [Mainardi (1995b); Mainardi (1996a); Mainardi (1996b)]. However, previous related works have been published in Proceedings of Conferences held in 1993, see [Mainardi (1994a); Mainardi (1995a)]. More complete treatments are found in author's CISM Lecture Notes, [Mainardi (1997)], and later in [Mainardi et al. (2001)], [Mainardi and Pagnini (2003)].

---

[3]Example of a report of an anonymous referee of the tuentieth century (June 1994), clearly against the use of fractional calculus in mathematical physics.

This paper is of insufficient interest to publish as a Letter; I believe that is sufficiently straightforward not to consider publishing it as a paper, either. The main drawback is that no application, either Physical or Mathematical, is really identified; what are these equations for? Anyone can write down some linear equations and then solve them, but that is not the point of doing Mathematical Physics. The paper should be rejected.

Appendix A

The Eulerian Functions

Here we consider the so-called Eulerian functions, namely the well known Gamma function and Beta function along with some special functions that turn out to be related to them, as the Psi function and the incomplete Gamma functions. We recall not only the main properties and representations of these functions, but we briefly consider also their applications in the evaluation of certain expressions relevant for the fractional calculus.

A.1 The Gamma function: $\Gamma(z)$

The *Gamma function* $\Gamma(z)$ is the most widely used of all the special functions: it is usually discussed first because it appears in almost every integral or series representation of other advanced mathematical functions. We take as its definition the *integral formula*

$$\Gamma(z) := \int_0^\infty u^{z-1} e^{-u}\, du, \quad \mathcal{R}e\,(z) > 0. \qquad (A.1)$$

This integral representation is the most common for $\Gamma(z)$, even if it is valid only in the right half-plane of $\mathbb{C}$.

The analytic continuation to the left half-plane is possible in different ways. As will be shown later, the *domain of analyticity* D_Γ of $\Gamma(z)$ turns out to be

$$D_\Gamma = \mathbb{C} - \{0, -1, -2, \dots, \}. \qquad (A.2)$$

Using integration by parts, (A.1) shows that, at least for $\mathcal{R}e\,(z) > 0$, $\Gamma(z)$ satisfies the simple *difference equation*

$$\Gamma(z+1) = z\,\Gamma(z), \qquad (A.3)$$

which can be iterated to yield
$$\Gamma(z+n) = z(z+1)\dots(z+n-1)\Gamma(z), \quad n \in \mathbb{N}. \qquad (A.4)$$
The recurrence formulas (A.3-4) can be extended to any $z \in D_\Gamma$. In particular, being $\Gamma(1) = 1$, we get for *non-negative integer values*
$$\Gamma(n+1) = n!, \quad n = 0, 1, 2, \dots. \qquad (A.5)$$
As a consequence $\Gamma(z)$ can be used to define the *Complex Factorial Function*
$$z! := \Gamma(z+1). \qquad (A.6)$$
By the substitution $u = v^2$ in (A.1) we get the *Gaussian Integral Representation*
$$\Gamma(z) = 2 \int_0^\infty e^{-v^2} v^{2z-1} dv, \quad \mathcal{R}e(z) > 0, \qquad (A.7)$$
which can be used to obtain $\Gamma(z)$ when z assumes *positive semi-integer values*. Starting from
$$\Gamma\left(\frac{1}{2}\right) = \int_{-\infty}^{+\infty} e^{-v^2} dv = \sqrt{\pi} \approx 1.77245, \qquad (A.8)$$
we obtain for $n \in \mathbb{N}$,
$$\Gamma\left(n+\frac{1}{2}\right) = \int_{-\infty}^{+\infty} e^{-v^2} v^{2n} dv = \Gamma\left(\frac{1}{2}\right) \frac{(2n-1)!!}{2^n} = \sqrt{\pi}\, \frac{(2n)!}{2^{2n}\, n!}. \qquad (A.9)$$

For the historical development of the Gamma function we refer the reader to the notable article [Davis (1959)]. It is surprising that the notation $\Gamma(z)$ and the name Gamma function were first used by Legendre in 1814 after that Euler had represented in 1729 his function through an infinite product, see Eq. (A.28). As a matter of fact Legendre introduced the representation (A.1) as a generalization of Euler's integral expression for $n!$,
$$n! = \int_0^1 (-\log t)^n\, dt.$$
In fact, changing variable $t \to u = -\log t$, we get
$$n! = \int_0^\infty e^{-u} u^n\, du = \Gamma(n+1).$$

Analytic continuation. The common way to derive the domain of analyticity (A.2) is to carry out the analytic continuation by the *mixed representation* due to Mittag-Leffler:

$$\Gamma(z) = \sum_{n=0}^{\infty} \frac{(-1)^n}{n!(z+n)} + \int_1^{\infty} e^{-u} u^{z-1} du, \quad z \in D_\Gamma. \quad (A.10)$$

This representation can be obtained from the so-called *Prym's decomposition*, namely by splitting the integral in (A.1) into 2 integrals, the one over the interval $0 \leq u \leq 1$ which is then developed in a series, the other over the interval $1 \leq u \leq \infty$, which, being uniformly convergent inside $\mathbb{C}$, provides an entire function. The terms of the series (uniformly convergent inside D_Γ) provide the *principal parts* of $\Gamma(z)$ at the corresponding poles $z_n = -n$. So we recognize that $\Gamma(z)$ is analytic in the entire complex plane except at the points $z_n = -n$ ($n = 0, 1, \ldots$), which turn out to be simple poles with residues $R_n = (-1)^n/n!$. The point at infinity, being an accumulation point of poles, is an essential non-isolated singularity. Thus $\Gamma(z)$ is a transcendental *meromorphic* function.

A formal way to obtain the domain of analyticity D_Γ is to carry out the required analytical continuation by the *Recurrence Formula*

$$\Gamma(z) = \frac{\Gamma(z+n)}{(z+n-1)(z+n-2)\ldots(z+1)z}, \quad (A.11)$$

that is obtained by iterating (A.3) written as $\Gamma(z) = \Gamma(z+1)/z$. In this way we can enter the left half-plane step by step. The numerator at the R.H.S of (A.11) is analytic for $\mathcal{R}e(z) > -n$; hence, the L.H.S. is analytic for $\mathcal{R}e(z) > -n$ except for simple poles at $z = 0, -1, \ldots, (-n+2), (-n+1)$. Since n can be arbitrarily large, we deduce the properties discussed above.

Another way to interpret the analytic continuation of the Gamma function is provided by the *Cauchy-Saalschütz representation*, which is obtained by iterated integration by parts in the basic representation (A.1). If $n \geq 0$ denotes any non-negative integer, we have

$$\Gamma(z) = \int_0^{\infty} u^{z-1} \left[e^{-u} - 1 + u + \frac{1}{2!}u^2 + \cdots + (-1)^{n+1}\frac{1}{n!}u^n \right] du, \quad (A.12)$$

in the strip $-(n+1) < \mathcal{R}e(z) < -n$.

To prove this representation the starting point is provided by the integral

$$\int_0^\infty u^{z-1} \left[e^{-u} - 1 \right] du, \quad -1 < \mathcal{R}e(z) < 0.$$

Integration by parts gives (the integrated terms vanish at both limits)

$$\int_0^\infty u^{z-1} \left[e^{-u} - 1 \right] du = \frac{1}{z} \int_0^\infty u^z e^{-u} du = \frac{1}{z} \Gamma(z+1) = \Gamma(z).$$

So, by iteration, we get (A.12).

Graph of the Gamma function on the real axis. Plots of $\Gamma(x)$ (continuous line) and $1/\Gamma(x)$ (dashed line) are shown for $-4 < x \leq 4$ in Fig. A.1 and for $0 < x \leq 3$ in Fig. A.2.

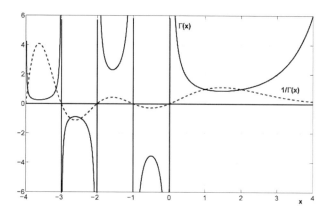

Fig. A.1 Plots of $\Gamma(x)$ (continuous line) and $1/\Gamma(x)$ (dashed line) for $-4 < x \leq 4$.

Hereafter we provide some analytical arguments that support the plots on the real axis. In fact, one can have an idea of the graph of the Gamma function on the real axis using the formulas

$$\Gamma(x+1) = x\Gamma(x), \quad \Gamma(x-1) = \frac{\Gamma(x)}{x-1},$$

to be iterated starting from the interval $0 < x \leq 1$, where $\Gamma(x) \to +\infty$ as $x \to 0^+$ and $\Gamma(1) = 1$.

For $x > 0$ the integral representation (A.1) yields $\Gamma(x) > 0$ and $\Gamma''(x) > 0$ since
$$\Gamma(x) = \int_0^\infty e^{-u}\, u^{x-1}\, du\,, \quad \Gamma''(x) = \int_0^\infty e^{-u}\, u^{x-1}\, (\log u)^2\, du\,.$$
As a consequence, on the positive real axis $\Gamma(x)$ turns out to be positive and *convex* so that it first decreases and then increases exhibiting a minimum value. Since $\Gamma(1) = \Gamma(2) = 1$, we must have a minimum at some x_0, $1 < x_0 < 2$. It turns out that $x_0 = 1.4616\ldots$ and $\Gamma(x_0) = 0.8856\ldots$; hence x_0 is quite close to the point $x = 1.5$ where Γ attains the value $\sqrt{\pi}/2 = 0.8862\ldots$.

On the negative real axis $\Gamma(x)$ exhibits vertical asymptotes at $x = -n$ $(n = 0, 1, 2, \ldots)$; it turns out to be positive for $-2 < x < -1$, $-4 < x < -3$, $\ldots$, and negative for $-1 < x < 0$, $-3 < x < -2$, $\ldots$.

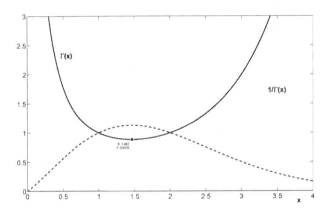

Fig. A.2 Plots of $\Gamma(x)$ (continuous line) and $1/\Gamma(x)$ (dashed line) for $0 < x \leq 3$.

The reflection or complementary formula.
$$\Gamma(z)\, \Gamma(1-z) = \frac{\pi}{\sin \pi z}\,. \qquad (A.13)$$
This formula, which shows the relationship between the Γ function and the trigonometric sin function, is of great importance together with the recurrence formula (A.3). It can be proven in several manners; the simplest proof consists in proving (A.13) for $0 < \mathcal{R}e\,(z) < 1$ and extend the result by analytic continuation to $\mathbb{C}$ except the points $0, \pm 1, \pm 2, \ldots$

The reflection formula shows that $\Gamma(z)$ has no zeros. In fact, the zeros cannot be in $z = 0, \pm 1, \pm 2, \ldots$ and, if $\Gamma(z)$ vanished for a non-integer z, because of (A.13), this zero would be a pole of $\Gamma(1-z)$, that cannot be true. This fact implies that $1/\Gamma(z)$ is an entire function.

The multiplication formulas. Gauss proved the following *Multiplication Formula*

$$\Gamma(nz) = (2\pi)^{(1-n)/2} \, n^{nz-1/2} \prod_{k=0}^{n-1} \Gamma(z + \frac{k}{n}), \quad n = 2, 3, \ldots, \quad (A.14)$$

which reduces, for $n = 2$, to *Legendre's Duplication Formula*

$$\Gamma(2z) = \frac{1}{\sqrt{2\pi}} \, 2^{2z-1/2} \, \Gamma(z) \, \Gamma(z + \frac{1}{2}), \quad (A.15)$$

and, for $n = 3$, to the *Triplication Formula*

$$\Gamma(3z) = \frac{1}{2\pi} \, 3^{3z-1/2} \, \Gamma(z) \, \Gamma(z + \frac{1}{3}) \, \Gamma(z + \frac{2}{3}). \quad (A.16)$$

Pochhammer's symbols. Pochhammer's symbols $(z)_n$ are defined for any non-negative integer n as

$$(z)_n := z \, (z+1) \, (z+2) \ldots (z+n-1) = \frac{\Gamma(z+n)}{\Gamma(z)}, \quad n \in \mathbb{N}. \quad (A.17)$$

with $(z)_0 = 1$. In particular, for $z = 1/2$, we obtain from (A.9)

$$\left(\frac{1}{2}\right)_n := \frac{\Gamma(n+1/2)}{\Gamma(1/2)} = \frac{(2n-1)!!}{2^n}.$$

We extend the above notation to negative integers, defining

$$(z)_{-n} := z \, (z-1) \, (z-2) \ldots (z-n+1) = \frac{\Gamma(z+1)}{\Gamma(z-n+1)}, \quad n \in \mathbb{N}. \quad (A.18)$$

Hankel integral representations. In 1864 Hankel provided a complex integral representation of the function $1/\Gamma(z)$ valid for *unrestricted* z; it reads:

$$\frac{1}{\Gamma(z)} = \frac{1}{2\pi i} \int_{Ha_-} \frac{e^t}{t^z} \, dt, \quad z \in \mathbb{C}, \quad (A.19a)$$

where Ha_- denotes the Hankel path defined as a contour that begins at $t = -\infty - ia$ ($a > 0$), encircles the branch cut that lies along the negative real axis, and ends up at $t = -\infty + ib$ $(b > 0)$. Of course, the branch cut is present when z is non-integer because t^{-z} is a multivalued function; in this case the contour can be chosen as in Fig. A.3 left, where

$$\arg(t) = \begin{cases} +\pi, & \text{above the cut,} \\ -\pi, & \text{below the cut.} \end{cases}$$

When z is an integer, the contour can be taken to be simply a circle around the origin, described in the counterclockwise direction.

An alternative representation is obtained assuming the branch cut along the positive real axis; in this case we get

$$\frac{1}{\Gamma(z)} = -\frac{1}{2\pi i}\int_{Ha_+}\frac{e^{-t}}{(-t)^z}\,dt\,,\quad z\in\mathbb{C}\,, \qquad (A.19b)$$

where Ha_+ denotes the Hankel path defined as a contour that begins at $t = +\infty + ib$ $(b > 0)$, encircles the branch cut that lies along the positive real axis, and ends up at $t = +\infty - ia$ $(a > 0)$. When z is non-integer the contour can be chosen as in Fig. A.3 left, where

$$\arg(t) = \begin{cases} 0, & \text{above the cut,} \\ 2\pi, & \text{below the cut.} \end{cases}$$

When z is an integer, the contour can be taken to be simply a circle around the origin, described in the counterclockwise direction.

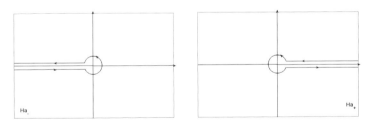

Fig. A.3 The left Hankel contour Ha_- (left); the right Hankel contour Ha_+(right).

We note that

$Ha_- \to Ha_+$ if $t \to t\,e^{-i\pi}$, and $Ha_+ \to Ha_-$ if $t \to t\,e^{+i\pi}$.

The advantage of Hankel representations (A.19a) and (A.19b) compared with the integral representation (A.1) is that they converge for *all* complex z and not just for $\mathcal{R}e\,(z) > 0$. As a consequence $1/\Gamma(z)$ is a transcendental *entire* function (of maximum exponential type); the point at infinity is an essential isolated singularity, which is an accumulation point of zeros ($z_n = -n$, $n = 0, 1, \ldots$). Since $1/\Gamma(z)$ is entire, $\Gamma(z)$ does not vanish in $\mathbb{C}$.

The formulas (A.19a) and (A.19b) are very useful for deriving integral representations in the complex plane for several special functions. Furthermore, using the reflection formula (A.13), we can get the integral representations of $\Gamma(z)$ itself in terms of the Hankel paths (referred to as *Hankel integral representations* for $\Gamma(z)$), which turn out to be valid in the whole domain of analyticity D_Γ.

The required Hankel integral representations that provide the analytical continuation of $\Gamma(z)$ turn out to be:

a) using the path Ha_-

$$\Gamma(z) = \frac{1}{2i\,\sin\,\pi z} \int_{Ha_-} e^t\, t^{z-1}\, dt, \quad z \in D_\Gamma; \qquad (A.20a)$$

b) using the path Ha_+

$$\Gamma(z) = -\frac{1}{2i\,\sin\,\pi z} \int_{Ha_+} e^{-t}\, (-t)^{z-1}\, dt, \quad z \in D_\Gamma. \qquad (A.20b)$$

Notable integrals via Gamma function.

$$\int_0^\infty e^{-st}\, t^\alpha\, dt = \frac{\Gamma(\alpha+1)}{s^{\alpha+1}}, \quad \mathcal{R}e\,(s) > 0, \quad \mathcal{R}e\,(\alpha) > -1. \qquad (A.21)$$

This formula provides the *Laplace transform of the power function* t^α.

$$\int_0^\infty e^{-at^\beta}\, dt = \frac{\Gamma(1+1/\beta)}{a^{1/\beta}}, \quad \mathcal{R}e\,(a) > 0, \quad \beta > 0. \qquad (A.22)$$

This integral for fixed $a > 0$ and $\beta = 2$ attains the well-known value $\sqrt{\pi/a}$ related to the *Gauss integral*. For fixed $a > 0$, the L.H.S. of (A.22) may be referred to as the *generalized Gauss integral*.

The function $I(\beta) := \Gamma(1+1/\beta)$ strongly decreases from infinity at $\beta = 0$ to a positive minimum (less than the unity) attained around $\beta = 2$ and then slowly increases to the asymptotic value 1 as $\beta \to \infty$,

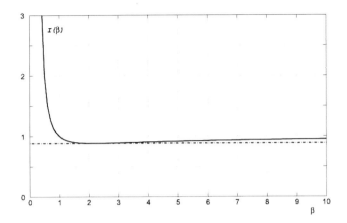

Fig. A.4   Plot of $I(\beta) := \Gamma(1+1/\beta)$ for $0 < \beta \leq 10$.

see Fig. A.4. The minimum value is attained at $\beta_0 = 2.16638\ldots$ and holds $I(\beta_0) = 0.8856\ldots$

A more general formula is

$$\int_0^\infty e^{-zt^\mu} t^{\nu-1}\, dt = \frac{1}{\mu} \frac{\Gamma(1+\nu/\mu)}{z^{\nu/\mu}}, \qquad (A.23)$$

where $\mathcal{R}e\,(z) > 0$, $\mu > 0$, $\mathcal{R}e\,(\nu) > 0$. This formula contains (A.21)-(A.22); it reduces to (A.21) for $z = s$, $\mu = 1$ and $\nu = \alpha + 1$, and to (A.22) for $z = a$, $\mu = \beta$ and $\nu = 1$.

Asymptotic formulas.

$$\Gamma(z) \simeq \sqrt{2\pi}\, e^{-z} z^{z-1/2} \left[1 + \frac{1}{12\,z} + \frac{1}{288\,z^2} + \ldots\right]; \qquad (A.24)$$

as $z \to \infty$ with $|\arg z| < \pi$. This asymptotic expression is usually referred as *Stirling formula*, originally given for $n!$. The accuracy of this formula is surprisingly very good on the positive real axis also for moderate values of $z = x > 0$, as it can be noted from the following exact formula,

$$x! = \sqrt{2\pi}\, e^{\left(-x + \frac{\theta}{12x}\right)} x^{x+1/2}; \quad x > 0, \qquad (A.25)$$

where θ is a suitable number in $(0, 1)$.

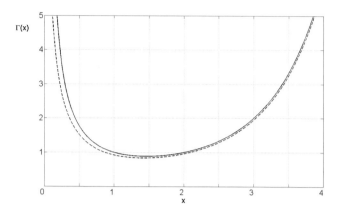

Fig. A.5 $\Gamma(x)$ (continuous line) compared with its first order Stirling approximation (dashed line).

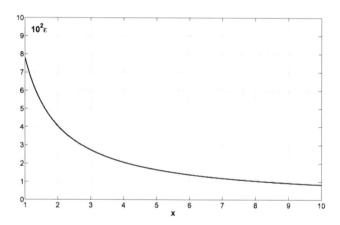

Fig. A.6 Relative error of the first order Stirling approximation to $\Gamma(x)$ for $1 \le x \le 10$.

In Fig. A.5 we show the comparison between the plot of the Gamma function (continuous line) with that provided by the first term of the Stirling approximation (in dashed line), in the range $0 \le x \le 4$.

In Fig. A.6 we show the relative error of the first term approximation with respect to the exact value in the range $1 \le x \le 10$; we note that this error decreases from less than 8% to less than 1%.

The two following asymptotic expressions provide a generalization of the Stirling formula.

If a, b denote two positive constants, we have

$$\Gamma(az+b) \simeq \sqrt{2\pi}\, e^{-az} (az)^{az+b-1/2}, \qquad (A.26)$$

as $z \to \infty$ with $|\arg z| < \pi$, and

$$\frac{\Gamma(z+a)}{\Gamma(z+b)} \simeq z^{a-b} \left[1 + \frac{(a-b)(a+b-1)}{2z} + \ldots \right], \qquad (A.27)$$

as $z \to \infty$ along any curve joining $z = 0$ and $z = \infty$ providing $z \neq -a, -a-1, \ldots$, and $z \neq -b, -b-1, \ldots$.

Infinite products. An alternative approach for introducing the Gamma function goes via infinite products provided by Euler in 1729 and Weierstrass in 1856. Let us start with the original formula given by Euler,

$$\Gamma(z) := \frac{1}{z} \prod_{n=1}^{\infty} \frac{\left(1 + \frac{1}{n}\right)^z}{\left(1 + \frac{z}{n}\right)} = \lim_{n \to \infty} \frac{n!\, n^z}{z(z+1)\ldots(z+n)}. \qquad (A.28)$$

The above limits exist for all $z \in D_\Gamma \subset \mathbb{C}$.

From Euler's formula (A.28) it is possible to derive Weierstrass' formula

$$\frac{1}{\Gamma(z)} = z\, e^{Cz} \prod_{n=1}^{\infty} \left[\left(1 + \frac{z}{n}\right) e^{-z/n} \right], \qquad (A.29)$$

where C, called the Euler-Mascheroni constant, is given by

$$C = 0.5772157\ldots = \begin{cases} \lim_{n \to \infty} \left(\sum_{k=1}^{n} \frac{1}{k} - \log n \right), \\ -\Gamma'(1) = -\int_0^\infty e^{-u} \log u\, du. \end{cases} \qquad (A.30)$$

A.2 The Beta function: $B(p, q)$

Euler's integral representation. The standard representation of the *Beta function* is

$$B(p, q) = \int_0^1 u^{p-1} (1-u)^{q-1} du, \quad \begin{cases} \mathcal{R}e\,(p) > 0, \\ \mathcal{R}e\,(q) > 0. \end{cases} \qquad (A.31)$$

Note that, from historical view-point, this representation for $B(p,q)$ is referred to as the *Euler integral of the first kind*, while the integral representation (A.1) for $\Gamma(z)$ is referred to as the *Euler integral of the second kind*.

The Beta function is a complex function of two complex variables whose analyticity properties will be deduced later, as soon as the relation with the Gamma function has been established.

Symmetry.
$$B(p,q) = B(q,p). \qquad (A.32)$$
This property is a simple consequence of the definition (A.31).

Trigonometric integral representation.
$$B(p,q) = 2 \int_0^{\pi/2} (\cos \vartheta)^{2p-1} (\sin \vartheta)^{2q-1} d\vartheta, \quad \begin{cases} \mathcal{R}e\,(p) > 0, \\ \mathcal{R}e\,(q) > 0. \end{cases} \qquad (A.33)$$
This noteworthy representation follows from (A.31) by setting $u = (\cos \vartheta)^2$.

Relation with the Gamma function.
$$B(p,q) = \frac{\Gamma(p)\,\Gamma(q)}{\Gamma(p+q)}. \qquad (A.34)$$
This relation is of fundamental importance. Furthermore, it allows us to obtain the analytical continuation of the Beta function.

The proof of (A.34) can be easily obtained by writing the product $\Gamma(p)\,\Gamma(q)$ as a double integral that is to be evaluated introducing polar coordinates. In this respect we must use the Gaussian representation (A.7) for the Gamma function and the trigonometric representation (A.33) for the Beta function. In fact,

$$\Gamma(p)\,\Gamma(q) = 4 \int_0^\infty \int_0^\infty e^{-(u^2+v^2)}\, u^{2p-1}\, v^{2q-1}\, du\, dv$$

$$= 4 \int_0^\infty e^{-\rho^2} \rho^{2(p+q)-1} d\rho \int_0^{\pi/2} (\cos \vartheta)^{2p-1} (\sin \vartheta)^{2q-1} d\vartheta$$

$$= \Gamma(p+q)\, B(p,q).$$

Henceforth, we shall exhibit other integral representations for $B(p,q)$, all valid for $\mathcal{R}e\,(p) > 0$, $\mathcal{R}e\,(q) > 0$.

Other integral representations. Integral representations on $[0, \infty)$ are

$$B(p,q) = \begin{cases} \int_0^\infty \dfrac{x^{p-1}}{(1+x)^{p+q}}\, dx\,, \\[2mm] \int_0^\infty \dfrac{x^{q-1}}{(1+x)^{p+q}}\, dx\,, \\[2mm] \dfrac{1}{2} \int_0^\infty \dfrac{x^{p-1} + x^{q-1}}{(1+x)^{p+q}}\, dx\,. \end{cases} \quad (A.35)$$

The first representation follows from (A.31) by setting $u = \dfrac{x}{1+x}$; the other two are easily obtained by using the symmetry property of $B(p,q)$.

A further integral representation on $[0, 1]$ is

$$B(p,q) = \int_0^1 \frac{y^{p-1} + y^{q-1}}{(1+y)^{p+q}}\, dy\,. \quad (A.36)$$

This representation is obtained from the first integral in (A.35) as a sum of the two contributions $[0, 1]$ and $[1, \infty)$.

Notable integrals via Beta function. The Beta function plays a fundamental role in the Laplace convolution of power functions. We recall that the Laplace convolution is the convolution between causal functions (i.e. vanishing for $t < 0$),

$$f(t) * g(t) = \int_{-\infty}^{+\infty} f(\tau)\, g(t - \tau)\, d\tau = \int_0^t f(\tau)\, g(t - \tau)\, d\tau\,.$$

The convolution satisfies both the commutative and associative properties;

$$f(t) * g(t) = g(t) * f(t)\,, \quad f(t) * [g(t) * h(t)] = [f(t) * g(t)] * h(t)\,.$$

It is straightforward to show, by setting in (A.31) $u = \tau/t$, the following identity

$$t^{p-1} * t^{q-1} = \int_0^t \tau^{p-1} (t - \tau)^{q-1}\, d\tau = t^{p+q-1}\, B(p,q)\,. \quad (A.37)$$

Introducing the causal Gel'fand-Shilov function

$$\Phi_\lambda(t) := \frac{t_+^{\lambda-1}}{\Gamma(\lambda)}, \quad \lambda \in \mathbb{C},$$

(where the suffix $+$ just denotes the causality property of vanishing for $t < 0$), we can write the previous result in the following interesting form:

$$\Phi_p(t) * \Phi_q(t) = \Phi_{p+q}(t). \qquad (A.38)$$

In fact, dividing by $\Gamma(p)\Gamma(q)$ the L.H.S of (A.37), and using (A.34), we just obtain (A.38).

In the following we show other relevant applications of the Beta function. The results (A.37-A.48) show that the convolution integral between two (causal) functions, which are absolutely integrable in any interval $[0,t]$ and bounded in every finite interval that does not include the origin, is not necessarily continuous at $t = 0$, even if a theorem ensures that this integral turns out to be continuous for any $t > 0$, see e.g. [Doetsch (1974)], pp. 47-48. In fact, considering two arbitrary real numbers α, β greater than -1, we have

$$I_{\alpha,\beta}(t) := t^\alpha * t^\beta = B(\alpha+1, \beta+1)\, t^{\alpha+\beta+1}, \qquad (A.39)$$

so that

$$\lim_{t \to 0^+} I_{\alpha,\beta}(t) = \begin{cases} +\infty & \text{if } -2 < \alpha+\beta < -1, \\ c(\alpha) & \text{if } \alpha+\beta = -1, \\ 0 & \text{if } \alpha+\beta > -1, \end{cases} \qquad (A.40)$$

where $c(\alpha) = B(\alpha+1, -\alpha) = \Gamma(\alpha+1)\Gamma(-\alpha) = \pi/\sin(-\alpha\pi)$.

We note that in the case $\alpha + \beta = -1$ the convolution integral attains for any $t > 0$ the constant value $c(\alpha) \geq \pi$. In particular, for $\alpha = \beta = -1/2$, we obtain the minimum value for $c(\alpha)$, i.e.

$$\int_0^t \frac{d\tau}{\sqrt{\tau}\sqrt{t-\tau}} = \pi. \qquad (A.41)$$

The Beta function is also used to prove some basic identities for the Gamma function, like the *complementary formula* (A.13) and the *duplication formula* (A.15). To prove the complementary formula we know that it is sufficient to prove it for real argument in the interval $(0,1)$, namely

$$\Gamma(\alpha)\,\Gamma(1-\alpha) = \frac{\pi}{\sin \pi \alpha}\,, \quad 0 < \alpha < 1\,.$$

We note from (A.34-A.35) that

$$\Gamma(\alpha)\,\Gamma(1-\alpha) = B(\alpha, 1-\alpha) = \int_0^\infty \frac{x^{\alpha-1}}{1+x}\,dx\,,$$

and from a classical exercise in complex analysis

$$\int_0^\infty \frac{x^{\alpha-1}}{1+x}\,dx = \frac{\pi}{\sin \pi \alpha}\,.$$

To prove the duplication formula we note that it is equivalent to

$$\Gamma(1/2)\,\Gamma(2z) = 2^{2z-1}\,\Gamma(z)\,\Gamma(z+1/2)\,,$$

and hence, after simple manipulations, to

$$B(z, 1/2) = 2^{2z-1}\,B(z, z)\,. \qquad (A.42)$$

This identity is easily verified for $\mathcal{R}e(z) > 0$, using the trigonometric representation (A.33) for the Beta function and noting that

$$\int_0^{\pi/2} (\cos \vartheta)^\alpha\,d\vartheta = \int_0^{\pi/2} (\sin \vartheta)^\alpha\,d\vartheta = 2^\alpha \int_0^{\pi/2} (\cos \vartheta)^\alpha\,(\sin \vartheta)^\alpha\,d\vartheta\,,$$

with $\mathcal{R}e(\alpha) > -1$, since $\sin 2\vartheta = 2\,\sin \vartheta\,\cos \vartheta$.

A.3 Logarithmic derivative of the Gamma function

The derivative of the Gamma function itself does not play an important role in the theory of special functions. It is not a very manageable function. Much more interesting is the logarithmic derivative of the Gamma function, called the ψ-function,

$$\psi(z) = \frac{d}{dz} \log \Gamma(z) = \frac{\Gamma'(z)}{\Gamma(z)}\,. \qquad (A.43)$$

By using the infinite product (A.28) it follows that

$$\psi(z) = -C + \sum_{n=0}^\infty \left(\frac{1}{n+1} - \frac{1}{z+n}\right)\,, \quad z \neq 0, -1, -2, \ldots \qquad (A.44)$$

Thus the ψ-function possesses simple poles at all non-positive integers. Like $\Gamma(z)$ it is a meromorphic transcendental function: the poles of $\psi(z)$ are thus identical with those of $\Gamma(z)$, but with different residues $R_n = -1$ (in agreement with the theorem of Logarithmic Index).

We have the *recursion relation*

$$\psi(z+1) = \psi(z) + \frac{1}{z}. \qquad (A.45)$$

Special values at positive integers at once follow from the series in (A.43):

$$\psi(1) = -C, \quad \psi(k+1) = -C + 1 + \frac{1}{2} + \frac{1}{3} + \cdots + \frac{1}{k}, \; k \in \mathbb{N}. \quad (A.46)$$

The derivative of $\psi(z)$ is also a meromorphic function and has double poles. This follows, for example, from

$$\psi'(z) = \sum_{n=0}^{\infty} \frac{1}{(z+n)^2}. \qquad (A.47)$$

Observe that the R.H.S. is positive on $(0, \infty)$ and that $\psi'(z)$ is the second derivative of $\log \Gamma(z)$. This shows that $\Gamma(x)$ is *log-convex* on $(0, \infty)$, a relevant property of the Gamma function on the positive real axis. This fact has been used in the legendary works of Bourbaki, according to which, for positive values of the argument, the Gamma function is defined uniquely by the following conditions,

$$f(x) > 0, \; f(x+1) = xf(x), \; f(1) = 1, \; f(x) \text{ is log-convex.} \quad (A.48)$$

Integral representations of the Psi function. For completeness let us recall the main integral representations for the *Psi function* valid in right half-planes of $\mathbb{C}$,

$$\psi(z+1) = -C + \int_0^1 \frac{1 - u^z}{1 - u} \, du, \quad \mathcal{R}e\,(z) > -1, \qquad (A.49)$$

and

$$\psi(z+1) = \log z + \int_0^\infty e^{zu} \left(\frac{1}{u} - \frac{1}{e^u - 1} \right) du, \quad \mathcal{R}e\,(z) > 0. \quad (A.50)$$

Graph of the Psi function on the real axis. A plot of the function $\psi(x)$ is shown in Fig. A.7 for $-4 < x \le 4$.

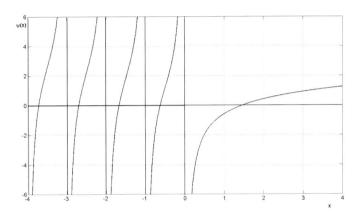

Fig. A.7 Plot of $\psi(x)$ for $-4 < x \le 4$.

A.4 The incomplete Gamma functions

The related *incomplete Gamma functions* are derived from splitting up the interval of integration in (A.1). The common definitions are

$$\gamma(\alpha,z) := \int_0^z u^{\alpha-1} e^{-u} \, du, \qquad (A.51)$$

$$\Gamma(\alpha,z) := \int_z^\infty u^{\alpha-1} e^{-u} \, du. \qquad (A.52)$$

For $\gamma(\alpha,z)$ we assume $\mathcal{R}e\,\alpha > 0$, with respect to z we assume $|\arg z| < \pi$. We thus have for $\mathcal{R}e\,\alpha > 0$ and $|\arg z| < \pi$:

$$\gamma(\alpha,z) = \Gamma(\alpha) - \Gamma(\alpha,z), \quad \Gamma(\alpha,z) = \Gamma(\alpha) - \gamma(\alpha,z). \qquad (A.53)$$

The importance of $\gamma(\alpha,z)$ in the framework of the special functions was pointed out in the fifties by Tricomi, who referred to it as the *Cinderella function*. Let us note that this function admits the alternative integral representation, obtained from (A.51) putting $u = vz$, i.e.,

$$\gamma(\alpha,z) := z^\alpha \int_0^1 v^{\alpha-1} e^{-vz} \, dv, \quad \mathcal{R}e\,\{\alpha\} > 0. \qquad (A.54)$$

We note that $\gamma(\alpha, z)$ is not generally uniform because of the factor z^α in (1.7) [like $\Gamma(\alpha, z)$]. Furthermore, even if analytically continued for $\mathcal{R}e\,\alpha \le 0$, it exhibits an infinite sequence of poles at $\alpha = 0, -1, -2, \dots$ [like $\Gamma(\alpha)$]. In order to get a uniform (*entire*) function in α and in z, one would consider the *modified* or *Tricomi* incomplete Gamma function

$$\gamma^*(\alpha, z) := \frac{\gamma(\alpha, z)}{\Gamma(\alpha)\, z^\alpha}, \qquad (A.55)$$

where we attribute to z^α its principal determination. Such function turns to be real valued for real α and real z. In particular we point out the noteworthy result

$$\gamma^*(-n, z) = z^n, \quad n = 0, 1, 2, \dots \qquad (A.56)$$

obtained by solving the indeterminate form $\frac{\infty}{\infty}$ for $\alpha = -n$, that are removable singularities.

The power series of $\gamma^*(\alpha, z)$, valid in all of $\mathbb{C} \times \mathbb{C}$, turns out to be

$$\gamma^*(\alpha, z) := \frac{1}{\Gamma(\alpha)} \sum_{n=0}^\infty \frac{(-z)^n}{(\alpha+n)\, n!} = e^{-z} \sum_{n=0}^\infty \frac{z^n}{\Gamma(\alpha+n+1)}. \qquad (A.57)$$

In terms of the function (A.55) both incomplete Gamma functions in (A.51)-(A.52) can be represented repectively as

$$\gamma(\alpha, z) = \Gamma(\alpha)\, z^\alpha\, \gamma^*(\alpha, z), \qquad (A.58)$$

$$\Gamma(\alpha, z) = \Gamma(\alpha)\, [1 - z^\alpha\, \gamma^*(\alpha, z)]. \qquad (A.59)$$

The asymptotic behaviour of the incomplete Gamma functions is elementary when only one of the two parameters α and z tends to infinity. Here we limit ourselves to provide the asymptotic expansion for $\Gamma(\alpha, z)$ for $z \to \infty$. From [Gatteschi (1973)] we have

$$\Gamma(\alpha, z) \sim e^{-z}\, z^\alpha \sum_{n=0}^\infty (-1)^n \frac{(1-\alpha)_n}{z^{n+1}}, \quad |z| \to \infty, \quad \arg z \ne \pi, \qquad (A.60)$$

where $(1-\alpha)_n := \Gamma(1-\alpha+n)/\Gamma(1-\alpha)$ denotes the Pochhammer symbol defined in (A.17). For more details on the incomplete Gamma functions we refer the reader to [Erdélyi *et al.* (1953-1955)] and to the more recent review [Gautschi (1998)].

Appendix B

The Bessel Functions

As Rainville pointed out in his classic booklet [Rainville 1960], no other special functions have received such detailed treatment in readily available treatises as the Bessel functions. Consequently, we here present only a brief introduction to the subject including the related Laplace transform pairs used in this book.

B.1 The standard Bessel functions

The Bessel functions of the first and second kind: J_ν, Y_ν.
The Bessel functions of the first kind $J_\nu(z)$ are defined from their power series representation:

$$J_\nu(z) := \sum_{k=0}^{\infty} \frac{(-1)^k}{\Gamma(k+1)\Gamma(k+\nu+1)} \left(\frac{z}{2}\right)^{2k+\nu}, \quad (B.1)$$

where z is a complex variable and ν is a parameter which can take arbitrary real or complex values. When ν is integer it turns out as an entire function; in this case

$$J_{-n}(z) = (-1)^n J_n(z), \quad n = 1, 2, \dots \quad (B.2)$$

In fact

$$J_n(z) = \sum_{k=0}^{\infty} \frac{(-1)^k}{k!(k+n)!} \left(\frac{z}{2}\right)^{2k+n},$$

$$J_{-n}(z) = \sum_{k=n}^{\infty} \frac{(-1)^k}{k!(k-n)!} \left(\frac{z}{2}\right)^{2k-n} = \sum_{s=0}^{\infty} \frac{(-1)^{n+s}}{(n+s)!s!} \left(\frac{z}{2}\right)^{2s+n}.$$

When ν is not integer the Bessel functions exhibit a branch point at $z = 0$ because of the factor $(z/2)^\nu$, so z is intended with $|\arg(z)| < \pi$ that is in the complex plane cut along the negative real semi-axis. Following a suggestion by Tricomi, see [Gatteschi (1973)], we can extract from the series in (B.1) that singular factor and set:

$$J_\nu^T(z) := (z/2)^{-\nu} J_\nu(z) = \sum_{k=0}^{\infty} \frac{(-1)^k}{k!\,\Gamma(k+\nu+1)} \left(\frac{z}{2}\right)^{2k}. \qquad (B.3)$$

The entire function $J_\nu^T(z)$ was referred to by Tricomi as the *uniform Bessel function*. In some textbooks on special functions, see e.g. [Kiryakova (1994)], p. 336, the related entire function

$$J_\nu^C(z) := z^{-\nu/2} J_\nu(2z^{1/2}) = \sum_{k=0}^{\infty} \frac{(-1)^k z^k}{k!\,\Gamma(k+\nu+1)} \qquad (B.4)$$

is introduced and named the *Bessel-Clifford function*.

Since for fixed z in the cut plane the terms of the series (B.1) are analytic function of the variable ν, the fact that the series is uniformly convergent implies that the Bessel function of the first kind $J_\nu(z)$ is an entire function of order ν.

The Bessel functions are usually introduced in the framework of the Fucks–Frobenius theory of the second order differential equations of the form

$$\frac{d^2}{dz^2} u(z) + p(z) \frac{d}{dz} u(z) + q(z)\, u(z) = 0, \qquad (B.5)$$

where $p(z)$ and $q(z)$ are assigned analytic functions. If we chose in (B.5)

$$p(z) = \frac{1}{z}, \quad q(z) = 1 - \frac{\nu^2}{z^2}, \qquad (B.6)$$

and solve by power series, we would just obtain the series in (B.1). As a consequence, we say that the Bessel function of the first kind satisfies the equation

$$u''(z) + \frac{1}{z} u'(z) + \left(1 - \frac{\nu^2}{z^2}\right) u(z) = 0, \qquad (B.7)$$

where, for shortness we have used the apices to denote differentiation with respect to z. It is customary to refer to Eq. (B.7) as the *Bessel differential equation*.

When ν is not integer the general integral of the Bessel equation is
$$u(z) = \gamma_1 J_\nu(z) + \gamma_2 J_{-\nu}(z), \quad \gamma_1, \gamma_2 \in \mathbb{C}, \tag{B.8}$$
since $J_{-\nu}(z)$ and $J_\nu(z)$ are in this case linearly independent with Wronskian
$$W\{J_\nu(z), J_{-\nu}(z)\} = -\frac{2}{\pi z} \sin(\pi\nu). \tag{B.9}$$
We have used the notation $W\{f(z), g(z)\} := f(z) g'(z) - f'(z) g(z)$.

In order to get a solution of Eq. (B.7) that is linearly independent from J_ν also when $\nu = n$ ($n = 0, \pm 1, \pm 2 \dots$) we introduce the Bessel function of the second kind
$$Y_\nu(z) := \frac{J_{-\nu}(z) \cos(\nu\pi) - J_{-\nu}(z)}{\sin(\nu\pi)}. \tag{B.10}$$
For integer ν the R.H.S of (B.10) becomes indeterminate so in this case we define $Y_n(z)$ as the limit
$$Y_n(z) := \lim_{\nu \to n} Y_\nu(z) = \frac{1}{\pi}\left[\frac{\partial J_\nu(z)}{\partial \nu}\bigg|_{\nu=n} - (-1)^n \frac{\partial J_{-\nu}(z)}{\partial \nu}\bigg|_{\nu=n}\right]. \tag{B.11}$$
We also note that (B.11) implies
$$Y_n(z) = (-1)^n Y_n(z). \tag{B.12}$$
Then, when ν is an arbitrary real number, the general integral of Eq. (B.7) is
$$u(z) = \gamma_1 J_\nu(z) + \gamma_2 Y_\nu(z), \quad \gamma_1, \gamma_2 \in \mathbb{C}, \tag{B.13}$$
and the corresponding Wronskian turns out to be
$$W\{J_\nu(z), Y_\nu(z)\} = \frac{2}{\pi z}. \tag{B.14}$$

The Bessel functions of the third kind: $H_\nu^{(1)}, H_\nu^{(2)}$. In addition to the Bessel functions of the first and second kind it is customary to consider the Bessel function of the third kind, or Hankel functions, defined as
$$H_\nu^{(1)}(z) := J_\nu(z) + iY_\nu(z), \quad H_\nu^{(2)}(z) := J_\nu(z) - iY_\nu(z). \tag{B.15}$$
These functions turn to be linearly independent with Wronskian
$$W\{H_\nu^{(1)}(z), H_\nu^{(2)}(z)\} = -\frac{4i}{\pi z}. \tag{B.16}$$

Using (B.10) to eliminate $Y_n(z)$ from (B.15), we obtain

$$\begin{cases} H_\nu^{(1)}(z) := \dfrac{J_{-\nu}(z) - e^{-i\nu\pi} J_\nu(z)}{i \sin(\nu\pi)}, \\ H_\nu^{(2)}(z) := \dfrac{e^{+i\nu\pi} J_\nu(z) - J_{-\nu}(z)}{i \sin(\nu\pi)}, \end{cases} \qquad (B.17)$$

which imply the important formulas

$$H_{-\nu}^{(1)}(z) = e^{+i\nu\pi} H_\nu^{(1)}(z), \quad H_{-\nu}^{(2)}(z) = e^{-i\nu\pi} H_\nu^{(2)}(z). \qquad (B.18)$$

The recurrence relations for the Bessel functions. The functions $J_\nu(z)$, $Y_\nu(z)$, $H_\nu^{(1)}(z)$, $H_\nu^{(2)}(z)$ satisfy simple *recurrence relations*. Denoting any one of them by $\mathcal{C}_\nu(z)$ we have:

$$\begin{cases} \mathcal{C}_\nu(z) = \dfrac{z}{2\nu} \left[\mathcal{C}_{\nu-1}(z) + \mathcal{C}_{\nu+1}(z) \right], \\ \mathcal{C}'_\nu(z) = \dfrac{1}{2} \left[\mathcal{C}_{\nu-1}(z) - \mathcal{C}_{\nu+1}(z) \right]. \end{cases} \qquad (B.19)$$

In particular we note

$$J'_0(z) = -J_1(z), \quad Y'_0(z) = -Y_1(z).$$

We note that $\mathcal{C}_\nu$ stands for *cylinder function*, as it is usual to call the different kinds of Bessel functions. The origin of the term *cylinder* is due to the fact that these functions are encountered in studying the boundary–value problems of potential theory for cylindrical coordinates.

A more general differential equation for the Bessel functions. The differential equation (B.7) can be generalized by introducing three additional complex parameters λ, p, q in such a way

$$z^2 w''(z) + (1 - 2p) z w'(z) + \left(\lambda^2 q^2 z^{2q} + p^2 - \nu^2 q^2 \right) w(z) = 0. \quad (B.20)$$

A particular integral of this equation is provided by

$$w(z) = z^p \mathcal{C}_\nu \left(\lambda z^q \right). \qquad (B.21)$$

We see that for $\lambda = 1$, $p = 0$, $q = 1$ we recover Eq. (B.7).

The asymptotic representations for the Bessel functions.

The asymptotic representations of the standard Bessel functions for $z \to 0$ and $z \to \infty$ are provided by the first term of the convergent series expansion around $z = 0$ and by the first term of the asymptotic series expansion for $z \to \infty$, respectively.

For $z \to 0$ (with $|\arg(z)| < \pi$ if ν is not integer) we have:

$$\begin{cases} J_{\pm n}(z) \sim (\pm 1)^n \dfrac{(z/2)^n}{n!}, & n = 0, 1, \dots, \\ J_\nu(z) \sim \dfrac{(z/2)^\nu}{\Gamma(\nu+1)}, & \nu \neq \pm 1, \pm 2 \dots. \end{cases} \quad (B.22)$$

$$\begin{cases} Y_0(z) \sim -iH_0^{(1)}(z) \sim iH_0^{(2)}(z) \sim \dfrac{2}{\pi} \log(z), \\ Y_\nu(z) \sim -iH_\nu^{(1)}(z) \sim iH_\nu^{(2)}(z) \sim -\dfrac{1}{\pi} \Gamma(\nu)(z/2)^{-\nu}, \ \nu > 0. \end{cases} \quad (B.23)$$

For $z \to \infty$ with $|\arg(z)| < \pi$ and for any ν we have:

$$\begin{cases} J_\nu(z) \sim \sqrt{\dfrac{2}{\pi z}} \cos\left(z - \nu\dfrac{\pi}{2} - \dfrac{\pi}{4}\right), \\ Y_\nu(z) \sim \sqrt{\dfrac{2}{\pi z}} \sin\left(z - \nu\dfrac{\pi}{2} - \dfrac{\pi}{4}\right), \\ H_\nu^{(1)}(z) \sim \sqrt{\dfrac{2}{\pi z}} e^{+i\left(z - \nu\frac{\pi}{2} - \frac{\pi}{4}\right)}, \\ H_\nu^{(2)}(z) \sim \sqrt{\dfrac{2}{\pi z}} e^{-i\left(z - \nu\frac{\pi}{2} - \frac{\pi}{4}\right)}. \end{cases} \quad (B.24)$$

The generating function of the Bessel functions of integer order.

The Bessel functions of the first kind $J_n(z)$ are simply related to the coefficients of the Laurent expansion of the function

$$w(z,t) = e^{z(t-1/t)/2} = \sum_{n=-\infty}^{+\infty} c_n(z) t^n, \quad 0 < |t| < \infty. \quad (B.25)$$

To this aim we multiply the power series of $e^{zt/2}$, $e^{-z/(2t)}$, and, after some manipulation, we get

$$w(z,t) = e^{z(t-1/t)/2} = \sum_{n=-\infty}^{+\infty} J_n(z) t^n, \quad 0 < |t| < \infty. \quad (B.26)$$

The function $w(z,t)$ is called the *generating function* of the Bessel functions of integer order, and formula (B.26) plays an important role in the theory of these functions.

Plots of the Bessel functions of integer order. Plots of the Bessel functions $J_\nu(x)$ and $Y_\nu(x)$ for integer orders $\nu = 0, 1, 2, 3, 4$ are shown in Fig. B.1 and in Fig. B.2, respectively.

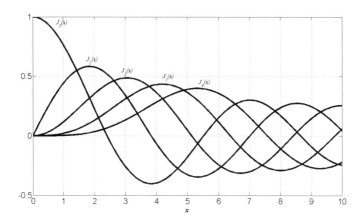

Fig. B.1   Plots of $J_\nu(x)$ with $\nu = 0, 1, 2, 3, 4$ for $0 \leq x \leq 10$.

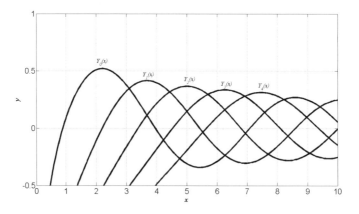

Fig. B.2   Plots of $Y_\nu(x)$ with $\nu = 0, 1, 2, 3, 4$ for $0 \leq x \leq 10$.

The Bessel functions of semi-integer order. We now consider the special cases when the order is a semi-integer number $\nu = n + 1/2$ ($n = 0, \pm 1, \pm 2, \pm 3, \dots$). In these cases the standard Bessel function can be expressed in terms of elementary functions.

In particular we have

$$J_{+1/2}(z) = \left(\frac{2}{\pi z}\right)^{1/2} \sin z, \quad J_{-1/2}(z) = \left(\frac{2}{\pi z}\right)^{1/2} \cos z. \quad (B.27)$$

The fact that any Bessel function of the first kind of half-integer order can be expressed in terms of elementary functions now follows from the first recurrence relation in (B.19), i.e.

$$J_{\nu-1} + J_{\nu+1} = \frac{2\nu}{z} J_\nu(z),$$

whose repeated applications gives

$$\begin{cases} J_{+3/2}(z) = \left(\frac{2}{\pi z}\right)^{1/2} \left[\frac{\sin z}{z} - \cos z\right], \\ J_{-3/2}(z) = -\left(\frac{2}{\pi z}\right)^{1/2} \left[\sin z - \frac{\cos z}{z}\right], \end{cases} \quad (B.28)$$

and so on.

To derive the corresponding formulas for Bessel functions of the second and third kind we start from the expressions (B.10) and (B.15) of these functions in terms of the Bessel functions of the first kind, and use (B.25). For example, we have:

$$Y_{1/2}(z) = -J_{-1/2}(z) = -\left(\frac{2}{\pi z}\right)^{1/2} \cos z, \quad (B.29)$$

$$H^{(1)}_{1/2}(z) = -i \left(\frac{2}{\pi z}\right)^{1/2} e^{+iz}, \quad H^{(2)}_{1/2}(z) = +i \left(\frac{2}{\pi z}\right)^{1/2} e^{-iz}. \quad (B.30)$$

It has been shown by Liouville that the case of half-integer order is the only case where the cylinder functions reduce to elementary functions.

It is worth noting that when $\nu = \pm 1/2$ the asymptotic representations (B.24) for $z \to \infty$ for all types of Bessel functions reduce to the exact expressions of the corresponding functions provided above. This could be verified by using the saddle-point method for the complex integral representation of the Bessel functions, that we will present in Subsection B.3.

B.2 The modified Bessel functions

The modified Bessel functions of the first and second kind: I_ν, K_ν. The modified Bessel functions of the first kind $J_\nu(z)$ with $\nu \in \mathbb{R}$ and $z \in \mathbb{C}$ are defined by the power series

$$I_\nu(z) := \sum_{k=0}^{\infty} \frac{1}{\Gamma(k+1)\Gamma(k+\nu+1)} \left(\frac{z}{2}\right)^{2k+\nu}. \qquad (B.31)$$

We also define the modified Bessel functions of the second kind $K_\nu(z)$:

$$K_\nu(z) := \frac{\pi}{2} \frac{I_{-\nu}(z) - I_\nu(z)}{\sin(\nu\pi)}. \qquad (B.32)$$

For integer ν the R.H.S of (B.32) becomes indeterminate so in this case we define $Y_n(z)$ as the limit

$$K_n(z) := \lim_{\nu \to n} K_\nu(z). \qquad (B.33)$$

Repeating the consideration of Section B.1, we find that $I_\nu(z)$ and $K_\nu(z)$ are analytic functions of z in the cut plane and entire function of the order ν. We recall that $K_\nu(z)$ is sometimes referred to as *Macdonald's function*. We note from the definitions (B.31) and (B.32) the useful formulas

$$I_{-n}(z) = I_n(z), \quad n = 0, \pm 1, \pm 2, \ldots$$
$$K_{-\nu}(z) = K_\nu(z), \quad \forall \nu.$$

The modified Bessel functions $I_\nu(z)$ and $K_\nu(z)$ are simply related to the standard Bessel function of argument $z\exp(\pm i\pi/2)$. If

$$-\pi < \arg(z) < \pi/2, \quad \text{i.e.,} \quad -\pi/2 < \arg(z\, e^{i\pi/2}) < \pi/2,$$

then (B.1) implies

$$I_\nu(z) = e^{-i\nu\pi/2} J_\nu(z\, e^{i\pi/2}). \qquad (B.34)$$

Similarly, according to (B.17), for the same value of z we have

$$K_\nu(z) = \frac{i\pi}{2} e^{i\nu\pi/2} H_\nu^1(z\, e^{i\pi/2}). \qquad (B.35)$$

On the other hand, if

$$-\pi/2 < \arg(z) < \pi, \quad \text{i.e.,} \quad -\pi < \arg(z\, e^{-i\pi/2}) < \pi/2,$$

then it is easily verified that

$$I_\nu(z) = e^{+i\nu\pi/2} J_\nu(z\, e^{-i\pi/2}), \qquad (B.36)$$

and

$$K_\nu(z) = -\frac{i\pi}{2} e^{-i\nu\pi/2} H_\nu^2(z\, e^{-i\pi/2}). \qquad (B.37)$$

The differential equation for the modified Bessel functions.
It is an immediate consequence of their definitions that $I_\nu(z)$ and $K_\nu(z)$ are linearly independent solutions of the differential equation

$$v''(z) + \frac{1}{z} v'(z) - \left(1 + \frac{\nu^2}{z^2}\right) v(z) = 0, \qquad (B.38)$$

which differs from the standard Bessel equation (B.7) only by the sign of one term, and reduces to Eq. (B.7) if in Eq. (B.38) we make the substitution $z = \pm it$. Like the standard Bessel equation, Eq. (B.38) is often encountered in Mathematical Physics and it is referred to as the *modified Bessel differential equation*. Its general solution, for arbitrary ν can be written in the form

$$v(z) = \gamma_1 I_\nu(z) + \gamma_2 K_\nu(z), \qquad \gamma_1, \gamma_2 \in \mathbb{C}. \qquad (B.39)$$

For the modified Bessel functions the corresponding Wronskian turns out to be

$$W\{I_\nu(z), K_\nu(z)\} = -\frac{1}{z}. \qquad (B.40)$$

The recurrence relations for the modified Bessel functions.
Like the cylinder functions, the modified Bessel functions $I_\nu(z)$ and $K_\nu(z)$ satisfy simple *recurrence relations*. However, at variance with the cylinder functions, we have to keep distinct the corresponding recurrence relations:

$$\begin{cases} I_\nu(z) = \dfrac{z}{2\nu}[I_{\nu-1}(z) - I_{\nu+1}(z)], \\[1em] I'_\nu(z) = \dfrac{1}{2}[I_{\nu-1}(z) + I_{\nu+1}(z)], \end{cases} \qquad (B.41)$$

and

$$\begin{cases} K_\nu(z) = -\dfrac{z}{2\nu}[K_{\nu-1}(z) - K_{\nu+1}(z)], \\[1em] K'_\nu(z) = -\dfrac{1}{2}[K_{\nu-1}(z) + K_{\nu+1}(z)]. \end{cases} \qquad (B.42)$$

Recurrence relations (B.41) and (B.42) can be written in a unified form if, following [Abramowitz and Stegun (1965)], we set

$$\mathcal{Z}_\nu(z) := \{I_\nu(z), e^{i\nu\pi} K_\nu(z)\}. \qquad (B.43)$$

In fact we get

$$\begin{cases} \mathcal{Z}_\nu(z) = \dfrac{z}{2\nu}\left[\mathcal{Z}_{\nu-1}(z) - \mathcal{Z}_{\nu+1}(z)\right], \\ \\ \mathcal{Z}'_\nu(z) = \dfrac{1}{2}\left[\mathcal{Z}_{\nu-1}(z) + \mathcal{Z}_{\nu+1}(z)\right], \end{cases} \quad (B.44)$$

that preserves the form of (B.41).

A more general differential equation for the modified Bessel functions. As for the standard Bessel functions we have provided the reader with a more general differential equation solved by related functions, see (B.20) and (B.21), so here we do it also for the modified Bessel functions. For this purpose it is sufficient to replace there λ^2 with $-\lambda^2$. Then, introducing three additional complex parameters λ, p, q in such a way that

$$z^2 w''(z) + (1-2p)z w'(z) + \left(-\lambda^2 q^2 z^{2q} + p^2 - \nu^2 q^2\right) w(z) = 0, \quad (B.45)$$

we get the required differential equation whose a particular integral is provided by

$$w(z) = z^p\, \mathcal{Z}_\nu\left(\lambda z^q\right). \quad (B.46)$$

Note that for $\lambda = 1$, $p = 0$, $q = 1$ in (B.45) we recover Eq. (B.38). Of course the constant $e^{i\nu\pi}$ multiplying the function $K_\nu(z)$ is not relevant for Eqs. (B.45)-(B.46), but it is essential to preserve the same form for the recurrence relations satisfied by the two functions denoted by $\mathcal{Z}_\nu(z)$, as shown in Eqs. (B.44).

The asymptotic representations for the modified Bessel functions. For the modified Bessel functions we have the following asymptotic representations as $z \to 0$ and as $z \to \infty$.

For $z \to 0$ (with $|\arg(z)| < \pi$ if ν is not integer) we have:

$$\begin{cases} I_{\pm n}(z) \sim \dfrac{(z/2)^n}{n!}, \quad n = 0, 1, \ldots, \\ \\ I_\nu(z) \sim \dfrac{(z/2)^\nu}{\Gamma(\nu+1)}, \quad \nu \neq \pm 1, \pm 2 \ldots. \end{cases} \quad (B.47)$$

and
$$\begin{cases} K_0(z) \sim \log(2/z), \\ K_\nu(z) \sim \dfrac{1}{2}\Gamma(\nu)(z/2)^{-\nu}, \ \nu > 0. \end{cases} \quad (B.48)$$

For $z \to \infty$ with $|\arg(z)| < \pi/2$ and for any ν we have:

$$I_\nu(z) \sim \frac{1}{\sqrt{2\pi}} z^{-1/2} e^z, \quad (B.49)$$

$$K_\nu(z) \sim \frac{1}{\sqrt{2\pi}} z^{-1/2} e^{-z}. \quad (B.50)$$

The generating function of the modified Bessel functions of integer order. For the modified Bessel functions of the first kind $I_n(z)$ of integer order we can establish a generating function following a procedure similar to that adopted for $J_n(z)$, see Eqs. (B.25)-(B.26). In fact, by considering the Laurent expansion of the function $\omega(z,t) = e^{z(t+1/t)/2}$ obtained by multiplying the power series of $e^{zt/2}$, $e^{z/(2t)}$, we get after some manipulation

$$\omega(z,t) = e^{z(t+1/t)/2} = \sum_{n=-\infty}^{+\infty} I_n(z)\, t^n, \quad 0 < |t| < \infty. \quad (B.51)$$

Plots of the modified Bessel functions of integer order. Plots of the Bessel functions $I_\nu(x)$ and $K_\nu(x)$ for integer orders $\nu = 0, 1, 2$ are shown in Fig. B.3.

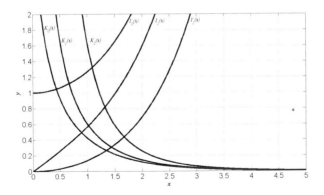

Fig. B.3 Plots of $I_\nu(x)$, $K_\nu(x)$ with $\nu = 0, 1, 2$ for $0 \leq x \leq 5$.

Since the modified Bessel functions exhibit an exponential behaviour for $x \to \infty$, see (B.49)-(B.50), we show the plots of $e^{-x} I_\nu(x)$ and $e^x K_\nu$ with $\nu = 0, 1, 2$ for $0 \le x \le 5$ in Fig. B.4.

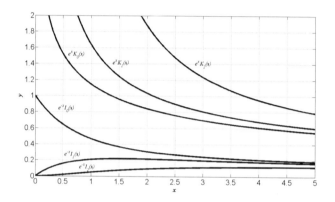

Fig. B.4 Plots of $e^{-x} I_\nu(x)$, $e^x K_\nu$ with $\nu = 0, 1, 2$ for $0 \le x \le 5$.

The modified Bessel functions of semi-integer order. Like the cylinder functions the modified Bessel functions of semi-integer order can be expressed in terms of elementary functions. Starting with the case $\nu = 1/2$ it is easy to recognize

$$I_{1/2}(z) = \left(\frac{2}{\pi z}\right)^{1/2} \sinh z, \quad I_{-1/2}(z) = \left(\frac{2}{\pi z}\right)^{1/2} \cosh z, \quad (B.52)$$

and

$$K_{1/2}(z) = K_{-1/2}(z) = \left(\frac{\pi}{2z}\right)^{1/2} e^{-z}. \quad (B.53)$$

For general index $\nu = n + 1/2$ the corresponding formulas are obtained from (B.48) and the recurrence relations (B.41) and (B.42).

B.3 Integral representations and Laplace transforms

Integral representations. The basic integral representation of the standard Bessel function $J_\nu(z)$ with $\mathcal{R}e\, z > 0$ is provided through the Hankel contour around the negative real axis, denoted by Ha_- and illustrated in Fig. A.3 left. We have, see e.g. [Davies (2002)],

$$J_\nu(z) = \frac{1}{2\pi i} \int_{Ha_-} p^{-\nu-1} \exp\left[\frac{z}{2}\left(p - \frac{1}{p}\right)\right] dp, \quad \mathcal{R}e\, z > 0, \quad (B.54)$$

where the restriction $\mathcal{R}e\, z > 0$ is necessary to make the integral converge. Then, we can split the integral in two contributions:
(i) from the circular path where $p = \exp(i\theta)$ $(-\pi < \theta < \pi)$;
(ii) from the straight paths where $p = \exp(s \pm i\pi)$ $(0 < s < \infty)$.
We have:

$$(i) \quad \frac{1}{2\pi} \int_{-\pi}^{+\pi} e^{(-i\nu\theta + iz\sin\theta)} d\theta = \frac{1}{\pi} \int_0^\pi \cos(\nu\theta - z\sin\theta)\, d\theta; \quad (B.55)$$

$$(ii) \quad \frac{1}{2\pi i} \int_\infty^0 e^{\nu s + i\pi\nu - z(e^s - e^{-s})/2}\, ds + \frac{1}{2\pi i} \int_\infty^0 e^{\nu s - i\pi\nu - z(e^s - e^{-s})/2}\, ds$$

$$= -\frac{\sin(\nu\pi)}{\pi} \int_0^\infty \exp(-z\sinh s - \nu s)\, ds. \quad (B.56)$$

Thus the final integral representation is

$$J_\nu(z) = \frac{1}{\pi} \int_0^\pi \cos(\nu\theta - z\sin\theta)\, d\theta$$
$$- \frac{\sin(\nu\pi)}{\pi} \int_0^\infty \exp(-z\sinh s - \nu s)\, ds, \quad \mathcal{R}e\, z > 0. \quad (B.57)$$

For integer ν the second integral gives no contribution. The first integral is known as Bessel's integral.

It is a simple matter to perform an analytic continuation of (B.54) to all the domain of analyticity of $J_\nu(z)$. If we temporarily restrict z to be real and positive, then the change of variables $u = pz/2$ yields

$$J_\nu(z) = \frac{(z/2)^\nu}{2\pi i} \int_{Ha_-} u^{-\nu-1} \exp\left[u - z^2/(4u)\right] du, \quad (B.58)$$

where the contour is unchanged since z is real. But the integral in (B.58) defines an entire function of z because it is single-valued and absolutely convergent for all z. We recognize from the pre-factor in (B.58) that $J_\nu(z)$ has a branch point in the origin, if ν is not integer.

In this case, according to the usual convention, we must introduce a branch cut on the negative real axis so that Eq. (B.58) is valid under the restriction $-\pi \arg(z) < \pi$.

We note that the series expansion of $J_\nu(z)$, Eq. (B.1) may be obtained from the integral representation (B.58) by replacing $\exp[-z^2/(4u)]$ by its Taylor series and integrating term by term and finally using Hankel's integral representation of the reciprocal of the Gamma function, Eq. (A.19a). Of course, the procedure can be inverted to yield the integral representation (B.58) from the series representation (B.1).

Other integral representations related to the class of Bessel functions can be found in any handbook of special functions.

Laplace transform pairs. Herewith we report a few of Laplace transform pairs related to Bessel functions extracted from [Ghizzetti and Ossicini (1971)], where the interested reader can find more formulas, all with the proof included. We first consider

$$J_\nu(\alpha t) \div \frac{\left(\sqrt{s^2+\alpha^2}-s\right)^\nu}{\alpha^\nu \sqrt{s^2+\alpha^2}}, \quad \mathcal{R}e\,\nu > -1, \ \mathcal{R}e\,s > |\mathcal{I}m\,\alpha|, \quad (B.59)$$

$$I_\nu(\alpha t) \div \frac{\left(s-\sqrt{s^2-\alpha^2}\right)^\nu}{\alpha^\nu \sqrt{s^2-\alpha^2}}, \quad \mathcal{R}e\,\nu > -1, \ \mathcal{R}e\,s > |\mathcal{R}e\,\alpha|. \quad (B.60)$$

Then, we consider the following transform pairs relevant for wave propagation problems:

$$J_0\left(\alpha\sqrt{t^2-a^2}\right)\Theta(t-a) \div \frac{e^{-a\sqrt{s^2+\alpha^2}}}{\sqrt{s^2+\alpha^2}}, \quad \mathcal{R}e\,s > |\mathcal{I}m\,\alpha|, \quad (B.61)$$

$$I_0\left(\alpha\sqrt{t^2-a^2}\right)\Theta(t-a) \div \frac{e^{-a\sqrt{s^2-\alpha^2}}}{\sqrt{s^2-\alpha^2}}, \quad \mathcal{R}e\,s > |\mathcal{R}e\,\alpha|, \quad (B.62)$$

$$a\alpha\,\frac{J_1\left(\alpha\sqrt{t^2-a^2}\right)}{\sqrt{t^2-a^2}}\Theta(t-a) \div e^{-as} - e^{-a\sqrt{s^2+\alpha^2}}, \quad (B.63)$$
$$\mathcal{R}e\,s > |\mathcal{I}m\,\alpha|,$$

$$a\alpha\,\frac{I_1\left(\alpha\sqrt{t^2-a^2}\right)}{\sqrt{t^2-a^2}}\Theta(t-a) \div e^{-as} - e^{-a\sqrt{s^2-\alpha^2}}, \quad (B.64)$$
$$\mathcal{R}e\,s > |\mathcal{R}e\,\alpha|.$$

B.4 The Airy functions

The Airy differential equation in the complex plane ($z \in \mathbb{C}$).
The Airy functions $Ai(z)$, $Bi(z)$ are usually introduced as the two linear independent solutions of the differential equation

$$\frac{d^2}{dz^2} u(z) - z\, u(z) = 0. \qquad (B.65)$$

The Wronskian turns

$$W\{Ai(z), Bi(z)\} = \frac{1}{\pi}.$$

Tatlor series.

$$Ai(z) = 3^{-2/3} \sum_{n=0}^{\infty} \frac{z^{3n}}{9\, n!\, \Gamma(n+2/3)} \\ - 3^{-4/3} \sum_{n=0}^{\infty} \frac{z^{3n+1}}{9\, n!\, \Gamma(n+4/3)}. \qquad (B.66a)$$

$$Bi(z) = 3^{-1/6} \sum_{n=0}^{\infty} \frac{z^{3n}}{9\, n!\, \Gamma(n+2/3)} \\ + 3^{-5/6} \sum_{n=0}^{\infty} \frac{z^{3n+1}}{9\, n!\, \Gamma(n+4/3)}. \qquad (B.66b)$$

We note

$$Ai(0) = Bi(0)/\sqrt{3} = 3^{-2/3}/\Gamma(2/3) \approx 0.355. \qquad (B.67)$$

Functional relations.

$$Ai(z) + \omega Ai(\omega z) + \omega^2 Ai(\omega^2 z) = 0, \quad \omega = e^{-2i\pi/3}. \qquad (B.68a)$$

$$Bi(z) = i\omega Ai(\omega z) - i\omega^2 Ai(\omega^2 z), \quad \omega = e^{-2i\pi/3}. \qquad (B.68b)$$

Relations with Bessel functions.

$$\begin{cases} Ai(z) = \frac{1}{3} z^{1/2} \left[I_{-1/3}(\zeta) - I_{1/3}(\zeta) \right], \\ Ai(-z) = \frac{1}{3} z^{1/2} \left[J_{1/3}(\zeta) + J_{-1/3}(\zeta) \right], \end{cases} \quad \zeta = \frac{2}{3} z^{3/2}. \qquad (B.69a)$$

$$\begin{cases} Bi(z) = \frac{1}{\sqrt{3}} z^{1/2} \left[I_{-1/3}(\zeta) + I_{1/3}(\zeta) \right], \\ Bi(-z) = \frac{1}{\sqrt{3}} z^{1/2} \left[J_{-1/3}(\zeta) - J_{1/3}(\zeta) \right], \end{cases} \quad \zeta = \frac{2}{3} z^{3/2}. \qquad (B.69b)$$

Asymptotic representations.

$$Ai(z) \sim \frac{1}{2\sqrt{\pi}} z^{-1/4} e^{-2z^{3/2}/3}, \quad z \to \infty, \quad |\arg z| < \pi. \quad (B.70a)$$

$$Bi(z) \sim \frac{1}{\sqrt{\pi}} z^{-1/4} e^{2z^{3/2}/3}, \quad z \to \infty, \quad |\arg z| < \pi/3. \quad (B.70b)$$

Integral representations for real variable ($z = x \in \mathbb{R}$).

$$\begin{aligned} Ai(x) &= \frac{1}{2\pi i} \int_{-i\infty}^{+i\infty} e^{\zeta x - \zeta^3/3} \, d\zeta \\ &= \frac{1}{\pi} \int_0^\infty \cos\left(ux + u^3/3\right) du. \end{aligned} \quad (B.71a)$$

$$Bi(x) = \frac{1}{\pi} \int_0^\infty \left[e^{ux - u^3/3} + \sin\left(ux + u^3/3\right) \right] du. \quad (B.71b)$$

Graphical representations for real variable ($z = x \in \mathbb{R}$).
We present the plots of the Airy functions with their derivatives on the real line in Figs. B.5 and B.6.

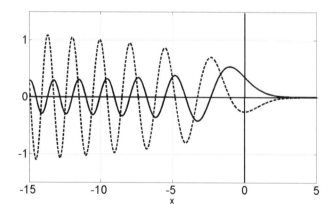

Fig. B.5 Plots of $Ai(x)$ (continuous line) and its derivative $Ai'(x)$ (dotted line) for $-15 \leq x \leq 5$.

As expected from their relations with the Bessel functions, see Eqs. (B.69a) and (B.69b), and from their asymptotic representations, see Eqs. (B.70a) and (B.70b), we note from the plots that the functions $Ai(x)$, $Bi(x)$ are monotonic for $x > 0$ ($Ai(x)$ is exponentially decreasing, $Bi(x)$ is exponentially increasing) whereas both of them are oscillating with an algebraic decay for $x < 0$.

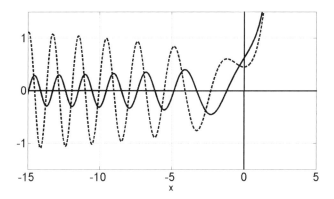

Fig. B.6 Plots of $Bi(x)$ (continuous line) and its derivative $Bi'(x)$ (dotted line) for $-15 \leq x \leq 5$.

These changes in behaviour along the real line are the most noteworthy characteristics of the Airy functions. For a survey on the applications of the Airy functions in physics we refer the interested reader to [Vallé and Soares (2004)].

Appendix C

The Error Functions

In this appendix we provide a survey on the class of so-called Error functions including some Laplace transform pairs related to them, which are relevant for the applications of fractional calculus.

C.1 The two standard Error functions

Basic definitions. The *Error function* is an entire function defined as

$$\operatorname{erf}(z) := \frac{2}{\sqrt{\pi}} \int_0^z e^{-\zeta^2}\, d\zeta, \quad z \in \mathbb{C}, \qquad (C.1)$$

where the integral is evaluated along an arbitrary path joining the origin to the point $\zeta = z$. The form of this path does not matter, since the integrand is an entire function of the complex variable ζ.

The *complementary Error function* is an entire function defined as

$$\operatorname{erfc}(z) := 1 - \operatorname{erf}(z) := \frac{2}{\sqrt{\pi}} \int_z^\infty e^{-\zeta^2}\, d\zeta, \quad z \in \mathbb{C}, \qquad (C.2)$$

where the path of integration is subjected to the restriction $\arg \zeta \to \vartheta$ with $|\vartheta| < \pi/4$ as $\zeta \to \infty$ along the path. In an obvious notation we can write the extreme of integration as $e^{i\vartheta} \infty$ and hence, in particular, $+\infty$ if we assume $\vartheta = 0$.

Recalling from Appendix A the definition of the incomplete Gamma functions, see (A.51)-(A.52), we recognize the identities:

$$\operatorname{erf}(z) = \frac{1}{\sqrt{\pi}} \gamma(1/2, z^2), \quad \operatorname{erfc}(z) = \frac{1}{\sqrt{\pi}} \Gamma(1/2, z^2). \qquad (C.3)$$

We note that, since

$$\int_0^{+\infty} e^{-u^2}\, du = \frac{1}{2} \int_{-\infty}^{+\infty} e^{-u^2}\, du = \frac{\sqrt{\pi}}{2}, \qquad (C.4)$$

then

$$\operatorname{erf}(+\infty) = 1, \quad \operatorname{erfc}(+\infty) = 0.$$

The factor $2/\sqrt{\pi}$ in front of the functions in (C.1) and (C.2) is kept to satisfy the above conditions at infinity. Some authors, however, do not put this pre-factor in their definitions of the Error functions, so the reader must be aware of the different definitions available in the literature.

Symmetry relations.

$$\operatorname{erf}(-z) = -\operatorname{erf}(z), \qquad \operatorname{erf}(\bar{z}) = \overline{\operatorname{erf}(z)}. \qquad (C.5)$$

Power series.

$$\operatorname{erf}(z) = \begin{cases} \dfrac{2}{\sqrt{\pi}} \displaystyle\sum_{n=0}^{\infty} \dfrac{(-1)^n}{n!\,(2n+1)} z^{2n+1}, \\[2ex] \dfrac{2}{\sqrt{\pi}} e^{-z^2} \displaystyle\sum_{n=0}^{\infty} \dfrac{2^n}{(2n+1)!!} z^{2n+1}, \end{cases} \quad z \in \mathbb{C}. \qquad (C.6)$$

Asymptotic expansions. For $\operatorname{erfc}(z)$ as $z \to \infty$ in the sector $|\arg z| < 3\pi/4$ we have:

$$\operatorname{erfc}(z) \sim \frac{1}{\sqrt{\pi}} \frac{e^{-z^2}}{z} \left(1 + \sum_{m=1}^{\infty} (-1)^m \frac{(2m-1)!!}{2^m} \frac{1}{z^{2m}} \right). \qquad (C.7)$$

Recalling that $(2m-1)!! = (2m)!/(2^m\, m!)$, $m = 1, 2, \ldots$, we get:

$$\operatorname{erfc}(z) \sim \frac{1}{\sqrt{\pi}} \frac{e^{-z^2}}{z} \left(1 + \sum_{m=1}^{\infty} (-1)^m \frac{(2m)!}{m!} \frac{1}{(2z)^{2m}} \right). \qquad (C.8)$$

Plots of the Error functions on the real axis. We easily recognize from (C.2) and (C.9) that *on the real line* $\operatorname{erfc}(x)$ is a decreasing function with limits $\operatorname{erfc}(-\infty) = 2$, $\operatorname{erfc}(+\infty) = 0$, whereas $\operatorname{erf}(x)$ is an increasing function with limits $\operatorname{erf}(-\infty) = -1$, $\operatorname{erf}(+\infty) = 1$.

To get more insight on the family of the Error functions we find it instructive to present in Fig. C.1 the plots of the functions $\mathrm{erf}(x)$ and $\mathrm{erfc}(x)$ along with the plot of the function

$$\mathrm{erf}'(x) := \frac{d}{dx}\mathrm{erf}(x) = \frac{2}{\sqrt{\pi}} e^{-x^2}, \qquad (C.9)$$

in the interval $-2 \le x \le +2$. In dashed line we enclose the plot of the leading term of the asymptotic expansion of $\mathrm{erfc}(x)$ as $x \to +\infty$, see (C.8). We recognize that a reasonably good matching of $\mathrm{erfc}(x)$ with this leading term is obtained already for $x \simeq 1.5$.

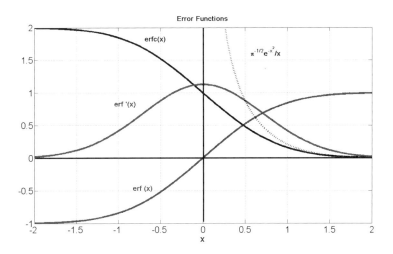

Fig. C.1 Plots of $\mathrm{erf}(x)$, $\mathrm{erf}'(x)$ and $\mathrm{erfc}(x)$ for $-2 \le x \le +2$.

C.2 Laplace transform pairs

Let us consider the most significant Laplace transform pairs related to the Error function. Let us start with functions that we denote by $\phi(a,t)$, $\psi(a,t)$ and $\chi(a,t)$, where $a > 0$ is a parameter and $t \ge 0$ is the time variable. These functions play a fundamental role in problems of diffusion where the parameter a is the space variable x; since they are interrelated in the Laplace domain as we later show, they will be referred to as the *three sisters functions of diffusion*.

$$\phi(a,t) := \mathrm{erfc}\left(\frac{a}{2\sqrt{t}}\right) \div \frac{e^{-a\,s^{1/2}}}{s} := \widetilde{\phi}(a,s)\,, \qquad (C.10)$$

$$\psi(a,t) := \frac{a}{2\sqrt{\pi}}\,t^{-3/2}\,e^{-a^2/(4t)} \div e^{-a\,s^{1/2}} := \widetilde{\psi}(a,s)\,, \qquad (C.11)$$

$$\chi(a,t) := \frac{1}{\sqrt{\pi}}\,t^{-1/2}\,e^{-a^2/(4t)} \div \frac{e^{-a\,s^{1/2}}}{s^{1/2}} := \widetilde{\chi}(a,s)\,, \qquad (C.12)$$

where $\mathcal{R}e\,s > 0$. We remind that

$$\phi(a,t) := \mathrm{erfc}\left(\frac{a}{2\sqrt{t}}\right) := 1 - \frac{2}{\sqrt{\pi}}\int_0^{a/(2\sqrt{t})} e^{-u^2}\,du\,, \quad t \ge 0.$$

All the three functions decay exponentially to 0 as $t \to 0^+$. Their Laplace transforms $\widetilde{\phi}(a,s)$, $\widetilde{\psi}(a,s)$, $\widetilde{\chi}(a,s)$ turn out to be interrelated via simple rules so that it is sufficient to prove only one transform pair to derive the remaining two pairs. For example,

$$\widetilde{\psi}(a,s) = s\,\widetilde{\phi}(a,s)\,, \quad \text{so} \quad \psi(a,t) = \frac{\partial}{\partial t}\phi(a,t)\,;$$

$$\widetilde{\psi}(a,s) = s^{1/2}\,\widetilde{\chi}(a,s)\,, \quad \text{so} \quad \psi(a,t) = -\frac{\partial}{\partial a}\chi(a,t)\,.$$

Using the Bromwich inversion formula for the Laplace transforms we obtain the following integral representations for the three sisters functions $\phi(a,t)$, $\psi(a,t)$, $\chi(a,t)$:

$$\phi(a,t) = 1 - \frac{1}{\pi}\int_0^\infty e^{-rt}\sin(a\sqrt{r})\,\frac{dr}{r} \div \widetilde{\phi}(a,s)\,, \qquad (C.10')$$

$$\psi(a,t) = \frac{1}{\pi}\int_0^\infty e^{-rt}\sin(a\sqrt{r})\,dr \div \widetilde{\psi}(a,s)\,, \qquad (C.11')$$

$$\chi(a,t) = \frac{1}{\pi}\int_0^\infty e^{-rt}\cos(a\sqrt{r})\,\frac{dr}{\sqrt{r}} \div \widetilde{\chi}(a,s)\,. \qquad (C.12')$$

In Fig. C.2 we show the plots of the sisters functions for $0 \le t \le 5$ assuming $a = 1$.

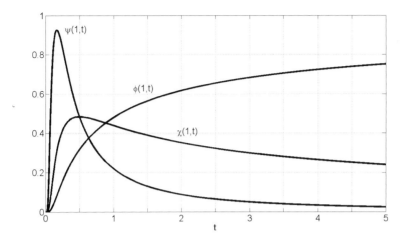

Fig. C.2 Plot of the three sisters functions $\phi(a,t)$, $\psi(a,t)$, $\chi(a,t)$ with $a = 1$ for $0 \leq t \leq 5$.

Another series of Laplace transform pairs involves Error functions in the Laplace domain. Assuming the original functions for $t \geq 0$ with a parameter $a > 0$, we have:

$$f_1(a,t) := e^{-a t^2} \div \frac{\sqrt{\pi}}{2\sqrt{a}} e^{s^2/(4a)} \operatorname{erfc}\left(\frac{s}{2\sqrt{a}}\right), \quad \forall s \in \mathbb{C}, \quad (C.13)$$

and, for $\mathcal{R}e\, s > 0$:

$$f_2(a,t) := \frac{1}{\sqrt{t+a}} \div \sqrt{\pi}\, s^{-1/2} e^{a s} \operatorname{erfc}\left[(a s)^{1/2}\right], \quad (C.14)$$

$$f_3(a,t) := \frac{1}{\sqrt{t(t+a)}} \div \pi\, a^{-1/2} e^{a s} \operatorname{erfc}\left[(a s)^{1/2}\right]. \quad (C.15)$$

C.3 Repeated integrals of the Error functions

Basic definitions and formulas.

$$I^n \operatorname{erfc}(z) := \int_z^\infty I^{n-1} \operatorname{erfc}(\zeta)\, d\zeta, \quad n = 0, 1, 2, \ldots, \quad (C.16)$$

with $\quad I^{-1} \operatorname{erfc}(z) = \frac{2}{\sqrt{\pi}} e^{-z^2}, \quad I^0 \operatorname{erfc}(z) = \operatorname{erfc}(z).$

The above definitions imply

$$I^n \operatorname{erfc}(z) = \frac{2}{\sqrt{\pi}} \int_z^\infty \frac{(u-z)^n}{n!} e^{-u^2} du, \qquad (C.17)$$

and

$$\frac{d}{dz} I^n \operatorname{erfc}(z) = -I^{n-1} \operatorname{erfc}(z). \qquad (C.18)$$

A relevant and useful formula is

$$\frac{d^n}{dz^n}\left(e^{z^2} \operatorname{erfc}(z)\right) = (-1)^n 2^n n! e^{z^2} I^n \operatorname{erfc}(z). \qquad (C.19)$$

Recurrence relations.

$$I^n \operatorname{erfc}(z) = -\frac{z}{n} I^{n-1} \operatorname{erfc}(z) - \frac{1}{2n} I^{n-2} \operatorname{erfc}(z). \qquad (C.20)$$

This formula is easily established by induction. In fact, integrating by part, we have

$$I \operatorname{erfc}(z) = \frac{1}{\sqrt{\pi}} e^{-z^2} - z \operatorname{erfc}(z),$$

and

$$I^2 \operatorname{erfc}(z) = \frac{1}{4}\left[(1 + 2z^2) \operatorname{erfc}(z) - \frac{2}{\sqrt{\pi}} z e^{-z^2}\right]$$

$$= \frac{1}{4}\left[\operatorname{erfc}(z) - 2z\, I \operatorname{erfc}(z)\right].$$

Differential equation. The second-order differential equation

$$\frac{d^2 y}{dz^2} + 2z \frac{dy}{dz} + 2n y = 0, \qquad (C.21)$$

admits as a general solution

$$y(z) = A\, I^n \operatorname{erfc}(z) + B\, I^n \operatorname{erfc}(-z), \qquad (C.22)$$

where A and B are arbitrary constants.

Power series. Integrating term by term the expansion in power series of $\operatorname{erf}(z)$ in (C.6) we get the following powers series

$$I^n \operatorname{erfc}(z) = \sum_{k=0}^\infty \frac{(-1)^n}{2^{n-k} k!\, \Gamma[1 + (n-k)/2]} z^k, \qquad (C.23)$$

where the terms corresponding to $k = n+2, n+4, \ldots$ are understood to be zero.

Asymptotic expansions. The following *asymptotic expansions* are given for $I^n \operatorname{erfc}(z)$, $n = 0, 1, 2\ldots$, as $z \to \infty$ in the sector $|\arg z| < \frac{3\pi}{4}$,

$$I^n \operatorname{erfc}(z) \sim \frac{2}{\sqrt{\pi}} \frac{e^{-z^2}}{(2z)^{n+1}} \left(1 + \sum_{m=1}^{\infty} \frac{(2m+n)!}{n!\, m!} \frac{(-1)^m}{(2z)^{2m}}\right). \quad (C.24)$$

C.4 The Erfi function and the Dawson integral

Definition of the Erfi function. In some applications we find the integral $\int_0^z e^{u^2} du$, that is related to the Error function of argument iz. In this case, at variance with the standard Error functions we prefer to not keep the pre-factor $2/\sqrt{\pi}$ and we define the entire function $\operatorname{Erfi}(z)$ as

$$\operatorname{Erfi}(z) := \int_0^z e^{u^2} du = \frac{\sqrt{\pi}}{2i} \operatorname{erf}(iz) = \frac{\sqrt{\pi}}{2i} \left[1 - \operatorname{erfc}(iz)\right]$$

$$= \begin{cases} \displaystyle\sum_{n=0}^{\infty} \frac{z^{2n+1}}{n!(2n+1)}, \\ \displaystyle e^{z^2} \sum_{n=0}^{\infty} \frac{2^n}{(2n+1)!!} z^{2n+1}, \end{cases} \quad z \in \mathbb{C}. \quad (C.25)$$

Asymptotic expansions of the Erfi function. The asymptotic expansion of $\operatorname{erfc}(z)$ as $z \to \infty$ in the sector $|\arg z| < 3\pi/4$, see (C.7), allows us to determine the asymptotic expansion of $\operatorname{Erfi}(z)$ in certain sectors of $\mathbb{C}$ excluding, however, the most relevant real axis.

In the particular case $z = x$ real the asymptotic expansion for $x \to +\infty$ reads, see [Gatteschi (1973)],

$$\operatorname{Erfi}(x) \sim \frac{e^{x^2}}{x} \left[1 + \sum_{m=1}^{\infty} \frac{(2m-1)!!}{(2x^2)^m}\right]. \quad (C.26)$$

Definition and plot of the Dawson integral. In Fig. C.3 we report on the positive real semi-axis the plot of the

$$\operatorname{Daw}(x) := \exp(-x^2) \operatorname{Erfi}(x), \quad (C.27)$$

known as *Dawson integral*, that is relevant in some problems of Mathematical Physics. This function exhibits its maximum at $x = 0.93143\ldots$ with value $0.54104\ldots$

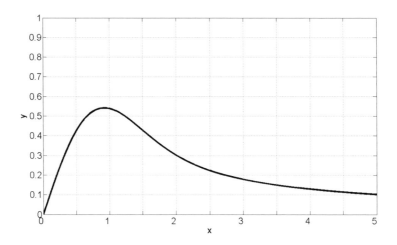

Fig. C.3 Plot of the Dawson integral Daw(x) for $0 \le x \le 5$.

C.5 The Fresnel integrals

Basic definitions. In some applications we find the integral

$$\int_0^x e^{\pm iu^2}\, du = \int_0^x \cos(u^2)\, du \pm i \int_0^x \sin(u^2)\, du, \qquad (C.28)$$

that is expected to be related to the Error function of argument

$$z = x\, e^{\pm i\pi/4} = x\, \frac{1 \pm i}{2}. \qquad (C.29)$$

The required relation is obtained comparing (C.28) with the definition (C.1) after setting in it the variable $\zeta = e^{\mp i\pi/4} u$. Then, we finally get

$$\int_0^x e^{\pm iu^2}\, du = \frac{\sqrt{\pi}}{2}\, e^{\pm i\pi/4}\, \text{erf}\left(e^{\mp i\pi/4} x\right). \qquad (C.30)$$

Appendix C: The Error Functions

After these notes, let us introduce the *Fresnel integrals* as the entire functions defined by

$$\begin{cases} C(z) := \int_0^z \cos\left(\frac{\pi}{2}\zeta^2\right) d\zeta, \\ \\ S(z) := \int_0^z \sin\left(\frac{\pi}{2}\zeta^2\right) d\zeta, \end{cases} \quad z \in \mathbb{C}, \qquad (C.31)$$

where the integrals are evaluated along an arbitrary path joining the origin to the point $\zeta = z$. The form of this path does not matter, since the integrand is an entire function of the complex variable ζ. In the literature there is in use another definition as follows,

$$\begin{cases} C_1(z) := \sqrt{\frac{2}{\pi}} \int_0^z \cos\left(\zeta^2\right) d\zeta = C\left(\sqrt{\frac{2}{\pi}}z\right), \\ \\ S_1(z) := \sqrt{\frac{2}{\pi}} \int_0^z \sin\left(\zeta^2\right) d\zeta = S\left(\sqrt{\frac{2}{\pi}}z\right), \end{cases} \quad z \in \mathbb{C}. \qquad (C.32)$$

As a matter of fact, the substantial difference between the two definitions (C.31) and (C.32) stands in a stretching in the independent variable. For both definitions, as $x \to +\infty$ all the functions approach to the limiting value $1/2$, namely

$$C(+\infty) = S(+\infty) = 1/2, \quad C_1(+\infty) = S_1(+\infty) = 1/2, \qquad (C.33)$$

as implied by the familiar formulas

$$\int_0^\infty \cos(u^2)\, du = \int_0^\infty \sin(u^2)\, du = \frac{1}{2}\sqrt{\frac{\pi}{2}}, \qquad (C.34)$$

derived from the complex integral

$$\int_0^\infty e^{\pm i u^2}\, du = \frac{\sqrt{\pi}}{2} e^{\pm i\pi/4} = \sqrt{\pi}\, \frac{1 \pm i}{4}. \qquad (C.35)$$

From now on we will consider the Fresnel integrals according to their definitions (C.31).

Power series.

$$\begin{cases} C(z) = \sum_{n=0}^\infty (-1)^n \frac{(\pi/2)^{2n}}{(4n+1)(2n)!} z^{4n+1}, \\ \\ S(z) = \sum_{n=0}^\infty (-1)^n \frac{(\pi/2)^{2n+1}}{(4n+3)(2n+1)!} z^{4n+3}, \end{cases} \quad z \in \mathbb{C}. \qquad (C.36)$$

Relation with the Error function. By using the considerations in Eqs. (C.28)-(C.30) we get:

$$C(z) \pm iS(z) = \frac{1}{\sqrt{2}} e^{\pm i\pi/4} \operatorname{erf}\left(\sqrt{\frac{2}{\pi}} e^{\mp i\pi/4} z\right), \qquad (C.37)$$

that implies

$$\begin{cases} C(z) = \frac{1}{2\sqrt{2}} \left[e^{-i\pi/4} \operatorname{erf}\left(\sqrt{\frac{2}{\pi}} e^{i\pi/4} z\right) + e^{i\pi/4} \operatorname{erf}\left(\sqrt{\frac{2}{\pi}} e^{-i\pi/4} z\right) \right], \\ \\ S(z) = \frac{1}{2\sqrt{2}} \left[e^{-i\pi/4} \operatorname{erf}\left(\sqrt{\frac{2}{\pi}} e^{i\pi/4} z\right) - e^{i\pi/4} \operatorname{erf}\left(\sqrt{\frac{2}{\pi}} e^{-i\pi/4} z\right) \right]. \end{cases}$$
$$(C.38)$$

Plots of the Fresnel integrals. For real $z = x$, the Fresnel integrals are real. Both $C(x)$ and $S(x)$ vanish for $x = 0$, and have oscillatory character, as follows from the formulas

$$C'(x) = \cos\left(\frac{\pi}{2}x\right), \quad S'(x) = \sin\left(\frac{\pi}{2}x\right), \qquad (C.39)$$

which show that $C(x)$ has extrema at $x = \pm\sqrt{2n+1}$, while $S(x)$ has extrema at $x = \pm\sqrt{2n}$, $n = 0, 1, 2 \ldots$. The largest maxima are $C(1) = 0,779893\ldots$ and $S(\sqrt{2}) = 0.713972\ldots$, respectively. Furthermore, as before noted, both the functions approach to the common value $1/2$ as $x \to +\infty$. In Fig. C.4 we show the plots of the Fresnel integrals in the interval $0 \le x \le 5$.

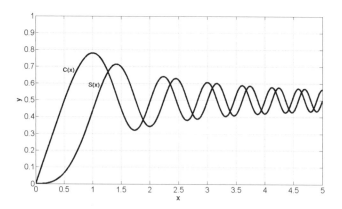

Fig. C.4 Plots of the Fresnel integrals for $0 \le x \le 5$.

Appendix C: The Error Functions

The Cornu spiral. The curve of the parametric equations
$$x(t) = C(t), \quad y(t) = S(t), \quad t \geq 0, \qquad (C.40)$$
is called the *Cornu spiral*, see Fig. C.5. It has the property to have the radius of curvature ρ proportional to the arc s measured from the origin. In fact we have

$$\begin{cases} \dfrac{dx}{dt} = \cos(\pi t^2/2), \\ \dfrac{dy}{dt} = \sin(\pi t^2/2), \end{cases} \quad \begin{cases} \dfrac{d^2 x}{dt^2} = -\pi t \sin(\pi t^2/2), \\ \dfrac{d^2 y}{dt^2} = \pi t \cos(\pi t^2/2), \end{cases} \qquad (C.41)$$

from which

$$\begin{cases} ds = \sqrt{(dx)^2 + (dy)^2} = dt, \quad s = t \\ \rho = \dfrac{dx\,(dy)^2 - dy\,(dx)^2}{[(dx)^2 + (dy)^2]^{3/2}} = \pi t \end{cases} \implies \rho = \pi s. \qquad (C.42)$$

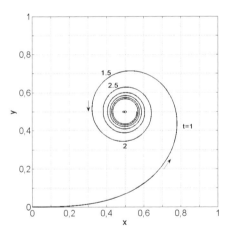

Fig. C.5   Plot of the Cornu spiral for $0 \leq x \leq 1$.

Appendix D

The Exponential Integral Functions

In this Appendix we provide a survey of the class of the Exponential integral functions including a list of Laplace transform pairs related to them, which are relevant for the applications of fractional calculus. Our main bibliographical sources have been the treatises [Abramowitz and Stegun (1965)], [Erdélyi et al. (1953-1955)], [Erdélyi (1956)], [Gatteschi (1973)], [Ghizzetti and Ossicini (1971)], [Jahnke and Emde (1943)], and our research papers [Mainardi et al. (2007); Mainardi et al. (2008)].

D.1   The classical Exponential integrals $\text{Ei}(z)$, $\mathcal{E}_1(z)$

The function $\text{Ei}(z)$.   A classical definition of the *Exponential integral* is

$$\text{Ei}(z) := -\int_{-z}^{\infty} \frac{e^{-u}}{u} \, du, \qquad (D.1)$$

where the Cauchy principal value of the integral is understood if $z = x > 0$. Some authors such as [Jahnke and Emde (1943)] adopt the following definition for $\text{Ei}(z)$,

$$\text{Ei}(z) := \int_{-\infty}^{z} \frac{e^{u}}{u} \, du, \qquad (D.2)$$

which is equivalent to (D.1). For this we note that

$$\int_{-z}^{\infty} \frac{e^{-u}}{u} \, du = -\int_{-\infty}^{z} \frac{e^{u}}{u} \, du,$$

where the Cauchy principal value is understood for $z = x > 0$.

Recalling the incomplete Gamma functions in (A.52), we note the identity

$$-\mathrm{Ei}\,(-z) = \Gamma(0,z) = \int_z^\infty \frac{e^{-u}}{u}\,du\,, \quad |\arg z| < \pi\,. \qquad (D.3)$$

The function $\mathcal{E}(z)$. In many texts on special functions the function $\Gamma(0,z)$ is usually taken as definition of Exponential integral and denoted by $\mathcal{E}_1(z)$ so that

$$\mathcal{E}_1(z) = -\mathrm{Ei}\,(-z) = \int_z^\infty \frac{e^{-u}}{u}\,du = \int_1^\infty \frac{e^{-zt}}{t}\,dt\,. \qquad (D.4)$$

This definition is then generalized to yield

$$\mathcal{E}_n(z) = \int_1^\infty \frac{e^{-zu}}{t^n}\,du\,, \quad n = 1, 2, \ldots\,, \qquad (D.5)$$

or, more generally

$$\mathcal{E}_\nu(z) = \int_1^\infty \frac{e^{-zu}}{u^\nu}\,du\,, \quad \nu \geq 1\,. \qquad (D.5a)$$

We note that, in contrast with the standard literature where the Exponential integrals are denoted by the letter E, we have used for them the letter $\mathcal{E}$: this choice is to avoid confusion with the standard notation for the Mittag-Leffler function $E_\alpha(z)$, treated in Appendix E, that plays a more relevant role in fractional calculus and hence in this book.

D.2 The modified Exponential integral Ein (z)

The basic definition. The whole subject matter can be greatly simplified if we agree to follow F.G. Tricomi who has proposed to consider the following *entire* function, formerly introduced by [Schelkunoff (1944)]

$$\mathrm{Ein}\,(z) := \int_0^z \frac{1 - e^{-u}}{u}\,du\,. \qquad (D.6)$$

Such a function, referred to as the *modified Exponential integral*, turns out to be entire, being the primitive of an entire function. The relation with the classical Exponential integrals will be given in (D.11).

On the real axis $\mathbb{R}$ the modified Exponential integral $\text{Ein}(x)$ is an *increasing* function because
$$\frac{d}{dx}\text{Ein}(x) = \frac{1-e^{-x}}{x} > 0, \quad \forall x \in \mathbb{R}. \tag{D.7}$$
In $\mathbb{R}^+$ the function $\text{Ein}(x)$ turns out to be a *Bernstein function*, which means that is positive, increasing, with the first derivative *completely monotonic*.

Power series. The *power series expansion* of $\text{Ein}(z)$, valid in all of $\mathbb{C}$, can be easily obtained by term-by-term integration and reads
$$\text{Ein}(z) := z - \frac{z^2}{2\cdot 2!} + \frac{z^3}{3\cdot 3!} - \frac{z^4}{4\cdot 4!} + \cdots = -\sum_{n=1}^{\infty}(-1)^n \frac{z^n}{n\, n!}. \tag{D.8}$$

Relation between $\text{Ein}(z)$ and $\mathcal{E}_1(z) = -\text{Ei}(-z)$. The relation between $\text{Ein}(z)$ and $\mathcal{E}_1(z) = -\text{Ei}(-z)$ can be obtained from the series expansion of $\Gamma(\alpha, z)$ in the limit as $\alpha \to 0$. For this purpose let us recall the relation (A.52) between the two incomplete gamma functions, which, using (A.53-A.54), allows us to write
$$\Gamma(\alpha, z) = \Gamma(\alpha) - \gamma(\alpha, z) = \Gamma(\alpha) - z^\alpha \sum_{n=0}^{\infty} \frac{(-z)^n}{(\alpha+n)n!}. \tag{D.9}$$
As a consequence of (D.7) and (D.9), we obtain
$$\Gamma(0, z) = \lim_{\alpha \to 0}\left[\Gamma(\alpha) - \frac{z^\alpha}{\alpha}\right] + \text{Ein}(z). \tag{D.10}$$
Since
$$\lim_{\alpha \to 0}\left[\Gamma(\alpha) - \frac{z^\alpha}{\alpha}\right] = \lim_{\alpha \to 0}\frac{\Gamma(\alpha+1) - z^\alpha}{\alpha}$$
$$= \lim_{\alpha \to 0}\frac{\Gamma(\alpha+1) - \Gamma(1)}{\alpha} + \lim_{\alpha \to 0}\frac{1 - z^\alpha}{\alpha} = -C - \log z,$$
where $C = -\Gamma'(1) = 0.577215\ldots$, see (A.30), denotes the Euler-Mascheroni constant, we finally obtain the required relation, i.e.
$$\mathcal{E}_1(z) = -\text{Ei}(-z) = \Gamma(0, z) = -C - \log z + \text{Ein}(z), \tag{D.11}$$
with $|\arg z| < \pi$. This relation is important for understanding the analytic properties of the classical Exponential integral functions in that it isolates the multi-valued part represented by the logarithmic function from the regular part represented by the entire function $\text{Ein}(z)$ given by the power series in (D.8), absolutely convergent in all of $\mathbb{C}$.

D.3 Asymptotics for the Exponential integrals

The asymptotic behaviour of the Exponential integrals $\text{Ei}(z)$, $\mathcal{E}_1(z)$ as $z \to \infty$ can be obtained from their integral representation (D.1), (D.4), respectively, noticing that

$$\mathcal{E}_1(z) := \int_z^\infty \frac{e^{-u}}{u}\,du = e^{-z}\int_0^\infty \frac{e^{-u}}{u+z}\,du\,. \qquad (D.12)$$

In fact, we can prove by repeated integrations by part that

$$G(z) := e^z\,\mathcal{E}_1(z) = \int_0^\infty \frac{e^{-u}}{u+z}\,du = \frac{1}{z}\left[\sum_{n=0}^{N-1}(-1)^n \frac{n!}{z^n} + R_N(z)\right], \qquad (D.13)$$

where it turns out $R_N(z) = O\left(z^{-N}\right)$ as $z \to \infty$ with $|\arg z| \leq \pi - \delta$. More precisely, we get from [Erdélyi (1956)]:

$$|R_N(z)| \leq \begin{cases} \dfrac{N!}{|z^N|}\dfrac{1}{\sin\delta}, & |\arg z| > \dfrac{\pi}{2}; \\ \dfrac{N!}{|z^N|}, & |\arg z| \leq \dfrac{\pi}{2}. \end{cases} \qquad (D.14)$$

Furthermore, if $z = x > 0$, it turns out $R_N(x) = (-1)^N \theta_N\, N!/x^N$, with $0 < \theta_N < 1$.

In conclusion, as $z \to \infty$ and $|\arg z| \leq \pi - \delta$, the required asymptotic expansion is derived from (D.13) and reads

$$\mathcal{E}_1(z) = -\text{Ei}(-z) = \Gamma(0,z) \sim \frac{e^{-z}}{z}\sum_{n=0}^\infty (-1)^n \frac{n!}{z^n}\,. \qquad (D.15)$$

By using the relation (D.10), we get the asymptotic expansion in the same sector for $\text{Ein}(z)$, which consequently includes a constant and a logarithmic term.

From (D.13)-(D.15), as $z \to 0^+$ with $|\arg z| \leq \pi - \delta$, we get the asymptotic expansion of the function

$$F(z) = G(1/z)/z := \int_0^\infty \frac{e^{-u}}{1+zu}\,du \sim \sum_{n=0}^\infty (-1)^n n!\,z^n\,. \qquad (D.16)$$

It is instructive to note that this series can be formally obtained by developing in geometric series the factor $1/(1+zu)$ (in positive powers of (zu)) and integrating term by term.

We note that the series in the R.H.S. of (D.16) for $z = x \in \mathbb{R}$ is the famous *Euler series*, which is divergent for any $x \neq 0$. It is usually treated in textbooks on asymptotics for historical reasons.

D.4 Laplace transform pairs for Exponential integrals

Let us consider the following three causal functions $f_1(t), f_2(t), f_3(t)$ related to Exponential integrals as follows:

$$f_1(t) := \mathcal{E}_1(t), \quad t > 0, \qquad (D.17a)$$

$$f_2(t) := \text{Ein}(t) = C + \log t + \mathcal{E}_1(t), \quad t > 0, \qquad (D.18a)$$

$$f_3(t) := C + \log t + e^t \mathcal{E}_1(t), \quad t > 0. \qquad (D.19a)$$

The corresponding Laplace transforms turn out to be:

$$\mathcal{L}\{f_1(t); s\} = \frac{1}{s} \log(s+1), \quad \mathcal{R}e\, s > 0, \qquad (D.17b)$$

$$\mathcal{L}\{f_2(t); s\} = -\frac{1}{s} \log\left(\frac{1}{s}+1\right), \quad \mathcal{R}e\, s > 0, \qquad (D.18b)$$

$$\mathcal{L}\{f_3(t); s\} = \frac{\log s}{s-1} - \frac{\log s}{s} = \frac{\log s}{s(s-1)}, \quad \mathcal{R}e\, s > 0. \qquad (D.19b)$$

The function $f_3(t)$ is found in problems of fractional relaxation, see [Mainardi et al. (2007)].

The proof of (D.17b) is found, for example, in [Ghizzetti and Ossicini (1971)], see Eq. [4.6.16] and pp. 104–105.

The proof of (D.18b) is hereafter provided in two ways, being in our opinion very instructive and useful for the applications in the text. The first proof is obtained as a consequence of the identity (D.10), i.e. $\text{Ein}(t) = \mathcal{E}_1(t) + C + \log t$, and the Laplace transform pair

$$\mathcal{L}\{\log t; s\} = -\frac{1}{s}[C + \log s], \quad \mathcal{R}e\, s > 0, \qquad (D.20a)$$

whose proof is found, for example, in [Ghizzetti and Ossicini (1971)], see Eq. [4.6.15] and p. 104. The second proof is direct and instructive. For this it is sufficient to compute the Laplace transform of the elementary function provided by the derivative of $\text{Ein}(t)$, that is, according to a standard exercise in the theory of Laplace transforms,

$$\mathcal{L}\left\{\frac{1-e^{-t}}{t}; s\right\} = \log\left(\frac{1}{s}+1\right), \quad \mathcal{R}e\, s > 0, \qquad (D.20b)$$

so that

$$f_2(t) := \text{Ein}(t) = \int_0^t f(t')\,dt' \div \frac{\widetilde{f}(s)}{s} = \frac{1}{s}\log\left(\frac{1}{s}+1\right), \quad \mathcal{R}e\,s > 0,$$

in agreement with (D.18b). After the previous proofs, the proof of (D.18c) is trivial.

In Fig. D.1 we report the plots of the functions $f_1(t)$, $f_2(t)$ and $f_3(t)$ for $0 \le t \le 10$.

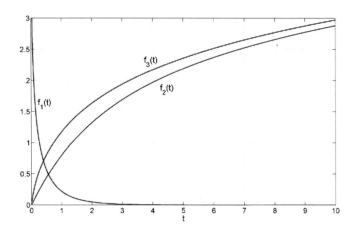

Fig. D.1 Plots of the functions $f_1(t)$, $f_2(t)$ and $f_3(t)$ for $0 \le t \le 10$.

We outline the different asymptotic behaviours of the three functions $f_1(t)$, $f_2(t)$ and $f_3(t)$ for small argument ($t \to 0^+$) and large argument ($t \to +\infty$) that can be easily obtained by using Eqs. (D.7), (D.10) and (D.15). However, it is instructive to derive the required asymptotic representations by using the Karamata Tauberian theory for Laplace transforms, see [Feller (1971)], Chapter XIII.5. We have

$$f_1(t) \sim \begin{cases} \log(1/t), & t \to 0^+, \\ e^{-t}/t, & t \to +\infty, \end{cases} \quad (D.21)$$

$$f_2(t) \sim \begin{cases} t, & t \to 0^+, \\ C + \log t, & t \to +\infty, \end{cases} \quad (D.22)$$

Appendix D: The Exponential Integral Functions   209

$$f_3(t) \sim \begin{cases} t \log(1/t), & t \to 0^+, \\ \log t, & t \to +\infty. \end{cases} \quad (D.23)$$

We conclude by pointing out the Laplace transform pair

$$\mathcal{L}\{\nu(t,a); s\} = \frac{1}{s^{a+1} \log s}, \quad \mathcal{R}e\, s > 0, \quad (D.24)$$

where

$$\nu(t,a) := \int_0^\infty \frac{t^{a+\tau}}{\Gamma(a+\tau+1)} d\tau, \quad a > -1. \quad (D.25)$$

For details on this transcendental function the reader is referred to the third volume of the Handbook of the Bateman Project [Erdélyi et al. (1953-1955)], see Chapter XVIII (devoted to the Miscellaneous functions) Section 18.3, pp. 217-224.

Appendix E

The Mittag-Leffler Functions

In this appendix we provide a survey of the high transcendental functions related to the Mittag-Leffler function, which are known to play fundamental roles in various applications of the fractional calculus. We simply refer to these as the class of functions of the Mittag-Leffler type. As usual we devote particular attention to their Laplace transforms. In Chapter 1 and Chapter 3 we have presented applications of these functions (with plots) in fractional relaxation phenomena. Here we present their applications for solving the Abel integral equations. At the end, we add some historical and bibliographical notes.

E.1 The classical Mittag-Leffler function $E_\alpha(z)$

The *Mittag-Leffler function*, that we denote by $E_\alpha(z)$ with $\alpha > 0$, is so named in honour of Gösta M. Mittag-Leffler, the great Swedish mathematician who introduced and investigated it at the beginning of the tuentieth century in a sequence of notes, see [Mittag-Leffler (1903a); (1903b); (1904); (1905)]. The function is defined by the following series representation, convergent in the whole complex plane,

$$E_\alpha(z) := \sum_{n=0}^{\infty} \frac{z^n}{\Gamma(\alpha n + 1)}, \quad \alpha > 0, \quad z \in \mathbb{C}, \qquad (E.1)$$

so $E_\alpha(z)$ is an *entire function*. In the limit for $\alpha \to 0^+$ the analyticity in the whole complex plane is lost since

$$E_0(z) := \sum_{n=0}^{\infty} z^n = \frac{1}{1-z}, \quad |z| < 1. \qquad (E.2)$$

The Mittag-Leffler function provides a simple generalization of the exponential function because of the substitution in the exponential series of $n! = \Gamma(n+1)$ with $(\alpha n)! = \Gamma(\alpha n + 1)$. So we recognize $E_1(z) = \exp(z)$. We also note that for the convergence of the power series in (E.1) the parameter α may be complex provided that $\mathcal{R}e\,(\alpha) > 0$, as pointed out in [Mittag-Leffler (1904)].

Other notable cases of definition (E.1), from which elementary functions are recovered, are

$$E_2\left(+z^2\right) = \cosh(z)\,, \qquad E_2\left(-z^2\right) = \cos(z)\,, \qquad z \in \mathbb{C}\,, \qquad (E.3)$$

and

$$E_{1/2}(\pm z^{1/2}) = e^z \left[1 + \mathrm{erf}\,(\pm z^{1/2})\right] = e^z\,\mathrm{erfc}(\mp z^{1/2})\,, \; z \in \mathbb{C}\,, \; (E.4)$$

where erf and erfc denote the error and the complementary error functions defined in Appendix C, see Eqs. (C.1)-(C.2), as

$$\mathrm{erf}\,(z) := \frac{2}{\sqrt{\pi}} \int_0^z e^{-u^2}\,du\,, \quad \mathrm{erfc}\,(z) := 1 - \mathrm{erf}\,(z)\,, \quad z \in \mathbb{C}\,.$$

In (E.4) for $z^{1/2}$ we mean the principal value of the square root of z in the complex plane cut along the negative real axis. With this choice $\pm z^{1/2}$ turns out to be positive/negative for $z \in \mathbb{R}^+$.

Since the identities in (E.3) are trivial, we present the proof only for (E.4). Avoiding the inessential polidromy with the substitution $\pm z^{1/2} \to z$, we write

$$E_{1/2}(z) = \sum_{m=0}^{\infty} \frac{z^{2m}}{\Gamma(m+1)} + \sum_{m=0}^{\infty} \frac{z^{2m+1}}{\Gamma(m+3/2)} = u(z) + v(z)\,. \quad (E.5)$$

Whereas the even part is easily recognized to be $u(z) = \exp(z^2)$, only after some manipulation can the odd part be proved to be $v(z) = \exp(z^2)\,\mathrm{erf}(z)$. To this end we need to recall from Appendix C the series representation of the error function, see (C.6),

$$\mathrm{erf}(z) = \frac{2}{\sqrt{\pi}}\,e^{-z^2} \sum_{m=0}^{\infty} \frac{2^m}{(2m+1)!!}\,z^{2m+1}\,, \quad z \in \mathbb{C}\,,$$

and note that $(2m+1)!! := 1 \cdot 3 \cdot 5 \cdots (2m+1) = 2^{m+1}\,\Gamma(m+3/2)/\sqrt{\pi}$.

An alternative proof is obtained by recognizing, after a term-wise differentiation of the series representation in (E.5), that $v(z)$ satisfies the following differential equation in $\mathbb{C}$,

$$v'(z) = 2\left[\frac{1}{\sqrt{\pi}} + z\, v(z)\right], \quad v(0) = 0,$$

whose solution can immediately be checked to be

$$v(z) = \frac{2}{\sqrt{\pi}}\, e^{z^2} \int_0^z e^{-u^2}\, du = e^{z^2}\, \mathrm{erf}(z).$$

The Mittag-Leffler functions of rational order. Let us now consider the Mittag-Leffler functions of rational order $\alpha = p/q$ with $p, q \in \mathbb{N}$ relatively prime. The relevant functional relations, that we quote from [Erdélyi et al. (1953-1955)], [Dzherbashyan (1966)], turn out to be

$$\left(\frac{d}{dz}\right)^p E_p(z^p) = E_p(z^p), \tag{E.6}$$

$$\frac{d^p}{dz^p} E_{p/q}\left(z^{p/q}\right) = E_{p/q}\left(z^{p/q}\right) + \sum_{k=1}^{q-1} \frac{z^{-k\,p/q}}{\Gamma(1 - k\,p/q)}, \quad q = 2, 3, \ldots, \tag{E.7}$$

$$E_{p/q}(z) = \frac{1}{p} \sum_{h=0}^{p-1} E_{1/q}\left(z^{1/p}\, e^{i2\pi h/p}\right), \tag{E.8}$$

and

$$E_{1/q}\left(z^{1/q}\right) = e^z \left[1 + \sum_{k=1}^{q-1} \frac{\gamma(1 - k/q, z)}{\Gamma(1 - k/q)}\right], \quad q = 2, 3, \ldots, \tag{E.9}$$

where $\gamma(a, z) := \int_0^z e^{-u}\, u^{a-1}\, du$ denotes the *incomplete gamma function* introduced at the end of Appendix A. Let us now sketch the proof for the above functional relations.

One easily recognizes that the relations (E.6) and (E.7) are immediate consequences of the definition (E.1). In order to prove the relation (E.8) we need to recall the identity

$$\sum_{h=0}^{p-1} e^{i2\pi hk/p} = \begin{cases} p \text{ if } k \equiv 0 \pmod{p}, \\ 0 \text{ if } k \not\equiv 0 \pmod{p}. \end{cases} \tag{E.10}$$

In fact, using this identity and the definition (E.1), we have

$$\sum_{h=0}^{p-1} E_\alpha(z\, e^{i2\pi h/p}) = p\, E_{\alpha p}(z^p)\,, \quad p \in \mathbb{N}\,. \qquad (E.11)$$

Substituting above α with α/p and z with $z^{1/p}$, we obtain

$$E_\alpha(z) = \frac{1}{p}\sum_{h=0}^{p-1} E_{\alpha/p}\left(z^{1/p}\, e^{i2\pi h/p}\right)\,, \quad p \in \mathbb{N}\,. \qquad (E.12)$$

Setting above $\alpha = p/q$, we finally obtain (E.8).

To prove the relation (E.9) we consider (E.7) for $p = 1$. Multiplying both sides by e^{-z}, we obtain

$$\frac{d}{dz}\left[e^{-z} E_{1/q}\left(z^{1/q}\right)\right] = e^{-z}\sum_{k=1}^{q-1}\frac{z^{-k/q}}{\Gamma(1-k/q)}\,. \qquad (E.13)$$

Then, upon integration of this and recalling the definition of the incomplete Gamma function, we obtain (E.9).

The relation (E.9) shows how the Mittag-Leffler functions of rational order can be expressed in terms of exponentials and incomplete gamma functions. In particular, taking in (E.9) $q = 2$, we now can verify again the relation (E.4). In fact, from (E.9) we obtain

$$E_{1/2}(z^{1/2}) = e^z\left[1 + \frac{1}{\sqrt{\pi}}\gamma(1/2,\,z)\right],$$

which is equivalent to (E.4) if we use the relation $\mathrm{erf}\,(z) = \gamma(1/2, z^2)/\sqrt{\pi}$, see (C.3).

The Mittag-Leffler integral representation. Many of the most important properties of $E_\alpha(z)$ follow from Mittag-Leffler's *integral representation*

$$E_\alpha(z) = \frac{1}{2\pi i}\int_{Ha}\frac{\zeta^{\alpha-1}\,e^\zeta}{\zeta^\alpha - z}\,d\zeta\,, \quad \alpha > 0,\ z \in \mathbb{C}\,, \qquad (E.14)$$

where the path of integration Ha (the *Hankel path*) is a loop which starts and ends at $-\infty$ and encircles the circular disk $|\zeta| \le |z|^{1/\alpha}$ in the positive sense: $-\pi \le \arg\zeta \le \pi$ on Ha. To prove (E.14), expand the integrand in powers of ζ, integrate term by term, and use Hankel's integral for the reciprocal of the Gamma function.

The integrand in (E.14) has a branch-point at $\zeta = 0$. The complex ζ-plane is cut along the negative real axis, and in the cut plane the integrand is single-valued: the principal branch of ζ^α is taken in the cut plane. The integrand has poles at the points $\zeta_m = z^{1/\alpha}\, e^{2\pi i m/\alpha}$, m integer, but only those of the poles lie in the cut plane for which $-\alpha\pi < \arg z + 2\pi m < \alpha\pi$. Thus, the number of the poles inside Ha is either $[\alpha]$ or $[\alpha + 1]$, according to the value of $\arg z$.

Asymptotic expansions. The most interesting properties of the Mittag-Leffler function are associated with its asymptotic expansions as $z \to \infty$ in various sectors of the complex plane.

For detailed asymptotic analysis, which includes the smooth transition across the Stokes lines, the interested reader is referred to [Wong and Zhao (2002)]. Here, we limit ourselves to report a few results available in [Erdélyi et al. (1953-1954)].

For the case $0 < \alpha < 2$ we have for $|z| \to \infty$:

$$E_\alpha(z) \sim \frac{1}{\alpha} \exp(z^{1/\alpha}) - \sum_{k=1}^{\infty} \frac{z^{-k}}{\Gamma(1 - \alpha k)}, \quad |\arg z| < \alpha\pi/2, \quad (E.15)$$

$$E_\alpha(z) \sim -\sum_{k=1}^{\infty} \frac{z^{-k}}{\Gamma(1 - \alpha k)}, \quad \alpha\pi/2 < \arg z < 2\pi - \alpha\pi/2. \quad (E.16)$$

For the case $\alpha \geq 2$ we have for $|z| \to \infty$:

$$E_\alpha(z) \sim \frac{1}{\alpha} \sum_m \exp\left(z^{1/\alpha} e^{2\pi i m/\alpha}\right) - \sum_{k=1}^{\infty} \frac{z^{-k}}{\Gamma(1 - \alpha k)}, \quad (E.17)$$

where $\arg z$ can assume any value between $-\pi$ and $+\pi$ inclusive, and m takes all integer values such that

$$-\alpha\pi/2 < \arg z + 2\pi m < \alpha\pi/2.$$

From the asymptotic properties (E.15)-(E.17) and the definition of the order (and type) of an entire function, we infer that the Mittag-Leffler function is an *entire function of order* $\rho = 1/\alpha$ (and type 1) for $\alpha > 0$; in a certain sense each $E_\alpha(z)$ is the simplest entire function of its order, see [Phragmén (1904)]. These properties are still valid but with $\rho = 1/\text{Re}\{\alpha\}$, if $\alpha \in \mathbb{C}$ with *positive real part*.

The Mittag-Leffler function also furnishes examples and counter-examples for the growth and other properties of entire functions of finite order, see [Buhl (1925)].

Complete monotonicity on the negative real axis. A relevant property of the classical Mittag-Leffler function is its *complete monotonicity* in the negative real axis, when its parameter α is less or equal 1. We write for $x > 0$:

$$E_\alpha(-x) \quad \text{CM} \quad \text{iff} \quad 0 < \alpha \le 1, \qquad (E.18)$$

where CM stands for *completely monotone* (function).

This property, formerly conjectured by Feller using probabilistic methods, was rigorously proven by [Pollard (1948)] based on the Bochner theorem, which provides a sufficient and necessary condition for the complete monotonicity. In other words, Pollard was able to give the representation of $E_\alpha(-x)$, for $x \in \mathbb{R}^+$, when $0 < \alpha < 1$, that reads

$$E_\alpha(-x) = \int_0^\infty e^{-ux} P_\alpha(u)\, du, \quad P_\alpha(u) \ge 0, \quad 0 < \alpha < 1. \quad (E.19)$$

Here $P_\alpha(u)$ is indeed a transcendental entire function with series representation

$$P_\alpha(u) = \frac{1}{\pi \alpha} \sum_{n=1}^\infty \frac{(-1)^{n-1}}{n!} \sin(\pi\alpha n)\, \Gamma(\alpha n + 1)]\, u^{n-1}, \qquad (E.20)$$

whose *non-negativity* was proven by Pollard being related in sign with the inverse Laplace transform of the CM function $\exp(-s^\alpha)$ derived in [Pollard (1946)].

Denoting this inverse Laplace transform by $P^*_\alpha(u)$, it turns out that

$$P_\alpha(u) = \frac{u^{-1-1/\alpha}}{\alpha} P^*_\alpha\left(u^{-1/\alpha}\right), \text{ with}$$

$$P^*_\alpha(u) = -\frac{1}{\pi} \sum_{n=0}^\infty \frac{(-1)^n}{n!} \sin(\pi\alpha n)\, \frac{\Gamma(\alpha n + 1)}{u^{\alpha n + 1}}. \qquad (E.21)$$

Later, in Appendix F, we will show that both the functions P and P^* introduced by Pollard are functions of the Wright type.

E.2 The Mittag-Leffler function with two parameters

A straightforward generalization of the Mittag-Leffler function is obtained by replacing the additive constant 1 in the argument of the

Gamma function in (E.1) by an arbitrary complex parameter β. It was formerly considered in 1905 by [Wiman (1905a)]; later, in the fifties, such generalization was investigated by Humbert and Agarwal, with respect to the Laplace transformation, see [Humbert (1953)], [Agarwal 1953], [Humbert and Agarwal (1953)]. Usually, when dealing with Laplace transform pairs, the parameter β is required to be positive as α.

For this function we agree to use the notation

$$E_{\alpha,\beta}(z) := \sum_{n=0}^{\infty} \frac{z^n}{\Gamma(\alpha n + \beta)}, \quad \mathcal{R}e\,\alpha > 0,\ \beta \in \mathbb{C},\ z \in \mathbb{C}. \quad (E.22)$$

Of course $E_{\alpha,1}(z) \equiv E_\alpha(z)$.

Particular simple cases (of trivial proof) are

$$E_{1,1}(z) = e^z,\ E_{1,2}(z) = \frac{e^z - 1}{z},\ E_{1,3}(z) = \frac{e^z - 1 - z}{z^2},$$

$$\ldots,\ E_{1,m}(z) = \frac{1}{z^{m-1}} \left(e^z - \sum_{n=0}^{m-2} \frac{z^n}{n!} \right), \quad (E.23)$$

and

$$E_{2,2}(+z^2) = \frac{\sinh(z)}{z},\quad E_{2,2}(-z^2) = \frac{\sin(z)}{z}. \quad (E.24)$$

Compare the identities in (E.24) with those in (E.3) concerning $E_{2,1}(\pm z^2) \equiv E_2(\pm z^2)$.

Recurrence relations. We list hereafter some general functional relations for the Mittag-Leffler function (E.22) of recursive kind, which involve both the two parameters α, β, see [Erdélyi et al. (1953-1955)], [Dzherbashyan (1966)],

$$E_{\alpha,\beta}(z) = \frac{1}{\Gamma(\beta)} + z\,E_{\alpha,\beta+\alpha}(z), \quad (E.25)$$

$$E_{\alpha,\beta}(z) = \beta E_{\alpha,\beta+1}(z) + \alpha z \frac{d}{dz} E_{\alpha,\beta+1}(z), \quad (E.26)$$

$$\left(\frac{d}{dz} \right)^p \left[z^{\beta-1} E_{\alpha,\beta}(z^\alpha) \right] = z^{\beta-p-1} E_{\alpha,\beta-p}(z^\alpha),\quad p \in \mathbb{N}. \quad (E.27)$$

Sometimes in the relevant literature we find functions expressed in terms of the Mittag-Leffler function of two parameters, when, with a minimum effort, they can be shown to be simply related to the classical Mittag-Leffler function of one parameter. An instructive example is offered by the function treated in Chapter 1, Eq. (1.45),

$$\Phi_\alpha(t) := t^{-(1-\alpha)} E_{\alpha,\alpha}(-t^\alpha) = -\frac{d}{dt} E_\alpha(-t^\alpha), \quad t \geq 0,$$

whose identity is valid for any $\alpha > 0$. This identity is easily proven by a direct calculation from the series representations, but it is not well noted in the literature.

The Mittag-Leffler integral representation. The integral representation of the Mittag-Leffler function with two parameters turns out to be

$$E_{\alpha,\beta}(z) = \frac{1}{2\pi i} \int_{Ha} \frac{\zeta^{\alpha-\beta} e^\zeta}{\zeta^\alpha - z} d\zeta, \quad \alpha, \beta > 0, \quad z \in \mathbb{C}, \qquad (E.28)$$

where the path of integration Ha is the usual *Hankel path* considered in (E.14) for the classical Mittag-Leffler function in one parameter.

Asymptotic expansions. As for the classical case, the most interesting properties of the Mittag-Leffler function in two parameters are associated with its asymptotic expansions as $z \to \infty$ in various sectors of the complex plane. These properties can be summarized as follows.

For the case $0 < \alpha < 2$ we have as $|z| \to \infty$:

$$E_{\alpha,\beta}(z) \sim \frac{1}{\alpha} \exp(z^{1/\alpha}) - \sum_{k=1}^\infty \frac{z^{-k}}{\Gamma(\beta - \alpha k)}, \quad |\arg z| < \alpha\pi/2, \qquad (E.29)$$

$$E_{\alpha,\beta}(z) \sim - \sum_{k=1}^\infty \frac{z^{-k}}{\Gamma(\beta - \alpha k)}, \quad \alpha\pi/2 < \arg z < 2\pi - \alpha\pi/2. \qquad (E.30)$$

For the case $\alpha \geq 2$ we have as $|z| \to \infty$:

$$E_{\alpha,\beta}(z) \sim \frac{1}{\alpha} \sum_m \exp\left(z^{1/\alpha} e^{2\pi i m/\alpha}\right) - \sum_{k=1}^\infty \frac{z^{-k}}{\Gamma(\beta - \alpha k)}, \qquad (E.31)$$

where $\arg z$ can assume any value between $-\pi$ and $+\pi$ inclusive and m takes all integer values such that
$$-\alpha\pi/2 < \arg z + 2\pi m < \alpha\pi/2\,.$$
We note that the additional parameter β has no influence on the character of the entire function $E_{\alpha,\beta}(z)$ with respect to $E_\alpha(z)$, so the Mittag-Leffler function in two parameters is still entire of order $\rho = 1/\mathcal{R}e\,(\alpha)$ and type 1.

The complete monotonicity on the negative real axis. Only recently the theorem by Pollard was extended to the complete monotonicity of the Mittag-Leffler function with two parameters on the negative real axis, see [Schneider (1996); Miller and Samko (1997); Miller and Samko (2001)]. Schneider proved that, for $x > 0$,

$$E_{\alpha,\beta}(-x) \quad \text{CM} \quad \text{iff} \quad \begin{cases} 0 < \alpha \leq 1, \\ \beta \geq \alpha\,. \end{cases} \qquad (E.32)$$

A part of the trivial case $\alpha = \beta = 1$, the statement, in virtue of Bernstein's theorem, is equivalent to prove that $E_{\alpha,\beta}(-x)$ is the Laplace transform of a *non-negative*, absolutely continuous function supported by $\mathbb{R}^+$, that we denote $S_{\alpha,\beta}$. We have, in our notation,

$$E_{\alpha,\beta}(-x) = \int_0^\infty e^{-ux} S_{\alpha,\beta}(u)\,du\,,\ S_{\alpha,\beta} \geq 0,\ \begin{cases} 0 < \alpha \leq 1, \\ \beta \geq \alpha, \end{cases} \qquad (E.33)$$

where $S_{\alpha,\beta}(r)$ is given by:
- for $0 < \alpha < 1$, $\beta \geq \alpha$:

$$S_{\alpha,\beta}(u) = \sum_{n=0}^\infty (-1)^n \frac{u^n}{n!\Gamma(\beta - \alpha - \alpha n)}\,,\ u \in \mathbb{R}^+\,, \qquad (E.34a)$$

- for $\alpha = 1$, $\beta > 1$:

$$S_{\alpha,\beta}(u) = \begin{cases} \dfrac{(1-u)^{\beta-2}}{\Gamma(\beta-1)}, & 0 \leq u \leq 1, \\ 0, & 1 < u < \infty\,. \end{cases} \qquad (E.34b)$$

For $\alpha = \beta = 1$ we recover the Delta generalized function centred in $u = 1$, namely $\delta(1-u) = (1-u)^{-1}/\Gamma(0)$. For the proof of the non-negativity we refer the reader to the original work by Schneider.

E.3 Other functions of the Mittag-Leffler type

There are various functions introduced in the literature that can be simply related to the Mittag-Leffler function with two parameters or are its generalizations by adding additional parameters. In this section we limit ourselves to introduce three functions, named after Miller and Ross, Rabotnov and Prabhakar; further generalizations are briefly mentioned in the notes at the end of this appendix along with their references.

The Miller-Ross function. In [Miller and Ross (1993)] the authors introduce the function $E_t(\nu, a)$ by the representation

$$E_t(\nu, a) := t^\nu \sum_{n=0}^\infty \frac{(at)^n}{\Gamma(n+1+\nu)}. \qquad (E.35)$$

Comparing this definition with (E.22) we recognize

$$E_t(-\nu, a) = t^\nu E_{1,\nu+1}(at). \qquad (E.36)$$

A part of the pre-factor t^ν the power series in (E.35) represents an entire function in $\mathbb{C}$ differing from the exponential series for the additive constant ν in the argument of $\Gamma(n+1) = n!$. According to [Wintner (1959)], this kind of generalization of the exponential series was formerly proposed by Heaviside, and contrasts with the subsequent generalization proposed by Mittag-Leffler.

Recalling the series representation of the incomplete Gamma functions in Section A.4, see Eqs. (A.55)-(A.57), we recognize the identities

$$E_t(\nu, a) = t^\nu \, \mathrm{e}^{at} \gamma^*(\nu, at) = \frac{a^{-\nu} \mathrm{e}^{at}}{\Gamma(\nu)} \gamma(\nu, at). \qquad (E.37)$$

For $\mathcal{R}e\,\nu > 0$, recalling the integral representation of the incomplete Gamma function in (E.37), we have

$$E_t(\nu, a) = \frac{a^{-\nu} \mathrm{e}^{at}}{\Gamma(\nu)} \int_0^{at} u^{\nu-1} \mathrm{e}^{-u}\, du = \frac{1}{\Gamma(\nu)} \int_0^t v^{\nu-1}\, \mathrm{e}^{a(t-v)}\, dv. \qquad (E.38)$$

For $\nu = -p$ ($p = 0, 1, \ldots$) and $a \neq 0$, using Eq. (A.55) $[\gamma^*(-p, at) \equiv (at)^p]$ in (E.37), we get

$$E_t(0, a) = \mathrm{e}^{at}, \; E_t(-1, a) = a\,\mathrm{e}^{at}, \; E_t(-2, a) = a^2\, \mathrm{e}^{at}, \ldots. \qquad (E.39)$$

In general we have for any $\nu > 0$:
$$E_t(-\nu, a) = t^{-\nu} \sum_{n=0}^{\infty} \frac{(at)^n}{\Gamma(n+1-\nu)} = {}_0D_t^\nu \left[e^{at} \right]. \qquad (E.40)$$

For $a = 0$ and arbitrary ν we recover the power law
$$E_t(\nu, 0) = \frac{t^\nu}{\Gamma(\nu+1)}, \qquad (E.41)$$

that degenerates for $\nu = -(p+1)$ ($p = 0, 1, \ldots$) into the derivatives of the generalized Dirac function $\delta^{(p)}(t) = t^{-p-1}/\Gamma(-p)$.

For $\nu = \pm 1/2$ and $a > 0$ a relationship with the error function is expected; we have
$$E_t(1/2, a) = a^{-1/2} e^{at} \operatorname{erf}\left[(at)^{1/2} \right], \qquad (E.42)$$

$$E_t(-1/2, a) = a\, E_t(1/2, a) + \frac{t^{-1/2}}{\sqrt{\pi}}. \qquad (E.43)$$

We also note the relationship
$$E_t(\nu, a) = a\, E_t(1+\nu, a) + \frac{t^\nu}{\Gamma(\nu+1)}. \qquad (E.44)$$

For more details on this function, including tables of values, recursion relations and applications, the interested reader is referred to [Miller and Ross (1993)].

The Rabotnov function. Yu N. Rabotnov, in his works on viscoelastity, see e.g. [Rabotnov (1948); (1969)] and [Rabotnov et al. (1969)] introduced the function of time t, depending on two parameters, that he denoted by $\alpha \in (-1, 0]$ (related to the type of viscoelasticity) and $\beta \in \mathbb{R}$;
$$\mathcal{E}_\alpha(\beta, t) := t^\alpha \sum_{n=0}^{\infty} \frac{\beta\, t^{n(\alpha+1)}}{\Gamma[(n+1)(\alpha+1)]}, \quad t \geq 0. \qquad (E.45)$$

Rabotnov, presumably unaware of the Mittag-Leffler function, referred to this function as the *fractional exponential function*, noting that for $\alpha = 0$ it reduces to the standard exponential $\exp(\beta t)$. In the literature such a function is mostly referred to as the *Rabotnov function*. The relation of this function with the Mittag-Leffler function in two parameters is thus obvious:
$$\mathcal{E}_\alpha(\beta, t) = t^\alpha E_{\alpha+1, \alpha+1}\left(\beta\, t^{\alpha+1} \right). \qquad (E.46)$$

However, even not yet noted in the literature, it is trivial to show, using (1.45), that
$$\mathcal{E}_\alpha(\beta, t) = \frac{1}{\beta} \frac{d}{dt} E_{\alpha+1}\left(\beta\, t^{\alpha+1}\right). \tag{E.47}$$
We have kept the notation of Rabotnov according to whom the main parameter α is shifted with respect to ours, so it appears confusing. However, in view of the fact that his constant β is negative, we have a complete equivalence with our theory of fractional relaxation discussed in Chapter 1 (Section 1.3), namely with Eq. (1.45), if we set in (E.47) $(\alpha+1) \to \alpha$ and $\beta \to -1$.

The Prabhakar function. In [Prabhakar (1971)] the author has introduced the function
$$E_{\alpha,\beta}^{\gamma}(z) := \sum_{n=0}^{\infty} \frac{(\gamma)_n}{n!\,\Gamma(\alpha n + \beta)}\, z^n\,, \quad \mathcal{R}e\,\alpha > 0,\ \beta \in \mathbb{C},\ \gamma > 0, \tag{E.48}$$
where $(\gamma)_n = \gamma(\gamma+1)\ldots(\gamma+n-1) = \Gamma(\gamma+n)/\Gamma(\gamma)$ denotes the Pochhammer symbol. For $\gamma = 1$ we recover the 2-parameter Mittag-Leffler function (E.22), i.e.
$$E_{\alpha,\beta}^{1}(z) := \sum_{n=0}^{\infty} \frac{z^n}{\Gamma(\alpha n + \beta)}, \tag{E.49}$$
and, for $\gamma = \beta = 1$, the classical Mittag-Leffler function (E.1),
$$E_{\alpha,1}^{1}(z) := \sum_{n=0}^{\infty} \frac{z^n}{\Gamma(\alpha n + 1)}. \tag{E.50}$$
The relevance of this function appears nowadays in certain fractional relaxation and diffusion phenomena, see e.g. [Hanyga and Seredyńska (2008a)] and [Figueiredo *et al.* (2009a); (2009b)].

E.4 The Laplace transform pairs

The Mittag-Leffler functions are connected to the Laplace integral through the relevant identities
$$\frac{1}{1-z} = \begin{cases} \displaystyle\int_0^\infty e^{-u}\, E_\alpha\left(u^\alpha z\right) du\,, \\[1em] \displaystyle\int_0^\infty e^{-u}\, u^{\beta-1}\, E_{\alpha,\beta}\left(u^\alpha z\right) du\,, \end{cases} \tag{E.51}$$

where $\alpha, \beta > 0$. The first integral was evaluated by Mittag-Leffler who showed that the region of its convergence contains the unit circle and is bounded by the line $\operatorname{Re} z^{1/\alpha} = 1$. This is also true for any $\beta > 0$.

The functions $e_\alpha(t;\lambda)$, $e_{\alpha,\beta}(t;\lambda)$. The above integrals are fundamental in the evaluation of the Laplace transform of $E_\alpha(-\lambda t^\alpha)$ and $E_{\alpha,\beta}(-\lambda t^\alpha)$ with $\alpha, \beta > 0$ and $\lambda \in \mathbb{C}$. Since these functions play a key role in problems of fractional calculus, we shall introduce a special notation for them.

Putting in (E.51) $u = st$ and $u^\alpha z = -\lambda t^\alpha$ with $t \geq 0$ and $\lambda \in \mathbb{C}$, and using the sign $\div$ for the juxtaposition of a function depending on t with its Laplace transform depending on s, we get the Laplace transform pairs for $\mathcal{R}e(s) > |\lambda|^{1/\alpha}$,

$$e_\alpha(t;\lambda) := E_\alpha(-\lambda t^\alpha) \div \frac{s^{\alpha-1}}{s^\alpha + \lambda} = \frac{s^{-1}}{1 + \lambda s^{-\alpha}}, \qquad (E.52)$$

and

$$e_{\alpha,\beta}(t;\lambda) := t^{\beta-1} E_{\alpha,\beta}(-\lambda t^\alpha) \div \frac{s^{\alpha-\beta}}{s^\alpha + \lambda} = \frac{s^{-\beta}}{1 + \lambda s^{-\alpha}}. \qquad (E.53)$$

Of course the results (E.52)-(E.53) can also be obtained formally by Laplace transforming term by term the series in (E.1) and (E.6) with $z = -\lambda t^\alpha$.

We note that, following the approach by Humbert and Agarwal, the Laplace transform pairs (E.52)-(E.53) allow us to obtain a number of functional relations satisfied by $e_\alpha(t;\lambda)$ and $e_{\alpha,\beta}(t;\lambda)$, for example,

$$e_{\alpha,\alpha}(t;\lambda) = -\frac{1}{\lambda}\frac{d}{dt}e_{\alpha,1}(t;\lambda) = \frac{t^{\alpha-1}}{\Gamma(\alpha)} - \lambda e_{\alpha,2\alpha}(t;\lambda), \qquad (E.54)$$

and

$$e_{\alpha,\beta}(t;\lambda) = \frac{d}{dt}e_{\alpha,\beta+1}(t;\lambda). \qquad (E.55)$$

To prove (E.54) we note with $\alpha > 0$:

$$e_{\alpha,\alpha}(t;\lambda) \div \frac{1}{s^\alpha + \lambda} = \begin{cases} -\dfrac{1}{\lambda}\left[\left(s\dfrac{s^{\alpha-1}}{s^\alpha + \lambda}\right) - 1\right], \\[2ex] \dfrac{1}{s^\alpha} - \dfrac{\lambda}{s^\alpha(s^\alpha + \lambda)}. \end{cases}$$

To prove (E.55) we note with $\alpha, \beta > 0$:
$$e_{\alpha,\beta}(t;\lambda) \div \frac{s^{\alpha-\beta}}{s^\alpha + \lambda} = s\frac{s^{\alpha-(\beta+1)}}{s^\alpha + \lambda} - 0.$$

If λ *is positive*, the functions $e_\alpha(t;\lambda)$ and $e_{\alpha,\beta}(t;\lambda)$ turn to be *completely monotone* for $t > 0$ when $0 < \alpha \leq 1$ and $0 < \alpha \leq \beta \leq 1$, respectively. These noteworthy properties can be regarded as a consequence of the theorems by Pollard and Schneider that we have previously stated, if we recall known results for CM functions[1] In fact, the CM properties (proved with respect the variables x) for the functions $E_\alpha(-x)$ and $E_{\alpha,\beta}(-x)$ still hold with respect to the variable t if we replace x by λt^α ($t \geq 0$), provided λ is a positive constant, since t^α is a Bernstein function just for $0 < \alpha \leq 1$. Furthermore for $0 < \beta < 1$ the function $e_{\alpha,\beta}(t;\lambda)$ turns out to be CM as a product of two CM functions.

It is instructive to prove the above properties independently from the theorems of Pollard and Scneider but directly from the Bochner-like integral representations of the functions $e_\alpha(t;\lambda)$ and $e_{\alpha,\beta}(t;\lambda)$ obtained from the inversion of their Laplace transforms (E.52), (E.53). In fact, excluding the trivial case $\alpha = \beta = 1$ for which $e_1(t;\lambda) = e_{1,1}(t;\lambda) = e^{-\lambda t}$, we can prove the existence of the corresponding spectral functions using the complex Bromwich formula for the inversion of Laplace transforms.

As an exercise in complex analysis (that we kindly invite the reader to carry out) we obtain the required integral representations

$$e_\alpha(t;\lambda) := \int_0^\infty e^{-rt} K_\alpha(r;\lambda)\, dr, \quad 0 < \alpha < 1, \qquad (E.56)$$

with *spectral function*

$$K_\alpha(r;\lambda) = \frac{1}{\pi} \frac{\lambda r^{\alpha-1} \sin(\alpha\pi)}{r^{2\alpha} + 2\lambda r^\alpha \cos(\alpha\pi) + \lambda^2} \geq 0, \qquad (E.57)$$

and

$$e_{\alpha,\beta}(t;\lambda) := \int_0^\infty e^{-rt} K_{\alpha,\beta}(r;\lambda)\, dr, \; 0 < \alpha \leq \beta < 1, \qquad (E.58)$$

---

[1] a) The composition of a CM function with a Bernstein function is still a CM function. b) The product of CM functions is a CM functions. For details see [Berg and Forst (1975)], [Gripenberg et al. (1990)].

with *spectral function*

$$K_{\alpha,\beta}(r;\lambda) = \frac{1}{\pi} \frac{\lambda \sin\left[(\beta-\alpha)\pi\right] + r^\alpha \sin(\beta\pi)}{r^{2\alpha} + 2\lambda r^\alpha \cos(\alpha\pi) + \lambda^2} r^{\alpha-\beta} \geq 0. \quad (E.59)$$

Since in these cases the contributions to the Bromwich inversion formula come only from the integration on the sides of the branch cut along the negative real axis (where $s = r \exp(\pm i\pi)$, $r \geq 0$), the spectral functions are simply derived by using the so-called Titchmarsh formula on Laplace inversion. In fact, recalling for a generic Laplace transform pair $f(t) \div \widetilde{f}(s)$ the Titchmarsh formula

$$f(t) = -\frac{1}{\pi} \int_0^\infty e^{-rt} \operatorname{Im}\left\{\widetilde{f}\left(re^{i\pi}\right)\right\} dr,$$

that requires the expressions of

$$-\operatorname{Im}\left\{\frac{s^{\alpha-1}}{s^\alpha + \lambda}\right\}, \quad -\operatorname{Im}\left\{\frac{s^{\alpha-\beta}}{s^\alpha + \lambda}\right\},$$

along the ray $s = r e^{i\pi}$ with $r \geq 0$ (the branch cut of the function s^α), we easily obtain the required expressions of the two spectral functions $K_\alpha(r;\lambda)$, $K_{\alpha,\beta}(r;\lambda)$, that, as a matter of fact, turn out to be non-negative.

The functions $e_{1/2}(t;\lambda)$, $e_{1/2,1/2}(t;\lambda)$. We note that in most handbooks containing tables for the Laplace transforms, the Mittag-Leffler function is ignored so that the transform pairs (E.52)-(E.53) do not appear if not in the special cases $\alpha = 1/2$ and $\beta = 1, 1/2$, written in terms of the error and complementary error functions, see e.g. [Abramowitz and Stegun (1965)]. In fact, in these cases we can use (E.4) and (E.28) and recover from (E.26)-(E.27) the two Laplace transform pairs

$$\frac{1}{s^{1/2}(s^{1/2} \pm \lambda)} \div e_{1/2}(t;\pm\lambda) = e^{\lambda^2 t} \operatorname{erfc}(\pm\lambda\sqrt{t}), \quad \lambda \in \mathbb{C}, \quad (E.60)$$

$$\frac{1}{s^{1/2} \pm \lambda} \div e_{1/2,1/2}(t;\pm\lambda) = \frac{1}{\sqrt{\pi t}} \mp \lambda e_{1/2}(t;\pm\lambda), \quad \lambda \in \mathbb{C}. \quad (E.61)$$

We also obtain the related pairs

$$\frac{1}{s^{1/2}(s^{1/2} \pm \lambda)^2} \div 2\sqrt{\frac{t}{\pi}} \mp 2\lambda t\, e_{1/2}(t;\pm\lambda), \quad \lambda \in \mathbb{C}, \quad (E.62)$$

$$\frac{1}{(s^{1/2} \pm \lambda)^2} \div \mp 2\lambda \sqrt{\frac{t}{\pi}} + (1 + 2\lambda^2 t)\, e_{1/2}(t; \pm\lambda), \quad \lambda \in \mathbb{C}. \quad (E.63)$$

In the pair (E.62) we have used the properties
$$\frac{1}{s^{1/2}(s^{1/2} \pm \lambda)^2} = -2\frac{1}{d}ds\left(\frac{1}{s^{1/2} \pm \lambda}\right), \qquad \frac{d^n}{ds^n}\widetilde{f}(s) \div (-t)^n\, f(t).$$

The pair (E.63) is easily obtained by noting that
$$\frac{1}{(s^{1/2} \pm \lambda)^2} = \frac{1}{s^{1/2}(s^{1/2} \pm \lambda)} \mp \frac{\lambda}{s^{1/2}(s^{1/2} \pm \lambda)^2}.$$

The function $e^{\gamma}_{\alpha,\beta}(t;\lambda)$. By using the Prabhakar function (E.48) we define the function
$$e^{\gamma}_{\alpha,\beta}(t;\lambda) := t^{\beta-1} E^{\gamma}_{\alpha,\beta}(-\lambda t^{\alpha}). \quad (E.64)$$

Substituting with $\alpha, \beta, \gamma > 0$ the series representation of the Prabhakar function in the Laplace transformation yields the identity
$$\int_0^{\infty} e^{-st}\, t^{\beta-1} E^{\gamma}_{\alpha,\beta}(at^{\alpha}) = \frac{1}{s^{\beta}} \sum_{n=0}^{\infty} (-1)^n \frac{\Gamma(\gamma+n)}{\Gamma(\gamma)} \left(\frac{\lambda}{s^{\alpha}}\right)^n. \quad (E.65)$$

On the other hand (using the binomial series)
$$(1+z)^{\gamma} = \sum_{n=0}^{\infty} \frac{\Gamma(1-\gamma)}{\Gamma(1-\gamma-n)n!} z^n = \sum_{n=0}^{\infty} (-1)^n \frac{\Gamma(\gamma+n)}{\Gamma(\gamma)n!} z^n. \quad (E.66)$$

Comparison between Eq. (E.65) and Eq. (E.66) yields the Laplace transform pair
$$e^{\gamma}_{\alpha,\beta}(t;\lambda) := t^{\beta-1} E^{\gamma}_{\alpha,\beta}(-\lambda t^{\alpha}) \div \frac{s^{\alpha\gamma-\beta}}{(s^{\alpha}+\lambda)^{\gamma}} = \frac{s^{-\beta}}{(1+\lambda s^{-\alpha})^{\gamma}}. \quad (E.67)$$

Eq. (E.67) holds (by analytic continuation) for $Re[\alpha] > 0$, $Re[\beta] > 0$. In particular we get the known Laplace transform pairs (E.53) for $e_{\alpha,\beta}(t;\lambda)$ and (E.52) for $e_{\alpha}(t;\lambda)$.

From the Laplace transform pair (E.67), using the rules stated in [Gripenberg et al. (1990)], [Hanyga and Seredyńska (2008a)] have shown that the Prabhakar function for $\lambda > 0$ and $\beta = 1$ is a CM function iff $0 < \alpha \le 1$ and $0 < \gamma \le 1$. More generally we can show that for $\lambda > 0$

$$e^{\gamma}_{\alpha,\beta}(t;\lambda) := t^{\beta-1} E^{\gamma}_{\alpha,\beta}(-\lambda t^{\alpha}) \text{ CM iff } \begin{cases} 0 < \alpha \le \beta \le 1, \\ 0 < \gamma \le 1. \end{cases} \quad (E.68)$$

E.5 Derivatives of the Mittag-Leffler functions

Let us denote the derivatives of order $k = 0, 1, 2\ldots$ of the function $E_{\alpha,\beta}(at^\alpha)$ for $a \in \mathbb{R}$ with respect its argument $z = at^\alpha$ by $E^{(k)}_{\alpha,\beta}(at^\alpha)$, where $\alpha, \beta > 0$. We then quote from [Podlubny (1999)] two interesting results concerning the fractional derivative and the Laplace transform of the function $t^{\alpha k + \beta - 1} E^{(k)}_{\alpha,\beta}(at^\alpha)$.

For the fractional derivative of order $\gamma > 0$, by differentiating term by term the series representation, we have:

$$_0D_t^\gamma \left\{ t^{\alpha k + \beta - 1} E^{(k)}_{\alpha,\beta}(at^\alpha) \right\} = t^{\alpha k + \beta - \gamma - 1} E_{\alpha,\beta}(at^\alpha). \tag{E.69}$$

For the Laplace transform we have for $\mathcal{R}e\, s > |a|^{1/\alpha}$:

$$\mathcal{L}\left\{ t^{\alpha k + \beta - 1} E^{(k)}_{\alpha,\beta}(at^\alpha); s \right\} = \frac{k!\, s^{\alpha - \beta}}{(s^\alpha - a)^{k+1}}, \quad k = 0, 1, 2, \ldots. \tag{E.70}$$

We find it instructive to prove the above Laplace transform formula for the classical Mittag-Leffler function (that is in the case $\beta = 1$), namely for $\mathcal{R}e\, s > |a|^{1/\alpha}$:

$$\mathcal{L}\left\{ t^{\alpha k} E^{(k)}_\alpha(at^\alpha); s \right\} = \frac{k!\, s^{\alpha - 1}}{(s^\alpha - a)^{k+1}}, \quad k = 0, 1, 2, \ldots. \tag{E.71}$$

In the particular case $\alpha = 1$ formula (E.71) reduces to

$$\mathcal{L}\{t^k e^{at}; s\} = \frac{k!}{(s - a)^{k+1}}, \quad k = 0, 1, 2, \ldots, \quad \mathcal{R}e\, s > |a|. \tag{E.72}$$

As a matter of fact (E.72) is known to be valid for $\mathcal{R}e\, s > \mathcal{R}e\, a$ and its proof is a consequence of the analyticity property of the Laplace transform, $\mathcal{L}\{t^k f(t); s\} = (-1)^k \widetilde{f}^{(k)}(s)$, applied to $f(t) = \exp(\pm at)$, for which

$$\mathcal{L}\{e^{\pm at}; s\} = \frac{1}{(s \mp a)}, \quad k = 0, 1, 2, \ldots, \quad \mathcal{R}e\, s > \pm \mathcal{R}e\, a.$$

However, Eq. (E.72) can be deduced for $\mathcal{R}e\, s > |a|$ by using the method of power series expansions as shown below. Indeed,

$$\int_0^\infty e^{-u} e^{\pm zu}\, du = \sum_{k=0}^\infty \frac{(\pm z)^k}{k!} \int_0^\infty e^{-u} u^k\, du = \sum_{k=0}^\infty (\pm z)^k = \frac{1}{1 \mp z},$$

from which, by k-iterated differentiation with respect to z,
$$\int_0^\infty e^{-u} u^k e^{\pm zu} \, du = \frac{k!}{(1 \mp z)^{k+1}}, \quad |z| < 1.$$
Now, by the substitutions $u = st$ and $z = a/s$ (for our purposes here we agree to take s real), we get after simple manipulations the identity in (E.73) for $s > |a|$. By analytic continuation the validity is extended to complex s with $\mathcal{R}e \, s > |a|$.

The above reasoning can be applied to the integral below in order to derive the more general Laplace transform formula (E.71). Indeed,
$$\int_0^\infty e^{-u} E_\alpha(\pm zu^\alpha) \, du = \sum_{k=0}^\infty \frac{(\pm z)^k}{\Gamma(\alpha k+1)} \int_0^\infty e^{-u} u^{\alpha k} \, du$$
$$= \sum_{k=0}^\infty (\pm z)^k = \frac{1}{1 \mp z},$$
from which, by k-iterated differentiating with respect to z,
$$\int_0^\infty e^{-u} u^{\alpha k} E_\alpha^{(k)}(\pm zu^\alpha) \, du = \frac{k!}{(1 \mp z)^{k+1}}, \quad |z| < 1.$$
Now, by the substitutions $u = st$ and $z = a/s^\alpha$, we get after simple manipulations the identity in (E.71) for $s > |a|^{1/\alpha}$, namely, by analytic continuation, for $\mathcal{R}e \, s > |a|^{1/\alpha}$.

E.6 Summation and integration of Mittag-Leffler functions

Hereafter we exhibit for completeness some formulas related to summation and integration of the Mittag-Leffler functions (in two parameters $\alpha, \beta > 0$), referring the interested reader to [Dzherbashyan (1966)], [Podlubny (1999)] for more formulas and details.

Concerning summation we outline,
$$E_{\alpha,\beta}(z) = \frac{1}{p} \sum_{h=0}^{p-1} E_{\alpha/p,\beta}\left(z^{1/p} e^{i2\pi h/p}\right), \quad p \in \mathbb{N}, \qquad (E.73)$$
from which we derive the *duplication formula*,
$$E_{\alpha,\beta}(z^2) = \frac{1}{2} \left[E_{\alpha/2,\beta}(+z) + E_{\alpha/2,\beta}(-z) \right]. \qquad (E.74)$$
As an example of this formula we can recover, for $\alpha = 2, \beta = 1$ the well-known expressions
$$\cosh(z) = \left(e^{+z} + e^{-z}\right)/2, \quad \cos(z) = \left(e^{+iz} + e^{-iz}\right)/2.$$

Concerning integration we outline another interesting formula in which the Gaussian probability density (the fundamental solution of the standard diffusion equation) enters with the Mittag-Leffler function:

$$\int_0^\infty e^{-x^2/(4t)} E_{\alpha,\beta}(-x^\alpha)\, x^{\beta-1}\, dx = \sqrt{\pi}\, t^{\beta/2}\, E_{\alpha/2,(\beta+1)/2}(t^{\alpha/2}), \quad (E.75)$$

where $\alpha, \beta > 0$ and $t > 0$. For $\beta = 1$ we get the interesting *duplication formula*

$$E_{\alpha/2}(-t^{\alpha/2}) = \frac{1}{\sqrt{\pi t}} \int_0^\infty e^{-x^2/(4t)} E_\alpha(-x^\alpha)\, dx. \quad (E.76)$$

Formula (E.75) can be obtained after some manipulations from term-by-term integration from 0 to ∞ of the series in

$$e^{-x^2/(4t)} E_{\alpha,\beta}(-x^\alpha)\, x^{\beta-1} = \sum_{n=0}^\infty \frac{x^{\alpha n+\beta-1}}{\Gamma(\alpha n + \beta)} e^{-x^2/(4t)}.$$

But, as pointed out in [Podlubny (1999)], it can be derived in a simpler way by applying a theorem on the Laplace transform theory (a corollary of the Efros theorem), according to which

$$\mathcal{L}\left\{ \frac{1}{\sqrt{\pi t}} \int_0^\infty e^{-x^2/(4t)} f(x)\, dx;\, s \right\} = s^{-1/2}\, \widetilde{f}(s^{1/2}).$$

For this purpose let us take

$$f(x) := x^{\beta-1}\, E_{\alpha,\beta}(-x^\alpha),$$

so that in view of the Laplace transform pair (E.53),

$$\widetilde{f}(s) = \frac{s^{\alpha-\beta}}{s^\alpha - 1}.$$

therefore

$$s^{-1/2}\, \widetilde{f}(s^{1/2}) = \frac{s^{\alpha/2-(\beta+1)/2}}{s^{\alpha/2} - 1} = \mathcal{L}\left\{ t^{(\beta+1)/2-1}\, E_{\alpha/2,(\beta+1)/2}(t^{\alpha/2});\, s \right\}.$$

E.7 Applications of the Mittag-Leffler functions to the Abel integral equations

In their CISM Lecture notes, [Gorenflo and Mainardi (1997)] have worked out the key role of the function of the Mittag-Leffler type $e_\alpha(t;\lambda) := E_\alpha(-\lambda t^\alpha)$ in treating Abel integral equations of the second kind and fractional differential equations, so improving the former results by [Hille and Tamarkin (1930)], [Barret (1954)], respectively. In particular, they have considered the differential equations of fractional order governing processes of *fractional relaxation* and *fractional oscillation*, where the functions $e_\alpha(t;\lambda)$ with $0 < \alpha < 1$ and $1 < \alpha < 2$ respectively, play fundamental roles, see also [Mainardi (1996b)], [Gorenflo and Mainardi (1996)]. We note that the fractional relaxation has been treated in Chapter 1 whereas for fractional oscillations we refer the interested reader to the mentioned papers by Gorenflo and Mainardi and to the most recent papers by Achar, Hanneken and collaborators, see [Achar et al. (2001); Achar et al. (2002); Achar et al. (2004)] and [Hanneken et al. (2005); Hanneken et al. (2007)].

In this section we recall the most relevant results for the Abel integral equations.

Abel integral equations of the second kind. Using the definition (1.2) of the Riemann-Liouville fractional integral of order $\alpha > 0$ the Abel integral equation of the second kind can be written as

$$u(t) + \frac{\lambda}{\Gamma(\alpha)} \int_0^t \frac{u(\tau)}{(t-\tau)^{1-\alpha}} \, d\tau := (1 + \lambda \, _0I_t^\alpha) \, u(t) = f(t), \quad (E.77)$$

where $t > 0$ and $\lambda \in \mathbb{C}$. As a consequence the equation can be formally solved as follows:

$$u(t) = (1 + \lambda \, _0I_t^\alpha)^{-1} f(t) = \left(1 + \sum_{n=1}^\infty (-\lambda)^n \, _0I_t^{\alpha n}\right) f(t). \quad (E.78)$$

Noting by (1.7-8) the convolution formulation

$$_0I_t^{\alpha n} f(t) = \Phi_{\alpha n}(t) * f(t) = \frac{t_+^{\alpha n - 1}}{\Gamma(\alpha n)} * f(t),$$

the formal solution reads

$$u(t) = f(t) + \left(\sum_{n=1}^{\infty}(-\lambda)^n \frac{t_+^{\alpha n -1}}{\Gamma(\alpha n)}\right) * f(t). \qquad (E.79)$$

Recalling definition (E.52) of the function $e_\alpha(t;\lambda)$ we note that

$$\sum_{n=1}^{\infty}(-\lambda)^n \frac{t_+^{\alpha n -1}}{\Gamma(\alpha n)} = \frac{d}{dt}E_\alpha(-\lambda t^\alpha) := \dot{e}_\alpha(t;\lambda), \quad t > 0. \qquad (E.80)$$

Finally, the solution reads

$$u(t) = f(t) + \dot{e}_\alpha(t;\lambda) * f(t). \qquad (E.81)$$

Of course the above formal proof can be made rigorous. Simply observe that because of the rapid growth of the Gamma function the infinite series in (E.79)-(E.80) are uniformly convergent in every bounded interval of the variable t so that term-wise integrations and differentiations are allowed. However, we prefer to use the alternative technique of Laplace transforms, which will allow us to obtain the solution in different forms, including result (E.81).

Applying Laplace transformation to (E.77) we obtain

$$\left[1 + \frac{\lambda}{s^\alpha}\right]\widetilde{u}(s) = \widetilde{f}(s) \implies \widetilde{u}(s) = \frac{s^\alpha}{s^\alpha + \lambda}\widetilde{f}(s). \qquad (E.82)$$

Now, let us proceed to obtain the inverse Laplace transform of (E.82) using Laplace transform pair (E.52) related to the Mittag-Leffler function $e_\alpha(t;\lambda)$.

We note that, to this aim, we can choose two different ways, by following the standard rules of the Laplace transformation. Firstly, writing (E.82) as

$$\widetilde{u}(s) = s\left[\frac{s^{\alpha-1}}{s^\alpha + \lambda}\widetilde{f}(s)\right], \qquad (E.82a)$$

we obtain

$$u(t) = \frac{d}{dt}\int_0^t f(t-\tau)\,e_\alpha(\tau;\lambda)\,d\tau. \qquad (E.83a)$$

Secondly, if we write (E.82) as

$$\widetilde{u}(s) = \frac{s^{\alpha-1}}{s^\alpha + \lambda}[s\widetilde{f}(s) - f(0^+)] + f(0^+)\frac{s^{\alpha-1}}{s^\alpha + \lambda}, \qquad (E.82b)$$

we obtain

$$u(t) = \int_0^t \dot{f}(t-\tau)\, e_\alpha(\tau;\lambda)\, d\tau + f(0^+)\, e_\alpha(t;\lambda)\,. \quad (E.83b)$$

We also note that, $e_\alpha(t;\lambda)$ being a function differentiable with respect to t with $e_\alpha(0^+;\lambda) = E_\alpha(0^+) = 1$, there exists another possibility to rewrite (E.82), namely

$$\widetilde{u}(s) = \left[s\, \frac{s^{\alpha-1}}{s^\alpha + \lambda} - 1 \right] \widetilde{f}(s) + \widetilde{f}(s)\,. \quad (E.82c)$$

Then we obtain

$$u(t) = \int_0^t f(t-\tau)\, \dot{e}_\alpha(\tau;\lambda)\, d\tau + f(t)\,, \quad (E.83c)$$

in agreement with (E.81). We see that the way b) is more restrictive than the ways a) and c) since it requires that $f(t)$ be differentiable with $\mathcal{L}$-transformable derivative.

E.8 Notes

We note that in the twentieth century the functions of the Mittag-Leffler type remained almost unknown to the majority of scientists because they have been unjustly ignored in many treatises on special functions, including the most common [Abramowitz and Stegun (1965)]. Furthermore, there appeared some relevant works where the authors arrived at series or integral representations of these functions without recognizing them, e.g. [Gnedenko and Kolmogorov (1968)], [Balakrishnan (1985)] and [Sanz-Serna (1988)].

Thanks to suggestion of Professor Rudolf Gorenflo, the *2000 Mathematics Subject Classification* has included these functions in their items, see *33E12: Mittag-Leffler functions and generalizations*.

From now on, let us consider a number of references where the functions of the Mittag-Leffler type have been dealt with sufficient detail.

A description of the most important properties of these functions is found in the third volume of the Handbook on *Higher Transcendental Functions* of the Bateman Project, [Erdélyi et al. (1953-1955)].

In it, the authors have included the Mittag-Leffler functions in the Chapter XVIII devoted to the so-called *miscellaneous functions*. The attribute of miscellaneous is due to the fact that only later, in the sixties, the Mittag-Leffler functions were recognized to belong to a more general class of higher transcendental functions, known as *Fox H-functions*[2]. In fact, this class was well established only after the seminal paper [Fox (1961)]. For more details on H-functions, see e.g. the specialized treatises [Kilbas and Saigo (2004)], [Mathai and Saxena (1978)], [Srivastava *et al.* (1982)].

Coming back to the classical Mittag-Leffler functions, we recommend the treatise on complex functions by [Sansone and Gerretsen (1960)], where a detailed account of these functions is given.

However, the specialized treatise, where more details on the functions of the Mittag-Leffler type are given, is surely [Dzherbashyan (1966)], in Russian. Unfortunately, no official English translation of this book is nowadays available. We can content ourselves for another book by the same author [Dzherbashyan (1993)] in English, where a brief description of these functions is given.

Details on Mittag-Leffler functions can also be found in some treatises devoted to the theory and/or applications of special functions, integral transforms and fractional calculus, e.g. [Davis (1936)], [Marichev (1983)], [Gorenflo and Vessella (1991)], [Samko *et al.* (1993)], [Kiryakova (1994)], [Carpinteri and Mainardi (1997)], [Podlubny (1999)], [Hilfer (2000a)], [West *et al.* (2003)], [Kilbas *et al.* (2006)], [Magin (2006)], [Debnath and Bhatta (2007)], [Mathai and Haubold (2008)].

As pioneering works of mathematical nature we refer to [Hille and Tamarkin (1930)] and [Barret (1954)]. The 1930 paper by Hille and Tamarkin was concerning the solution of the Abel integral equation of the second kind (a particular fractional integral equation). The 1956 paper by Barret was concerning the general solution of the linear fractional differential equation with constant coefficients.

---

[2]It is opinion of the present author that the analysis of the Mittag-Leffler functions as particular cases of the general and cumbersome class of the Fox H-functions is not suitable for applied scientists. In fact they are accustomed to deal with an essential number of parameters and without much generality.

Concerning earlier applications of the Mittag-Leffler function in physics, we refer to the contributions by Kenneth S. Cole, see [Cole (1933)] (mentioned in [Davis (1936)], p. 287), in connection with nerve conduction, and by F.M. de Oliveira Castro, [De Oliveira Castro (1939)], and Bertram Gross, see [Gross (1947a)], in connection with dielectrical and mechanical relaxation, respectively. Subsequently, [Caputo and Mainardi (1971a); (1971b)] have proved that Mittag-Leffler functions appear whenever derivatives of fractional order are introduced in the constitutive equations of a linear viscoelastic body. Since then, several other authors have pointed out the relevance of the Mittag-Leffler function for fractional viscoelastic models, as pointed out in Chapter 3.

In recent times the attention of mathematicians and applied scientists towards the functions of the Mittag-Leffler type has increased, overall because of their relation with the fractional calculus and its applications. In addition to the books and papers already quoted, here we would like to draw the reader's attention to some relevant papers on the functions of the Mittag-Leffler type, in alphabetic order of the first author, [Al Saqabi and Tuan (1996)], [Brankov and Tonchev (1992)], [Berberan-Santos (2005a); (2005b); (2005c)] [Gorenflo et al (1997)], [Gorenflo and Mainardi (2008)], [Haubold et al. (2009)], [Hilfer and Anton (1995)], [Hilfer (2008)], [Jayakamur (2003)], [Kilbas and Saigo (1996)], [Lin (1998)], [Mainardi and Gorenflo (2000)], [Mainardi et al. (2000); (2004); (2005)], [Mathai et al. (2006)] [Metzler and Klafter (2002)], [Pillai (1990)], [Saigo and Kilbas (1998)], [Scalas et al. (2004)], [Saxena et al. (2006d)] [Sedletski (2004)], [Srivastava and Saxena (2001)], [Weron and Kotulski (1996)], [Wong and Zhao (2002)], and references therein. This list, however, is not exhaustive. More references can be found in the huge bibliography at the end of the book.

To the author's knowledge, earlier plots of the Mittag-Leffler functions are found (presumably for the first time in the literature of fractional calculus and special functions) in [Caputo and Mainardi (1971a)][3]. Precisely, these authors have provided plots of the func-

---

[3]Recently, this paper has been reprinted in *Fractional Calculus and Applied Analysis* under the kind permission of Birkhäuser Verlag AG.

tion $E_\nu(-t^\nu)$ for some values of $\nu \in (0, 1]$ adopting linear–logarithmic scales, in the framework of fractional relaxation for viscoelastic media. In those years not only such a function was still almost ignored, but also fractional calculus was not yet well accepted by the community of physicists.

Recently, numerical routines for functions of the Mittag-Leffler type have been provided, see e.g. [Gorenflo et al. (2002)] (with *MATHEMATICA*), [Podlubny (2006)] (with *MATLAB*) and [Seybold and Hilfer (2005)]. Furthermore, in [Freed et al. (2002)], an appendix is devoted to the table of Padè approximants for the Mittag-Leffler function $E_\alpha(-x)$.

Because the fractional calculus has actually attracted a wide interest in different areas of applied sciences, we think that the Mittag-Leffler function is nowadays exiting from its isolated life as *Cinderella*. (using the term coined by F.G. Tricomi in the fifties for the incomplete Gamma function). We like to refer to the classical Mittag-Leffler function as the *Queen function of fractional calculus*, and to consider all the related functions as her court, see [Mainardi and Gorenflo (2007)].

In this appendix, we have limited ourselves to functions of the Mittag-Leffler type in one variable with 1, 2 or 3 parameters. A treatment of the Mittag-Leffler functions containing more parameters and more variables is outside the aim of this book: for this see the recent survey papers by [Kiryakova (2008); (2009a)] and references therein.

We finally point out that the analytical continuation of the classical Mittag-Leffler function when the parameter α is negative has been recently considered by [Hanneken et al. (2009)].

Appendix F

The Wright Functions

In this appendix we provide a survey of the high transcendental functions known in the literature as Wright functions. We devote particular attention for two functions of the Wright type, which, in virtue of their role in applications of fractional calculus, we have called auxiliary functions. We also discuss their relevance in probability theory showing their connections with Lévy stable distributions. At the end, we add some historical and bibliographical notes.

F.1 The Wright function $W_{\lambda,\mu}(z)$

The *Wright function*, that we denote by $W_{\lambda,\mu}(z)$, is so named in honour of E. Maitland Wright, the eminent British mathematician, who introduced and investigated this function in a series of notes starting from 1933 in the framework of the theory of partitions, see [Wright (1933); (1935a); (1935b)]. The function is defined by the series representation, convergent in the whole complex plane,

$$W_{\lambda,\mu}(z) := \sum_{n=0}^{\infty} \frac{z^n}{n!\,\Gamma(\lambda n + \mu)}, \quad \lambda > -1, \quad \mu \in \mathbb{C}, \qquad (F.1)$$

so $W_{\lambda,\mu}(z)$ is an *entire function*. Originally, Wright assumed $\lambda > 0$, and, only in 1940, he considered $-1 < \lambda < 0$, see [Wright (1940)]. We note that in the handbook of the Bateman Project [Erdélyi *et al.* (1953-1955)], Vol. 3, Ch. 18, presumably for a misprint, λ is restricted to be non-negative. We distinguish the Wright functions in *first kind* ($\lambda \geq 0$) and *second kind* ($-1 < \lambda < 0$).

The integral representation. The *integral representation* reads
$$W_{\lambda,\mu}(z) = \frac{1}{2\pi i} \int_{Ha} e^{\sigma + z\sigma^{-\lambda}} \frac{d\sigma}{\sigma^\mu}, \quad \lambda > -1, \quad \mu \in \mathbb{C}, \qquad (F.2)$$
where Ha denotes the Hankel path. The equivalence between the series and integral representations is easily proven by using the Hankel formula for the Gamma function, see (A.19),
$$\frac{1}{\Gamma(\zeta)} = \int_{Ha} e^u u^{-\zeta} du, \quad \zeta \in \mathbb{C},$$
and performing a term-by-term integration. The exchange between series and integral is legitimate by the uniform convergence of the series, being $W_{\lambda,\mu}(z)$ an entire function. We have:
$$W_{\lambda,\mu}(z) = \frac{1}{2\pi i} \int_{Ha} e^{\sigma + z\sigma^{-\lambda}} \frac{d\sigma}{\sigma^\mu} = \frac{1}{2\pi i} \int_{Ha} e^{\sigma} \left[\sum_{n=0}^{\infty} \frac{z^n}{n!} \sigma^{-\lambda n} \right] \frac{d\sigma}{\sigma^\mu}$$
$$= \sum_{n=0}^{\infty} \frac{z^n}{n!} \left[\frac{1}{2\pi i} \int_{Ha} e^{\sigma} \sigma^{-\lambda n - \mu} d\sigma \right] = \sum_{n=0}^{\infty} \frac{z^n}{n!\, \Gamma[\lambda n + \mu]}.$$

Furthermore, it is possible to prove that the Wright function is entire of order $1/(1+\lambda)$ hence of exponential type only if $\lambda \geq 0$. The case $\lambda = 0$ is trivial since $W_{0,\mu}(z) = e^z/\Gamma(\mu)$.

Asymptotic expansions. For the detailed asymptotic analysis in the whole complex plane for the Wright functions, the interested reader is referred to [Wong and Zhao (1999a); (1999b)]. These authors have provided asymptotic expansions of the Wright functions of the first and second kind following a new method for smoothing Stokes' discontinuities.

As a matter of fact, the second kind is the most interesting for us. By setting $\lambda = -\nu \in (-1, 0)$, we recall the asymptotic expansion originally obtained by Wright himself, that is valid in a suitable sector about the negative real axis as $|z| \to \infty$,
$$W_{-\nu,\mu}(z) = Y^{1/2-\mu} e^{-Y} \left[\sum_{m=0}^{M-1} A_m Y^{-m} + O(|Y|^{-M}) \right], \qquad (F.3)$$
$$Y = Y(z) = (1-\nu)(-\nu^\nu z)^{1/(1-\nu)},$$
where the A_m are certain real numbers.

Generalization of the Bessel functions. The Wright functions turn out to be related to the well-known Bessel functions J_ν and I_ν for $\lambda = 1$ and $\mu = \nu + 1$. In fact, by using the series definitions (B.1) and (B.31) for the Bessel functions and the series definitions (F.1) for the Wright functions, we easily recognize the identities:

$$J_\nu(z) := \left(\frac{z}{2}\right)^\nu \sum_{n=0}^\infty \frac{(-1)^n (z/2)^{2n}}{n!\,\Gamma(n + +\nu + 1)} = \left(\frac{z}{2}\right)^\nu W_{1,\nu+1}\left(-\frac{z^2}{4}\right),$$
$$W_{1,\nu+1}(-z) := \sum_{n=0}^\infty \frac{(-1)^n z^n}{n!\,\Gamma(n + \nu + 1)} = z^{-\nu/2} J_\nu(2z^{1/2}),$$
(F.4)

and

$$I_\nu(z) := \left(\frac{z}{2}\right)^\nu \sum_{n=0}^\infty \frac{(z/2)^{2n}}{n!\,\Gamma(n + +\nu + 1)} = \left(\frac{z}{2}\right)^\nu W_{1,\nu+1}\left(\frac{z^2}{4}\right),$$
$$W_{1,\nu+1}(z) := \sum_{n=0}^\infty \frac{z^n}{n!\,\Gamma(n + \nu + 1)} = z^{-\nu/2} I_\nu(2z^{1/2}).$$
(F.5)

As far as the standard Bessel functions J_ν are concerned, the following observations are worth noting. We first note that the Wright function $W_{1,\nu+1}(-z)$ reduces to the entire function $\mathcal{C}_\nu(z)$ known as *Bessel-Clifford function* introduced Eq. (B.4). Then, in view of the first equation in (F.4) some authors refer to the Wright function as the *Wright generalized Bessel function* (misnamed also as the *Bessel-Maitland function*) and introduce the notation for $\lambda \geq 0$, see e.g. [Kiryakova (1994)], p. 336,

$$J_\nu^{(\lambda)}(z) := \left(\frac{z}{2}\right)^\nu \sum_{n=0}^\infty \frac{(-1)^n (z/2)^{2n}}{n!\,\Gamma(\lambda n + \nu + 1)} = \left(\frac{z}{2}\right)^\nu W_{\lambda,\nu+1}\left(-\frac{z^2}{4}\right). \quad (F.6)$$

Similar remarks can be extended to the modified Bessel functions I_ν.

Recurrence relations. Hereafter, we quote some relevant recurrence relations from [Erdélyi *et al.* (1953-1954)], Vol. 3, Ch. 18:

$$\lambda z\, W_{\lambda,\lambda+\mu}(z) = W_{\lambda,\mu-1}(z) + (1-\mu)\, W_{\lambda,\mu}(z), \quad (F.7)$$

$$\frac{d}{dz} W_{\lambda,\mu}(z) = W_{\lambda,\lambda+\mu}(z). \quad (F.8)$$

We note that these relations can easily be derived from (F.1).

F.2 The auxiliary functions $F_\nu(z)$ and $M_\nu(z)$ in $\mathbb{C}$

In his earliest analysis of the time-fractional diffusion-wave equation [Mainardi (1994a)], the author introduced the two *auxiliary functions* of the Wright type:

$$F_\nu(z) := W_{-\nu,0}(-z), \quad 0 < \nu < 1, \qquad (F.9)$$

and

$$M_\nu(z) := W_{-\nu,1-\nu}(-z), \quad 0 < \nu < 1, \qquad (F.10)$$

interrelated through

$$F_\nu(z) = \nu z M_\nu(z). \qquad (F.11)$$

As it is shown in Chapter 6, the motivation was based on the inversion of certain Laplace transforms in order to obtain the fundamental solutions of the fractional diffusion-wave equation in the space-time domain. Here we will devote particular attention to the mathematical properties of these functions limiting at the essential the discussion for the general Wright functions. The reader is referred to the Notes for some historical and bibliographical details.

Series representations. The *series representations* of our auxiliary functions are derived from those of $W_{\lambda,\mu}(z)$. We have:

$$\begin{aligned} F_\nu(z) &:= \sum_{n=1}^\infty \frac{(-z)^n}{n!\,\Gamma(-\nu n)} \\ &= \frac{1}{\pi} \sum_{n=1}^\infty \frac{(-z)^{n-1}}{n!} \Gamma(\nu n + 1)\,\sin(\pi \nu n), \end{aligned} \qquad (F.12)$$

and

$$\begin{aligned} M_\nu(z) &:= \sum_{n=0}^\infty \frac{(-z)^n}{n!\,\Gamma[-\nu n + (1-\nu)]} \\ &= \frac{1}{\pi} \sum_{n=1}^\infty \frac{(-z)^{n-1}}{(n-1)!} \Gamma(\nu n)\,\sin(\pi \nu n), \end{aligned} \qquad (F.13)$$

where we have used the well-known reflection formula for the Gamma function, see (A.13),

$$\Gamma(\zeta)\,\Gamma(1-\zeta) = \pi/\sin \pi \zeta.$$

We note that $F_\nu(0) = 0$, $M_\nu(0) = 1/\Gamma(1-\nu)$ and that the relation (F.11), consistent with the recurrence relation (F.7), can be derived from (F.12)-(F.13) arranging the terms of the series.

The integral representations. The *integral representations* of our auxiliary functions are derived from those of $W_{\lambda,\mu}(z)$. We have:

$$F_\nu(z) := \frac{1}{2\pi i} \int_{Ha} e^{\sigma - z\sigma^\nu} d\sigma, \quad z \in \mathbb{C}, \quad 0 < \nu < 1, \quad (F.14)$$

$$M_\nu(z) := \frac{1}{2\pi i} \int_{Ha} e^{\sigma - z\sigma^\nu} \frac{d\sigma}{\sigma^{1-\nu}}, \quad z \in \mathbb{C}, \quad 0 < \nu < 1. \quad (F.15)$$

We note that the relation (F.11) can be obtained directly from (F.14) and (F.15) with an integration by parts, i.e.

$$\int_{Ha} e^{\sigma - z\sigma^\nu} \frac{d\sigma}{\sigma^{1-\nu}} = \int_{Ha} e^\sigma \left(-\frac{1}{\nu z} \frac{d}{d\sigma} e^{-z\sigma^\nu} \right) d\sigma$$
$$= \frac{1}{\nu z} \int_{Ha} e^{\sigma - z\sigma^\nu} d\sigma.$$

The passage from the series representation to the integral representation and vice-versa for our auxiliary functions can be derived in a way similar to that adopted for the general Wright function, that is by expanding in positive powers of z the exponential function $\exp(-z\sigma^\nu)$, exchanging the order between the series and the integral and using the Hankel representation of the reciprocal of the Gamma function, see (A.19a).

Since the radius of convergence of the power series in (F.12)-(F.13) can be proven to be infinite for $0 < \nu < 1$, our auxiliary functions turn out to be entire in z and therefore the exchange between the series and the integral is legitimate[1].

Special cases. Explicit expressions of $F_\nu(z)$ and $M_\nu(z)$ in terms of known functions are expected for some particular values of ν.

In [Mainardi and Tomirotti (1995)] the authors have shown that for $\nu = 1/q$, where $q \geq 2$ is a positive integer, the auxiliary functions can be expressed as a sum of $(q-1)$ simpler entire functions.

In the particular cases $q = 2$ and $q = 3$ we find from (F.13),

$$M_{1/2}(z) = \frac{1}{\sqrt{\pi}} \sum_{m=0}^{\infty} (-1)^m \left(\frac{1}{2} \right)_m \frac{z^{2m}}{(2m)!} = \frac{1}{\sqrt{\pi}} \exp\left(-z^2/4 \right), \quad (F.16)$$

---

[1] The author in [Mainardi (1994a)] proved these properties independently from [Wright (1940)], because at that time he was aware only of [Erdélyi et al. (1953-1955)] where λ was restricted to be non-negative.

and

$$M_{1/3}(z) = \frac{1}{\Gamma(2/3)} \sum_{m=0}^{\infty} \left(\frac{1}{3}\right)_m \frac{z^{3m}}{(3m)!} - \frac{1}{\Gamma(1/3)} \sum_{m=0}^{\infty} \left(\frac{2}{3}\right)_m \frac{z^{3m+1}}{(3m+1)!} \quad (F.17)$$

$$= 3^{2/3} \operatorname{Ai}\left(z/3^{1/3}\right),$$

where Ai denotes the *Airy function* defined in Appendix B (Section B.4).

Furthermore, it can be proved that $M_{1/q}(z)$ satisfies the differential equation of order $q-1$

$$\frac{d^{q-1}}{dz^{q-1}} M_{1/q}(z) + \frac{(-1)^q}{q} z\, M_{1/q}(z) = 0, \quad (F.18)$$

subjected to the $q-1$ initial conditions at $z = 0$, derived from (F.13),

$$M_{1/q}^{(h)}(0) = \frac{(-1)^h}{\pi} \Gamma[(h+1)/q] \sin[\pi\,(h+1)/q], \quad (F.19)$$

with $h = 0, 1, \ldots q-2$.

We note that, for $q \geq 4$, Eq. (F.18) is akin to the *hyper-Airy* differential equation of order $q-1$, see e.g. [Bender and Orszag (1987)]. Consequently, in view of the above considerations, the auxiliary function $M_\nu(z)$ could be referred to as the *generalized hyper-Airy function*.

F.3 The auxiliary functions $F_\nu(x)$ and $M_\nu(x)$ in $\mathbb{R}$

We point out that the most relevant applications of Wright functions, especially our auxiliary functions, are when the independent variable is real. More precisely, in this Section we will consider functions of the variable x with $x \in \mathbb{R}^+$ or $x \in \mathbb{R}$.

When the support is all of $\mathbb{R}$, we agree to consider *even functions*, that is, functions defined in a symmetric way. In this case, to stress the symmetry property of the function, the independent variable may be denoted by $|x|$.

We point out that in the limit $\nu \to 1^-$ the function $M_\nu(x)$, for $x \in \mathbb{R}^+$, tends to the Dirac generalized function $\delta(x-1)$.

The asymptotic representation of $M_\nu(x)$. Let us first point out the asymptotic behaviour of the function $M_\nu(x)$ as $x \to +\infty$. Choosing as a variable x/ν rather than x, the computation of the asymptotic representation by the saddle-point approximation yields, see [Mainardi and Tomirotti (1995)],

$$M_\nu(x/\nu) \sim a(\nu)\, x^{(\nu - 1/2)/(1 - \nu)} \exp\left[-b(\nu)\, x^{1/(1 - \nu)}\right], \quad (F.20)$$

where

$$a(\nu) = \frac{1}{\sqrt{2\pi(1 - \nu)}} > 0, \quad b(\nu) = \frac{1 - \nu}{\nu} > 0. \quad (F.21)$$

The above evaluation is consistent with the first term in Wright's asymptotic expansion (F.3) after having used the definition (F.10).

Plots of $M_\nu(x)$. We show the plots of our auxiliary functions on the real axis for some rational values of the parameter ν.

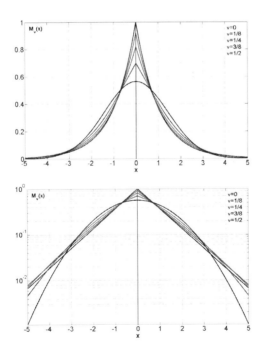

Fig. F.1 Plots of the Wright type function $M_\nu(x)$ with $\nu = 0, 1/8, 1/4, 3/8, 1/2$ for $-5 \leq x \leq 5$; top: linear scale, bottom: logarithmic scale.

To gain more insight of the effect of the parameter itself on the behaviour close to and far from the origin, we will adopt both linear and logarithmic scale for the ordinates.

In Figs. F.1 and F.2 we compare the plots of the $M_\nu(x)$ Wright auxiliary functions in $-5 \le x \le 5$ for some rational values in the ranges $\nu \in [0, 1/2]$ and $\nu \in [1/2, 1]$, respectively. Thus in Fig. F.1 we see the transition from $\exp(-|x|)$ for $\nu = 0$ to $1/\sqrt{\pi}\exp(-x^2)$ for $\nu = 1/2$, whereas in Fig. F.2 we see the transition from $1/\sqrt{\pi}\exp(-x^2)$ to the delta function $\delta(1 - |x|)$ for $\nu = 1$.

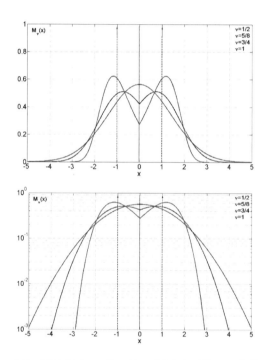

Fig. F.2 Plots of the Wright type function $M_\nu(x)$ with $\nu = 1/2, 5/8, 3/4, 1$ for $-5 \le x \le 5$: top: linear scale; bottom: logarithmic scale.

In plotting $M_\nu(x)$ at fixed ν for sufficiently large x the asymptotic representation (F.20)-(F.21) is very useful because, as x increases, the numerical convergence of the series in (F.13) becomes poor and poor up to being completely inefficient. Henceforth, the matching between the series and the asymptotic representation is relevant.

However, as $\nu \to 1^-$, the plotting remains a very difficult task because of the high peak arising around $x = \pm 1$. In this case the saddle-point method, improved as in [Kreis and Pipkin (1986)], can successfully be used to visualize some structure in the peak while it tends to the Dirac delta function, see also [Mainardi and Tomirotti (1997)] and Chapter 6 for a related wave-propagation problem. With Pipkin's method we are able to get the desired matching with the series representation just in the region around the maximum $x \approx 1$, as shown in Fig. F.3. Here we exhibit the significant plots of the auxiliary function $M_\nu(x)$ with $\nu = 1 - \epsilon$ for $\epsilon = 0.01$ and $\epsilon = 0.001$ and we compare the series representation (100 terms, dashed line), the saddle-point representation (dashed-dotted line), and the Pipkin representation (continuous line).

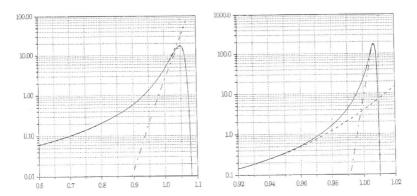

Fig. F.3 Comparison of the representations of $M_\nu(x)$ with $\nu = 1 - \epsilon$ around the maximum $x \approx 1$ obtained by Pipkin's method (continuous line), 100 terms-series (dashed line) and the saddle-point method (dashed-dotted line). Left: $\epsilon = 0.01$; Right: $\epsilon = 0.001$.

F.4 The Laplace transform pairs

Let us write the Laplace transform of the Wright function as

$$W_{\lambda,\mu}(\pm r) \div \mathcal{L}\left[W_{\lambda,\mu}(\pm r); s\right] := \int_0^\infty e^{-sr} W_{\lambda,\mu}(\pm r)\, dr,$$

where r denotes a non-negative real variable, i.e. $0 \le r < +\infty$, and s is the Laplace complex parameter.

When $\lambda > 0$ the series representation of the Wright function can be transformed term-by-term. In fact, for a known theorem of the theory of the Laplace transforms, see e.g. [Doetsch (1974)], the Laplace transform of an entire function of exponential type can be obtained by transforming term-by-term the Taylor expansion of the original function around the origin. In this case the resulting Laplace transform turns out to be analytic and vanishing at infinity. As a consequence, we obtain *the Laplace transform pair for the Wright function of the first kind* as

$$W_{\lambda,\mu}(\pm r) \div \frac{1}{s} E_{\lambda,\mu}\left(\pm \frac{1}{s}\right), \quad \lambda > 0, \quad |s| > \rho > 0, \qquad (F.22)$$

where $E_{\lambda,\mu}$ denotes the generalized Mittag-Leffler function in two parameters, and ρ is an arbitrary positive number. The proof is straightforward, noting that

$$\sum_{n=0}^{\infty} \frac{(\pm r)^n}{n!\,\Gamma(\lambda n + \mu)} \div \frac{1}{s}\sum_{n=0}^{\infty} \frac{(\pm 1/s)^n}{\Gamma(\lambda n + \mu)},$$

and recalling the series representation (E.22) of the generalized Mittag-Leffler function,

$$E_{\alpha,\beta}(z) := \sum_{n=0}^{\infty} \frac{z^n}{\Gamma(\alpha n + \beta)}, \quad \alpha > 0, \quad z \in \mathbb{C}.$$

For $\lambda \to 0^+$ Eq. (F.22) provides the Laplace transform pair

$$W_{0^+,\mu}(\pm r) := \frac{e^{\pm r}}{\Gamma(\mu)} \div \frac{1}{\Gamma(\mu)} \frac{1}{s \mp 1}.$$

This means

$$W_{0^+,\mu}(\pm r) \div \frac{1}{s} E_{0,\mu}\left(\pm \frac{1}{s}\right) = \frac{1}{\Gamma(\mu)s} E_0\left(\pm \frac{1}{s}\right), \quad |s| > 1, \qquad (F.23)$$

where, in order to be consistent with (F.22), we have formally put, according to (E.2),

$$E_{0,\mu}(z) := \sum_{n=0}^{\infty} \frac{z^n}{\Gamma(\mu)} = \frac{1}{\Gamma(\mu)} E_0(z) = \frac{1}{\Gamma(\mu)}\frac{1}{1-z}, \quad |z| < 1.$$

We recognize that in this limitig case the Laplace transform exhibits a simple pole at $s = \pm 1$ while for $\lambda > 0$ it exhibits an essential singularity at $s = 0$.

Appendix F: The Wright Functions

For $-1 < \lambda < 0$ the Wright function turns out to be an entire function of order greater than 1, so that the term-by-term transformation representation is no longer legitimate. Thus, for Wright functions of the second kind, care is required in establishing the existence of the Laplace transform, which necessarily must tend to zero as $s \to \infty$ in its half-plane of convergence.

For the sake of convenience we first derive the Laplace transform for the special case of $M_\nu(r)$; the exponential decay as $r \to \infty$ of the *original* function provided by (F.20) ensures the existence of the *image* function. From the integral representation (F.13) of the M_ν function we obtain

$$M_\nu(r) \div \frac{1}{2\pi i} \int_0^\infty e^{-sr} \left[\int_{Ha} e^{\sigma - r\sigma^\nu} \frac{d\sigma}{\sigma^{1-\nu}} \right] dr$$

$$= \frac{1}{2\pi i} \int_{Ha} e^\sigma \sigma^{\nu-1} \left[\int_0^\infty e^{-r(s+\sigma^\nu)} dr \right] d\sigma = \frac{1}{2\pi i} \int_{Ha} \frac{e^\sigma \sigma^{\nu-1}}{\sigma^\nu + s} d\sigma.$$

Then, by recalling the integral representation (E.14) of the Mittag-Leffler function,

$$E_\alpha(z) = \frac{1}{2\pi i} \int_{Ha} \frac{\zeta^{\alpha-1} e^\zeta}{\zeta^\alpha - z} d\zeta, \quad \alpha > 0,$$

we obtain the Laplace transform pair

$$M_\nu(r) \div E_\nu(-s), \quad 0 < \nu < 1. \tag{F.24}$$

Although transforming the Taylor series of $M_\nu(r)$ term-by-term is not legitimate, this procedure yields a series of negative powers of s that represents the asymptotic expansion of the correct Laplace transform, $E_\nu(-s)$, as $s \to \infty$ in a sector around the positive real axis. Indeed we get

$$\sum_{n=0}^\infty \frac{\int_0^\infty e^{-sr}(-r)^n dr}{n! \Gamma(-\nu n + (1-\nu))} = \sum_{n=0}^\infty \frac{(-1)^n}{\Gamma(-\nu n + 1 - \nu)} \frac{1}{s^{n+1}}$$

$$= \sum_{m=1}^\infty \frac{(-1)^{m-1}}{\Gamma(-\nu m + 1)} \frac{1}{s^m} \sim E_\nu(-s), \quad s \to \infty,$$

consistently with the asymptotic expansion (E.16).

We note that (F.24) contains the well-known Laplace transform pair, see e.g. [Doetsch (1974)],

$$M_{1/2}(r) := \frac{1}{\sqrt{\pi}} \exp\left(-r^2/4\right) \div E_{1/2}(-s) := \exp\left(s^2\right) \operatorname{erfc}(s),$$

that is valid for all $s \in \mathbb{C}$.

Analogously, using the more general integral representation (F.2) of the Wright function, we can get *the Laplace transform pair for the Wright function of the second kind*. For the case $\lambda = -\nu \in (-1, 0)$, with $\mu > 0$ for simplicity, we obtain,

$$W_{-\nu,\mu}(-r) \div E_{\nu,\mu+\nu}(-s), \quad 0 < \nu < 1. \quad (F.25)$$

We note the minus sign in the argument in order to ensure the the existence of the Laplace transform thanks to the Wright asymptotic formula (F.3) valid in a sector about the negative real axis.

In the limit as $\lambda \to 0^-$ we formally obtain the Laplace transform pair

$$W_{0^-,\mu}(-r) := \frac{e^{-r}}{\Gamma(\mu)} \div \frac{1}{\Gamma(\mu)} \frac{1}{s+1}.$$

In order to be consistent with (F.24) we rewrite

$$W_{0^-,\mu}(-r) \div E_{0,\mu}(-s) = \frac{1}{\Gamma(\mu)} E_0(-s), \quad |s| < 1. \quad (F.26)$$

Therefore, as $\lambda \to 0^\pm$, we note a sort of continuity in the formal results (F.23) and (F.26) because

$$\frac{1}{(s+1)} = \begin{cases} (1/s) E_0(-1/s), & |s| > 1; \\ E_0(-s), & |s| < 1. \end{cases} \quad (F.27)$$

We now point out the relevant Laplace transform pair related to the *auxiliary* functions of argument $r^{-\nu}$ proved in [Mainardi (1994a); (1996a); (1996b)]:

$$\frac{1}{r} F_\nu(1/r^\nu) = \frac{\nu}{r^{\nu+1}} M_\nu(1/r^\nu) \div e^{-s^\nu}, \quad 0 < \nu < 1. \quad (F.28)$$

$$\frac{1}{\nu} F_\nu(1/r^\nu) = \frac{1}{r^\nu} M_\nu(1/r^\nu) \div \frac{e^{-s^\nu}}{s^{1-\nu}}, \quad 0 < \nu < 1. \quad (F.29)$$

We recall that the Laplace transform pairs in (F.28) were formerly considered by [Pollard (1946)], who provided a rigorous proof based on a formal result by [Humbert (1945)]. Later [Mikusiński (1959a)] achieved a similar result based on his theory of operational calculus, and finally, albeit unaware of the previous results, [Buchen and Mainardi (1975)] derived the result in a formal way, as stressed in

Chapter 5. We note, however, that all these authors were not informed about the Wright functions. To our actual knowledge, the former author who derived the Laplace transforms pairs (F.28)-(F.29) in terms of Wright functions of the second kind was [Stankovich (1970)].

Hereafter, we will provide two independent proofs of (F.28) by carrying out the inversion of $\exp(-s^\nu)$, either by the complex Bromwich integral formula, see [Mainardi (1994a); Mainardi (1996a)], or by the formal series method, see [Buchen and Mainardi (1975)]. Similarly, we can act for the Laplace transform pair (F.29).

For the complex integral approach we deform the Bromwich path Br into the Hankel path Ha, that is equivalent to the original path, and we set $\sigma = sr$. Recalling (F.14)-(F.15), we get

$$\mathcal{L}^{-1}\left[\exp\left(-s^\nu\right)\right] = \frac{1}{2\pi i}\int_{Br} e^{sr - s^\nu} ds = \frac{1}{2\pi i r}\int_{Ha} e^{\sigma - (\sigma/r)^\nu} d\sigma$$

$$= \frac{1}{r} F_\nu\left(1/r^\nu\right) = \frac{\nu}{r^{\nu+1}} M_\nu\left(1/r^\nu\right).$$

For the series approach, let us expand the Laplace transform in series of negative powers and invert term by term. Then, after recalling (F.12)-(F.13), we obtain:

$$\mathcal{L}^{-1}\left[\exp\left(-s^\nu\right)\right] = \sum_{n=0}^\infty \frac{(-1)^n}{n!}\mathcal{L}^{-1}\left[s^{\nu n}\right] = \sum_{n=1}^\infty \frac{(-1)^n}{n!}\frac{r^{-\nu n -1}}{\Gamma(-\nu n)}$$

$$= \frac{1}{r} F_\nu\left(1/r^\nu\right) = \frac{\nu}{r^{\nu+1}} M_\nu\left(1/r^\nu\right).$$

We note the relevance of Laplace transforms (F.24) and (F.28) in pointing out the non-negativity of the Wright function $M_\nu(x)$ and the complete monotonicity of the Mittag-Leffler functions $E_\nu(-x)$ for $x > 0$ and $0 < \nu < 1$. In fact, since $\exp\left(-s^\nu\right)$ denotes the Laplace transform of a probability density (precisely, the extremal Lévy stable density of index ν, see [Feller (1971)]) the L.H.S. of (F.28) must be non-negative, and so also must the L.H.S of (F.24). As a matter of fact the Laplace transform pair (F.24) shows, replacing s by x, that the spectral representation of the Mittag-Leffler function $E_\nu(-x)$ is expressed in terms of the Wright M-function $M_\nu(r)$, that is:

$$E_\nu(-x) = \int_0^\infty e^{-rx} M_\nu(r)\, dr, \ 0 < \nu < 1, \ x \geq 0. \qquad (F.30)$$

We now recognize that Eq. (F.30) is consistent with Eqs. (E.19)-(E.21) derived by [Pollard (1948)].

It is instructive to compare the spectral representation of $E_\nu(-x)$ with that of the function $E_\nu(-t^\nu)$. From Eqs. (E.56)-(E.57) we can write

$$E_\nu(-t^\nu) = \int_0^\infty e^{-rt} K_\nu(r) \, dr, \quad 0 < \nu < 1, \, t \geq 0, \qquad (F.31)$$

where the *spectral function* reads

$$K_\nu(r) = \frac{1}{\pi} \frac{r^{\nu-1} \sin(\nu\pi)}{r^{2\nu} + 2 r^\nu \cos(\nu\pi) + 1}. \qquad (F.32)$$

The relationship between $M_\nu(r)$ and $K_\nu(r)$ is worth exploring. Both functions are non-negative, integrable and normalized in $\mathbb{R}^+$, so they can be adopted in probability theory as density functions. The normalization conditions derive from Eqs. (F.30) and (F.31) since

$$\int_0^{+\infty} M_\nu(r) \, dr = \int_0^{+\infty} K_\nu(r) \, dr = E_\nu(0) = 1.$$

In the following section we will discuss the probability interpretation of the M_ν function with support both in $\mathbb{R}^+$ and in $\mathbb{R}$ whereas for K_ν we note that it has been interpreted as spectral distribution of relaxation/retardation times in the fractional Zener viscoelastic model, see Chapter 3, Section 3.2, Fig. 3.3.

We also note that for certain renewal processes, functions of Mittag-Leffler and Wright type can be adopted as probability distributions of waiting times, as shown in [Mainardi et al. (2005)], where such distributions are compared. We refer the interested reader to that paper for details.

F.5 The Wright M-functions in probability

We have already recognized that the Wright M-function with support in $\mathbb{R}^+$ can be interpreted as probability density function (*pdf*). Consequently, extending the function in a symmetric way to all of $\mathbb{R}$ and dividing by 2 we have a *symmetric pdf* with support in $\mathbb{R}$. In the former case the variable is usually a time coordinate whereas

Appendix F: The Wright Functions

in the latter the variable is the absolute value of a space coordinate. We now provide more details on these densities in the framework of the theory of probability. As in Section F.3, we agree to denote by x and $|x|$ the variables in $\mathbb{R}^+$ and $\mathbb{R}$, respectively.

The absolute moments of order δ. The *absolute moments* of order $\delta > -1$ in $\mathbb{R}^+$ of the Wright M-function *pdf* in $\mathbb{R}^+$ are finite and turn out to be

$$\int_0^\infty x^\delta M_\nu(x)\,dx = \frac{\Gamma(\delta+1)}{\Gamma(\nu\delta+1)}, \quad \delta > -1, \quad 0 \le \nu < 1. \quad (F.33)$$

In order to derive this fundamental result we proceed as follows, based on the integral representation (F.15).

$$\int_0^\infty x^\delta M_\nu(x)\,dx = \int_0^\infty x^\delta \left[\frac{1}{2\pi i}\int_{Ha} e^{\sigma - x\sigma^\nu}\frac{d\sigma}{\sigma^{1-\nu}}\right] dx$$

$$= \frac{1}{2\pi i}\int_{Ha} e^\sigma \left[\int_0^\infty e^{-x\sigma^\nu} x^\delta\,dx\right] \frac{d\sigma}{\sigma^{1-\nu}}$$

$$= \frac{\Gamma(\delta+1)}{2\pi i}\int_{Ha} \frac{e^\sigma}{\sigma^{\nu\delta+1}}\,d\sigma = \frac{\Gamma(\delta+1)}{\Gamma(\nu\delta+1)}.$$

Above we have legitimated the exchange between the two integrals and we have used the identity

$$\int_0^\infty e^{-x\sigma^\nu} x^\delta\,dx = \frac{\Gamma(\delta+1)}{(\sigma^\nu)^{\delta+1}},$$

derived from (A.23) along with the Hankel formula (A.19a).

In particular, for $\delta = n \in \mathbb{N}$, the above formula provides the moments of integer order that can also be computed from the Laplace transform pair (F.24) as follows:

$$\int_0^{+\infty} x^n M_\nu(x)\,dx = \lim_{s\to 0}(-1)^n \frac{d^n}{ds^n} E_\nu(-s) = \frac{\Gamma(n+1)}{\Gamma(\nu n+1)}.$$

Incidentally, we note that the Laplace transform pair (F.24) could be obtained using the fundamental result (F.33) by developing in power series the exponential kernel of the Laplace transform and then transforming the series term-by-term.

The characteristic function. As well-known in probability theory the Fourier transform of a density provides the so-called *characteristic function*. In our case we have:

$$\mathcal{F}\left[\tfrac{1}{2}M_\nu(|x|)\right] := \frac{1}{2}\int_{-\infty}^{+\infty} M_\nu(|x|)\,dx \\ = \int_0^\infty \cos(\kappa x)\,M_\nu(x)\,dx = E_{2\nu}(-\kappa^2). \qquad (F.34)$$

For this prove it is sufficient to develop in series the cosine function and use formula (F.33),

$$\int_0^\infty \cos(\kappa x)\,M_\nu(x)\,dx = \sum_{n=0}^\infty (-1)^n \frac{\kappa^{2n}}{(2n)!}\int_0^\infty x^{2n}\,M_\nu(x)\,dx \\ = \sum_{n=0}^\infty (-1)^n \frac{\kappa^{2n}}{\Gamma(2\nu n+1)} = E_{2\nu}(-\kappa^2).$$

Relations with Lévy stable distributions. We find it worthwhile to discuss the relations between the Wright M-functions and the so-called *Lévy stable distributions*. The term stable has been assigned by the French mathematician Paul Lévy, who, in the twenties of the last century, started a systematic research in order to generalize the celebrated *Central Limit Theorem* to probability distributions with infinite variance. For stable distributions we can assume the following DEFINITION: *If two independent real random variables with the same shape or type of distribution are combined linearly and the distribution of the resulting random variable has the same shape, the common distribution (or its type, more precisely) is said to be stable.*

The restrictive condition of stability enabled Lévy (and then other authors) to derive the *canonic form* for the characteristic function of the densities of these distributions. Here we follow the parameterization in [Feller (1952); (1971)] revisited in [Gorenflo and Mainardi (1998b)] and in [Mainardi et al. (2001)]. Denoting by $L_\alpha^\theta(x)$ a generic stable density in $\mathbb{R}$, where α is the *index of stability* and and θ the asymmetry parameter, improperly called *skewness*, its characteristic function reads:

$$L_\alpha^\theta(x) \div \widehat{L}_\alpha^\theta(\kappa) = \exp\left[-\psi_\alpha^\theta(\kappa)\right],\quad \psi_\alpha^\theta(\kappa) = |\kappa|^\alpha\,e^{i(\operatorname{sign}\kappa)\theta\pi/2}, \qquad (F.35)$$

$$0 < \alpha \le 2,\ |\theta| \le \min\{\alpha, 2-\alpha\}.$$

We note that the allowed region for the parameters α and θ turns out to be a diamond in the plane $\{\alpha, \theta\}$ with vertices in the points $(0,0)$, $(1,1)$, $(1,-1)$, $(2,0)$, that we call the *Feller-Takayasu diamond*, see Fig. F.4. For values of θ on the border of the diamond (that is $\theta = \pm\alpha$ if $0 < \alpha < 1$, and $\theta = \pm(2-\alpha)$ if $1 < \alpha < 2$) we obtain the so-called *extremal stable densities*.

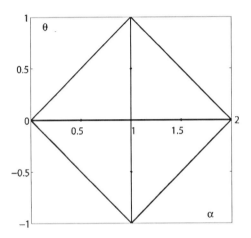

Fig. F.4 The Feller-Takayasu diamond for Lévy stable densities.

We note the *symmetry relation* $L_\alpha^\theta(-x) = L_\alpha^{-\theta}(x)$, so that a stable density with $\theta = 0$ is symmetric.

Stable distributions have noteworthy properties of which the interested reader can be informed from the relevant existing literature. Here-after we recall some peculiar PROPERTIES:
- *The class of stable distributions possesses its own domain of attraction*, see e.g. [Feller (1971)].
- *Any stable density is unimodal and indeed bell-shaped*, i.e. its n-th derivative has exactly n zeros in $\mathbb{R}$, see [Gawronski (1984)].
- *The stable distributions are self-similar and infinitely divisible.*

These properties derive from the canonic form (F.35) through the scaling property of the Fourier transform.
Self-similarity means
$$L_\alpha^\theta(x,t) \div \exp\left[-t\psi_\alpha^\theta(\kappa)\right] \iff L_\alpha^\theta(x,t) = t^{-\alpha} L_\alpha^\theta(x/t^\alpha)], \quad (F.36)$$

where t is a positive parameter. If t is time, then $L_\alpha^\theta(x,t)$ is a spatial density evolving on time with self-similarity.

Infinite divisibility means that for every positive integer n, the characteristic function can be expressed as the nth power of some characteristic function, so that any stable distribution can be expressed as the n-fold convolution of a stable distribution of the same type. Indeed, taking in (F.35) $\theta = 0$, without loss of generality, we have

$$\mathrm{e}^{-t|\kappa|^\alpha} = \left[\mathrm{e}^{-(t/n)|\kappa|^\alpha}\right]^n \Longleftrightarrow L_\alpha^0(x,t) = \left[L_\alpha^0(x,t/n)\right]^{*n}, \quad (F.37)$$

where

$$\left[L_\alpha^0(x,t/n)\right]^{*n} := L_\alpha^0(x,t/n) * L_\alpha^0(x,t/n) * \cdots * L_\alpha^0(x,/nt)$$

is the multiple Fourier convolution in $\mathbb{R}$ with n identical terms.

Only for a few particular cases, the inversion of the Fourier transform in (F.35) can be carried out using standard tables, and well-known probability distributions are obtained.

For $\alpha = 2$ (so $\theta = 0$), we recover the *Gaussian pdf*, that turns out to be the only stable density with finite variance, and more generally with finite moments of any order $\delta \geq 0$. In fact

$$L_2^0(x) = \frac{1}{2\sqrt{\pi}} \mathrm{e}^{-x^2/4}. \quad (F.38)$$

All the other stable densities have finite absolute moments of order $\delta \in [-1, \alpha)$ as we will later show.

For $\alpha = 1$ and $|\theta| < 1$, we get

$$L_1^\theta(x) = \frac{1}{\pi} \frac{\cos(\theta\pi/2)}{[x + \sin(\theta\pi/2)]^2 + [\cos(\theta\pi/2)]^2}, \quad (F.39)$$

which for $\theta = 0$ includes the *Cauchy-Lorentz pdf*,

$$L_1^0(x) = \frac{1}{\pi} \frac{1}{1+x^2}. \quad (F.40)$$

In the limiting cases $\theta = \pm 1$ for $\alpha = 1$ we obtain the *singular Dirac pdf's*

$$L_1^{\pm 1}(x) = \delta(x \pm 1). \quad (F.41)$$

In general, we must recall the power series expansions provided in [Feller (1971)]. We restrict our attention to $x > 0$ since the evaluations for $x < 0$ can be obtained using the symmetry relation. The convergent expansions of $L_\alpha^\theta(x)$ ($x > 0$) turn out to be;

for $0 < \alpha < 1$, $|\theta| \le \alpha$:

$$L_\alpha^\theta(x) = \frac{1}{\pi x} \sum_{n=1}^{\infty} (-x^{-\alpha})^n \frac{\Gamma(1+n\alpha)}{n!} \sin\left[\frac{n\pi}{2}(\theta - \alpha)\right]; \quad (F.42)$$

for $1 < \alpha \le 2$, $|\theta| \le 2 - \alpha$:

$$L_\alpha^\theta(x) = \frac{1}{\pi x} \sum_{n=1}^{\infty} (-x)^n \frac{\Gamma(1+n/\alpha)}{n!} \sin\left[\frac{n\pi}{2\alpha}(\theta - \alpha)\right]. \quad (F.43)$$

From the series in (F.42) and the symmetry relation we note that *the extremal stable densities for $0 < \alpha < 1$ are unilateral*, precisely vanishing for $x > 0$ if $\theta = \alpha$, vanishing for $x < 0$ if $\theta = -\alpha$. In particular the unilateral extremal densities $L_\alpha^{-\alpha}(x)$ with $0 < \alpha < 1$ have support in $\mathbb{R}^+$ and Laplace transform $\exp(-s^\alpha)$. For $\alpha = 1/2$ we obtain the so-called *Lévy-Smirnov pdf*:

$$L_{1/2}^{-1/2}(x) = \frac{x^{-3/2}}{2\sqrt{\pi}} e^{-1/(4x)}, \quad x \ge 0. \quad (F.44)$$

As a consequence of the convergence of the series in (F.42)-(F.43) and of the symmetry relation we recognize that the stable *pdf*'s with $1 < \alpha \le 2$ are entire functions, whereas with $0 < \alpha < 1$ have the form

$$L_\alpha^\theta(x) = \begin{cases} (1/x)\, \Phi_1(x^{-\alpha}) & \text{for } x > 0, \\ (1/|x|)\, \Phi_2(|x|^{-\alpha}) & \text{for } x < 0, \end{cases} \quad (F.45)$$

where $\Phi_1(z)$ and $\Phi_2(z)$ are distinct entire functions. The case $\alpha = 1$ ($|\theta| < 1$) must be considered in the limit for $\alpha \to 1$ of (F.42)-(F.43), because the corresponding series reduce to power series akin with geometric series in $1/x$ and x, respectively, with a finite radius of convergence. The corresponding stable *pdf*'s are no longer represented by entire functions, as can be noted directly from their explicit expressions (F.39)-(F.40).

We omit to provide the asymptotic representations of the stable densities referring the interested reader to [Mainardi et al. (2001)]. However, based on asymptotic representations, we can state as follows; for $0 < \alpha < 2$ the stable *pdf*'s exhibit *fat tails* in such a way that their absolute moment of order δ is finite only if $-1 < \delta < \alpha$.

More precisely, one can show that for non-Gaussian, not extremal, stable densities the asymptotic decay of the tails is
$$L_\alpha^\theta(x) = O\left(|x|^{-(\alpha+1)}\right), \quad x \to \pm\infty. \quad (F.46)$$
For the extremal densities with $\alpha \neq 1$ this is valid only for one tail (as $|x| \to \infty$), the other (as $|x| \to \infty$) being of exponential order. For $1 < \alpha < 2$ the extremal *pdf*'s are two-sided and exhibit an exponential left tail (as $x \to -\infty$) if $\theta = +(2-\alpha)$, or an exponential right tail (as $x \to +\infty$) if $\theta = -(2-\alpha)$. Consequently, the Gaussian *pdf* is the unique stable density with finite variance. Furthermore, when $0 < \alpha \leq 1$, the first absolute moment is infinite so we should use the median instead of the non-existent expected value in order to characterize the corresponding *pdf*.

Let us also recall a relevant identity between stable densities with index α and $1/\alpha$ (a sort of reciprocity relation) pointed out in [Feller (1971)], that is, assuming $x > 0$,
$$\frac{1}{x^{\alpha+1}} L_{1/\alpha}^\theta(x^{-\alpha}) = L_\alpha^{\theta^*}(x), \quad 1/2 \leq \alpha \leq 1, \quad \theta^* = \alpha(\theta+1) - 1. \quad (F.47)$$
The condition $1/2 \leq \alpha \leq 1$ implies $1 \leq 1/\alpha \leq 2$. A check shows that θ^* falls within the prescribed range $|\theta^*| \leq \alpha$ if $|\theta| \leq 2 - 1/\alpha$. We leave as an exercise for the interested reader the verification of this reciprocity relation in the limiting cases $\alpha = 1/2$ and $\alpha = 1$.

From a comparison between the series expansions in (F.42)-(F.43) and in (F.14)-(F.15), we recognize that for $x > 0$ our *auxiliary functions of the Wright type are related to the extremal stable densities as follows*, see [Mainardi and Tomirotti (1997)],
$$L_\alpha^{-\alpha}(x) = \frac{1}{x} F_\alpha(x^{-\alpha}) = \frac{\alpha}{x^{\alpha+1}} M_\alpha(x^{-\alpha}), \quad 0 < \alpha < 1, \quad (F.48)$$
$$L_\alpha^{\alpha-2}(x) = \frac{1}{x} F_{1/\alpha}(x) = \frac{1}{\alpha} M_{1/\alpha}(x), \quad 1 < \alpha \leq 2. \quad (F.49)$$
In Eqs. (F.48)-(F.49), for $\alpha = 1$, the skewness parameter turns out to be $\theta = -1$, so we get the singular limit
$$L_1^{-1}(x) = M_1(x) = \delta(x-1). \quad (F.50)$$
More generally, all (regular) stable densities, given in Eqs. (F.42)-(F.43), were recognized to belong to the class of Fox *H*-functions, as formerly shown by [Schneider (1986)], see also [Mainardi et al. (2005)]. This general class of high transcendental functions is out of the scope of this book.

The Wright IM-function in two variables. In view of time-fractional diffusion processes related to time-fractional diffusion equations it is worthwhile to introduce the function in two variables

$$\mathbb{M}_\nu(x,t) := t^{-\nu} M_\nu(xt^{-\nu}), \quad 0 < \nu < 1, \quad x,t \in \mathbb{R}^+, \quad (F.51)$$

which defines a spatial probability density in x evolving in time t with self-similarity exponent $H = \nu$. Of course for $x \in \mathbb{R}$ we have to consider the symmetric version obtained from (F.51) multiplying by $1/2$ and replacing x by $|x|$.

Hereafter we provide a list of the main properties of this function, which can be derived from the Laplace and Fourier transforms for the corresponding Wright M-function in one variable.

From Eq. (F.29) we derive the Laplace transform of $\mathbb{M}_\nu(x,t)$ with respect to $t \in \mathbb{R}^+$,

$$\mathcal{L}\{\mathbb{M}_\nu(x,t); t \to s\} = s^{\nu-1} e^{-xs^\nu}. \quad (F.52)$$

From Eq. (F.24) we derive the Laplace transform of $\mathbb{M}_\nu(x,t)$ with respect to $x \in \mathbb{R}^+$,

$$\mathcal{L}\{\mathbb{M}_\nu(x,t); x \to s\} = E_\nu(-st^\nu). \quad (F.53)$$

From Eq. (F.34) we derive the Fourier transform of $\mathbb{M}_\nu(|x|,t)$ with respect to $x \in \mathbb{R}$,

$$\mathcal{F}\{\mathbb{M}_\nu(|x|,t); x \to \kappa\} = 2E_{2\nu}(-\kappa^2 t^\nu). \quad (F.54)$$

Using the Mellin transforms [Mainardi *et al.* (2003)] derived the following integral formula,

$$\mathbb{M}_\nu(x,t) = \int_0^\infty \mathbb{M}_\lambda(x,\tau)\,\mathbb{M}_\mu(\tau,t)\,d\tau, \quad \nu = \lambda\mu. \quad (F.55)$$

Special cases of the Wright IM-function are simply derived for $\nu = 1/2$ and $\nu = 1/3$ from the corresponding ones in the complex domain, see Eqs. (F.16)-(F.17). We devote particular attention to the case $\nu = 1/2$ for which we get from (F.16) the Gaussian density in $\mathbb{R}$,

$$\mathbb{M}_{1/2}(|x|,t) = \frac{1}{2\sqrt{\pi}t^{1/2}} e^{-x^2/(4t)}. \quad (F.56)$$

For the limiting case $\nu = 1$ we obtain

$$\mathbb{M}_1(|x|,t) = \frac{1}{2}[\delta(x-t) + \delta(x+t)]. \quad (F.57)$$

F.6 Notes

In the early nineties, in his former analysis of fractional equations interpolating diffusion and wave-propagation, the present author, see e.g. [Mainardi (1994a)], introduced the functions of the Wright type $F_\nu(z) := W_{-\nu,0}(-z)$ and $M_\nu(z) := W_{-\nu,1-\nu}(-z)$ with $0 < \nu < 1$, in order to characterize the fundamental solutions for typical boundary value problems, as it is shown in Chapter 6.

Being then only aware of the Handbook of the Bateman project, where the parameter λ of the Wright function $W_{\lambda,\mu}(z)$ was erroneously restricted to non-negative values, the author thought to have originally extended the analyticity property of the original Wright function by taking $\nu = -\lambda$ with $\nu \in (0,1)$. So he introduced the entire functions F_ν and M_ν as *auxiliary functions* for his purposes. Presumably for this reason, the function M_ν is referred to as the *Mainardi function* in the treatise by [Podlubny (1999)] and in some research papers including [Balescu (2007a)], [Chechkin et al. (2008)], [Germano et al. (2009)], [Gorenflo et al. (1999); (2000)], [Hanyga (2002b)], [Kiryakova (2009a); (2009b)].

It was Professor B. Stanković, during the presentation of the paper [Mainardi and Tomirotti (1995)] at the Conference *Transform Methods and Special Functions, Sofia 1994*, who informed the author that this extension for $-1 < \lambda < 0$ had been already made by Wright himself in 1940 (following his previous papers in the thirties), see [Wright (1940)]. In his paper [Mainardi et al. (2005)], devoted to the 80th birthday of Professor Stanković, the author used the occasion to renew his personal gratitude to Professor Stanković for this earlier information that led him to study the original papers by Wright and to work (also in collaboration) on the functions of the Wright type for further applications.

For more mathematical details on the functions of the Wright type, the reader may be referred to [Kilbas et al. (2002)] and the references therein. For the numerical point of view we like to highlight the recent paper by [Luchko (2008)], where algorithms are provided for computation of the Wright function on the real axis with prescribed accuracy.

Furthermore, from the stochastic point of view, the Wright M-function emerges as a natural generalization of the Gaussian density for time-fractional diffusion processes. In fact, when these self-similar non-Markovian processes are characterized by stationary increments, so that they are defined only through their first and second moments, which indeed is a property of Gaussian processes, the Wright $M-pdf$ plays the main role as the Gaussian. Thus, such a class of processes, denoted as *generalized grey Brownian motion*, generalizes the Gaussian class of the *fractional Brownian motion* and covers stochastic models of anomalous diffusion, both of slow and fast type. See for details [Mura and Mainardi (2008)], [Mura and Pagnini (2008)], [Mura *et al.* (2008)] and the recent tutorial survey by [Mainardi *et al.* (2009)].

Bibliography

Abel, N.H. (1823). Oplösning af et Par Opgaver ved Hjelp af bestemie Integraler *Magazin for Naturvidenskaberne*, Aargang 1, Bind 2, 11–27, in Norwegian. French translation "Solution de quelques problèmes à l'aide d'intégrales définies" in: L. Sylov and S. Lie (Editors), *Oeuvres Complètes de Niels Henrik Abel* (Deuxième Edition), Vol I, Christiania (1881), pp. 11–27. Reprinted by Éditions Jacques Gabay, Sceaux, 1992.

Abel, N.H. (1826). Aufloesung einer mechanischen Aufgabe, *Journal für die reine und angewandte Mathematik* (Crelle), **1**, 153–157, in German. French translation "Reolution d'un problème de mecanique" in: L. Sylov and S. Lie (Editors), *Oeuvres Complètes de Niels Henrik Abel* (Deuxième Edition), Vol I, Christiania (1881), pp. 97–101, reprinted by Éditions Jacques Gabay, Sceaux, 1992. English translation "Solution of a mechanical problem" with comments by J.D. Tamarkin in D.E. Smith (Editor), *A Source Book in Mathematics*, Dover, New York, 1959, pp. 656–662.

Abramowitz, M. and Stegun, I. A. (1965). *Handbook of Mathematical Functions*, Dover, New York.

Achar, B.N.N and Hanneken, J.W. (2007). Dynamic response of the fractional relaxor-oscillator to a harmonic driving force, in [Sabatier *et al.* (2007)], pp. 243–256.

Achar, B.N.N. and Hanneken, J.W. (2009). Microscopic formulation of fractional calculus theory of viscoelasticity based on lattice dynamics, *Physica Scripta* **T136**, 014011/1–7.

Achar, B.N.N, Hanneken, J.W. and Clarke, T. (2001). Dynamics of the fractional oscillator, *Physica A* **297**, 361–367.

Achar, B.N.N, Hanneken, J.W. and Clarke, T (2002). Response characteristics of a fractional oscillator, *Physica A* **309**, 275–288.

Achar, B.N.N, Hanneken, J.W. and Clarke, T (2004). Damping characteristics of a fractional oscillator, *Physica A* **339**, 311–319.

Achar, B.N.N, Lorenzo, C.F. and Hartley, T.T. (2007). The Caputo fractional derivative: initialization issues relative to fractional differential equations, in [Sabatier *et al.* (2007)], pp. 27–42.

Achenbach, J. D. (1973). *Wave Propagation in Elastic Solids*, North-Holland, Amsterdam.

Achenbach, J. D. and Reddy, D. P. (1967). A note on wave propagation in linearly viscoelastic media, *Z. Angew. Math. Phys.* **18**, 141–144.

Adams, E. P. (1939). Smithsonian Mathematical Formulae and Tables of Elliptic Functions, *Smithon misc. Collns*, **74**, No 1, Smithsonian Insitution, Washington.

Adolfsson, K. (2004). Nonlinear fractional order viscoelasticity at large strains, *Nonlinear Dynamics* **38** No 1-2, 233–246.

Adolfsson, K., Enelund, M. and Larsson, S. (2003). Adaptive discretization of an integro-differential equation with a weakly singular convolution kernel, *Comput. Methods Appl. Mech. Engrg.* **192**, 5285–5304.

Adolfsson, K. and Enelund, M. (2003). Fractional derivative viscoelasticity at large deformations, *Nonlinear Dynamics* **33** No 3, 301–321.

Adolfsson, K., Enelund, M. and Olsson, P. (2005). On the fractional order model of viscoelasticity, *Mechanics of Time-Dependent Materials* **9** No 1, 15–34.

Adolfsson, K., Enelund, M. and Larsson, S. (2008). Space-time discretization of an integro-differential equation modeling quasi-static fractional-order viscoelasticity, *Journal of Vibration and Control* **14** No. 9-10, 1631–1649.

Agarwal, R. P. (1953). A propos d'une note de M. Pierre Humbert, *C.R. Acad. Sci. Paris* **236**, 2031–2032.

Agrarwal, O.P. (2000). A general solution for the fourth-order fractional diffusion-wave equation, *Fractional Calculus and Applied Analysis* **3**, 1–12.

Agrarwal, O.P. (2001). A general solution for a fourth-order fractional diffusion-wave equation defined in a bounded domain, *Computers and Structures* **79**, 1497–1501.

Agrarwal, O.P. (2002). Solution for a fractional diffusion-wave equation defined in a bounded domain, *Nonlinear Dynamics* **29**, 145–155.

Agrarwal, O.P. (2003). Response of a diffusion-wave system subjected to deterministic and stochastic fields, *Z. angew. Math. Mech. (ZAMM)* **83**, 265–274.

Agrawal, O.P. (2008). A general finite element formulation for fractional variational problems, *J. Math. Anal. Appl.* **337**, 1–12.

Aki, K. and Richards, P.G. (1980). *Quantitative Seismology*, Freeman, San Francisco.

Alcoutlabi, M. and Martinez-Vega, J.J. (1998). Application of fractional calculus to viscoelastic behaviour modelling and to the physical ageing phenomenon in glassy amorphous polymers *Polymer* **39** No 25, D 6269–6277.

Alcoutlabi, M. and Martinez-Vega, J.J. (1999). The effect of physical ageing on the time relaxation spectrum of amorphous polymers: the fractional calculus approach. *J. Material Science* **34**, 2361–2369.

Alcoutlabi, M. and Martinez-Vega, J.J. (2003). Modeling of the viscoelastic behavior of amorphous polymers by the differential and integration fractional method: the relaxation spectrum $H(\tau)$, *Polymer* **44**, 7199–7208.

Alemany P.A. (1995). Fractional diffusion equation for fractal-time-continuous-time random walks, *Chaos, Solitons and Fractals* **6**, 7–10.

Alfrey, T. (1948). *Mechanical Behavior of High Polymers*, Interscience, New York.

Alfrey, T. and Doty, P.M. (1945). Methods of specifying the properties of viscoelastic materials, *J. Appl. Phys.* **16**, 700–713.

Al Saqabi, B. N. and Tuan, V. K. (1996). Solution of a fractional differintegral equation, *Integral Transforms and Special Functions* **4**, 321–326.

Amelung, F. and Wolf, D (1994). Viscoelastic perturbations of the Earth: significance of the incremental gravitational force in models of glacial isostasy, *Geophys. J. Int.*, **117**, 864–879.

Anastassiou, G.A. (2009). *Fractional Differentiation Inequalities*, Springer, Dordrecht, Heidelberg.

Anderson, D.L. (1967). The anelasticity of the mantle, *Geophys. J. R. Astr. Soc.* **14**, 135–164.

Anderssen, R.S. and Davies, A.R. (2001). Simple moving-average formulae for the direct recovery of the relaxation spectrum, *J. Rheol.* **45**, 1–27.

Anderssen, R.S., Husain, S.A. and Loy, R.J. (2004). The Kohlrausch function: properties and applications, *ANZIAM J.* **45 E**, C800-C816.

Anderssen, R.S. and Loy, R.J. (2001). On the scaling of molecular weight distribution functionals, *J. Rheol.* **45**, 891–901.

Anderssen, R.S. and Loy, R.J. (2002a). Completely monotone fading memory relaxation modulii, *Bull. Aust. Math. Soc.* **65**, 449–460.

Anderssen, R.S. and Loy, R.J. (2002b). Rheological implications of completely monotone fading memory, *J. Rheol.* **46**, 1459–1472.

Anderssen, R.S. and Mead, D.W. (1998). Theoretical derivation of molecular weight scaling for rheological parameters, *J. Non-Newtonian Fluid Mech.* **76**, 299–306.

Anderssen, R. S. and Westcott, M. (2000). The molecular weight distribution problem and reptation mixing rules, *ANZIAM J.* **42**, 26–40.

Andrade, E.N. (1910). On the viscous flow of metals and allied phenomena. *Proc. R. Soc. London A* **84**, 1–12.

Andrade, E.N. (1962). On the validity of the $t^{1/3}$ law of flow of metals, *Phil. Mag.* **7** No 84, 2003–2014.

André, S., Meshaka, Y. and Cunat, C. (2003). Rheological constitutive equation of solids: a link between models of irreversible thermodynamics and on fractional order derivative equations, *Rheol. Acta* **42** No 6, 500–515.

Angulo, J.M., Ruiz-Medina, M.D., Anh, V.V. and Grecksch, W. (2000). Fractional diffusion and fractional heat equation, *Adv. in Appl. Probab.* **32**, 1077–1099.

Anh, V.V. and Leonenko, N.N. (2000). Scaling laws for fractional diffusion-wave equations with singular data, *Statist. Probab. Lett.* **48**, 239–252.

Anh, V.V. and Leonenko, N.N. (2001). Spectral analysis of fractional kinetic equations with random data, *J. Statistical Physics* **104**, 1349–1387.

Annin, B.D. (1961). Asymptotic expansion of an exponential function of fractional order, *Prikl. Matem. i Mech. (PMM)* **25** No 4, 796–798.

Aprile, A. (1995). *Su un Modello Costitutivo Isteretico Evolutivo per Dispositivi di Protezione Sismica a Smorzamento Viscoelastico: Aspetti Teorici, Sperimentali ed Applicativi*, PhD Thesis, University of Bologna, Faculty of Engineering. [In Italian]

Arenz, R.J. (1964). Uniaxial wave propagation in realistic viscoelastic materials, *J. Appl. Mech.* **Vol 31** No. 1, 17–21. [Trans. ASME, E]

Artin, E. (1964). *The Gamma Function*, Holt, Rinehart and Winston, New York.

Atanackovic, T.M. (2002a). A modified Zener model of a viscoelastic body, *Continuum Mechanics and Thermodynamics* **14**, 137–148.

Atanackovic, T.M. (2002b). A generalized model for the uniaxial isothermal deformation of a viscoelastic body, *Acta Mechanica* **159**, 77–86.

Atanackovic, T.M. (2003). On a distributed derivative model of a viscoelastic body, *Comptes Rendus Academie des Sciences, Paris, Mechanics* **331**, 687–692.

Atanackovic, T.M. (2004). *Applications of Fractional Calculus in Mechanics*, Lecture Notes at the National Technical University of Athens, June 2004, pp. 100.

Atanackovic, T.M., Budincevic, M. and Pilipovic, S. (2005). On a fractional distributed-order oscillator, *J. Phys. A: Math. Gen.* **38**, 1–11.

Atanackovic, T. and Novakovic, B. N. (2002). On a fractional derivative type of a viscoelastic body, *Theoretical and Applied Mechanics (Belgrade)* **28-29**, 27–37.

Atanackovic, T.M., Oparnica, LJ. and Pilipovic, S. (2006). On a model of viscoelastic rod in unilateral contact with a rigid wall, *IMA Journal of Applied Mathematics* **71**, 1–13.

Atanackovic, T.M., Oparnica, L. and Pilipovic, S. (2007a). On a nonlinear distributed order fractional differential equation, *J. Math. Anal. Appl.* **328**, 590–608.

Atanackovic, T.M. and Pilipovic, S. (2005). On a class of equations arising in linear viscoelasticity theory, *Zeitschrift angew. Math. Mech. (ZAMM)* **85**, 748–754.

Atanackovic, T.M., Pilipovic, S. and Zorica, D. (2007). A diffusion wave equation with two fractional derivatives of different order, *J. Physics A* **40**, 5319–5333.

Atanackovic, T.M., and Spasic, D.T. (2004). On a viscoelastic compliant contact impact model, *Journal of Applied Mechanics (Transactions of ASME)* **71**, 134–138.

Atanackovic, T.M. and Stankovic, B. (2002). Dynamics of a viscoelastic rod of fractional derivative type, *Z. angew. Math. Mech. (ZAMM)* **82**, 377–386.

Atanackovic, T.M. and Stankovic, B. (2004). On a system of differential equations with fractional derivatives arising in rod theory, *J. Phys. A: Math. Gen.* **37**, 1241–1250.

Atanackovic, T.M., and Stankovic, B. (2004). Stability of an elastic rod on a fractional derivative type of a foundation, *Journal of Sound and Vibration* **227**, 149–161.

Atanackovic, T.M., and Stankovic, B. (2004). An expansion formula for fractional derivatives and its application, *Fractional Calculus and Applied Analysis* **7** No 3, 365–378.

Atanackovic, T.M., and Stankovic, B. (2007). On a differential equation with left and right fractional derivatives, *Fractional Calculus and Applied Analysis* **10** No 2, 139–150.

Attewell, P.B. and Brentnall, D. (1964). Internal friction: some considerations of the frequency response of rocks and other metallic and non-metallic materials, *Int. J. Rock Mech. Mining Sci.* **1**, 231–254.

Attewell, P.B. and Ramana Y.V. (1966). Wave attenuation and internal friction as functions of the frequency, *Geophysics* **31**, 1049–1056.

Babenko, Yu. I. (1986). *Heat and Mass Transfer*, Chimia, Leningrad. [in Russian]

Baclic, B.S., and Atanackovic, T.M. (2000). Stability and creep of a fractional derivative order viscoelastic rod, *Bulletin de la Classe de Sciences, Mathmatiques et Naturelles Serbian Academy of Sciences and Arts* **25**, 115–131.

Baerwald, H.G. (1930). Uber die Fortpflanzung von Signalen in dispergierenden Systemen. Zweiter Teil: Verlustarme kontinuierliche Systems, *Ann. Physik* (Ser. V) **7**, 731–760.

Bagley, R.L. (1979). *Applications of Generalized Derivatives to Viscoelasticity*, Ph. D. Dissertation, Air Force Institute of Technology.

Bagley, R.L. (1989). Power law and fractional calculus of viscoelasticity, *AIAA Journal* **27**, 1412–1417.

Bagley, R.L. (1990). On the fractional order initial value problem and its engineering applications, in K. Nishimoto (Editor), *Fractional Calculus and Its Application*, College of Engineering, Nihon University, pp. 12–20. [Proc. Int. Conf. held at Nihon University, Tokyo 1989]

Bagley, R.L. (1991). The thermorheologically complex material, *Int. J. Engng. Sc.* **29**, 797–806.

Bagley, R.L. (2007). On the equivalence of the Riemann-Liouville and the Caputo fractional order derivatives in modeling of linear viscoelastic materials, *Fractional Calculus and Applied Analysis* **10** No 2, 123–126.

Bagley, R.L. and Calico, R.A. (1991). Fractional order state equations for the control of viscoelastically damped structures, *Journal of Guidance, Control and Dynamics* **14**, 304–311.

Bagley, R.L. and Torvik, P.J. (1979). A generalized derivative model for an elastomer damper, *Shock Vib. Bull.* **49**, 135–143.

Bagley, R.L. and Torvik, P.J. (1983a). A theoretical basis for the application of fractional calculus, *J. Rheology* **27**, 201–210.

Bagley, R.L. and Torvik, P.J. (1983b). Fractional calculus - A different approach to the finite element analysis of viscoelastically damped structures, *AIAA Journal* **21**, 741–748.

Bagley, R.L. and Torvik, P.J. (1985). Fractional calculus in the transient analysis of viscoelastically damped structures, *AIAA Journal* **23**, 918–925.

Bagley, R.L. and Torvik, P.J. (1986). On the fractional calculus model of viscoelastic behavior, *J. Rheology* **30**, 133–155.

Bagley, R.L. and Torvik, P.J. (2000a). On the existence of the order domain and the solution of distributed order equations, Part I, *Int. J. Appl. Math.* **2**, 865–882.

Bagley, R.L. and Torvik, P.J. (2000b). On the existence of the order domain and the solution of distributed order equations, Part II, *Int. J. Appl. Math.* **2**, 965–987.

Baillie, R.T. and King, M.L. (1996). Fractional differencing and long memory processes, *J. Econometrics* **73**, 1–3.

Baker, G.A. (1975). *Essentials of Padè Approximants*, Academic Press, New York.

Baker, G.A. and Gammel, G.L. (1970). *The Padè Approximants in Theoretical Physics*, Academic Press, New York.

Balakrishnan, V. (1985) Anomalous diffusion in one dimension, *Physica A* **132**, 569–580.

Balescu, R. (1997). *Statistical Mechanics - Matter out of Equilibrium*, Imperial College Press, London.
Balescu, R. (2000). Memory effects in plasma transport theory, *Plasma Phys. Contr. Fusion* **42**, B1–B13.
Balescu, R. (2005). *Aspects of anomalous transport in plasmas*, Taylor and Francis, London.
Balescu, R. (2007a). V-Langevin equations, continuous time random walks and fractional diffusion, *Chaos, Solitons and Fractals* **34**, 62–80.
Balescu, R. (2007b). V-Langevin equations, continuous time random walks and fractional diffusion, Extended Version, pp. 69. [E-print: http://arxiv.org/abs/0704.2517]
Baldock, G.R. and Bridgeman, T. (1981). *Mathematical Theory of Wave Motion*, Ellis Horwood, Chichester.
Bano, M. (2004). Modelling of GPR waves for lossy media obeying a complex power law of frequency for dielectric permittivity, *Geophys. Prosp.* **52**, 11–26.
Barkai, E. (2001). Fractional Fokker-Planck equation, solution, and applications, *Phys. Rev. E* **63**, 046118/1–17.
Barkai, E. (2002). CTRW pathways to the fractional diffusion equation, *Chem. Phys.* **284**, 13–27.
Barkai, E. (2003). Aging in subdiffusion generated by a deterministic dynamical system, *Phys. Rev. Lett.* **90**, 104101/1–4.
Barkai, E. Metzler, R. and Klafter, J. (2000) From continuous-time random walks to the fractional Fokker-Planck equation, *Physical Review E* **61**, 132–138.
Barkai, E. and Sokolov, I.M. (2007). On Hilfer's objection to the fractional time diffusion equation, *Physica A* **373**, 231–236.
Barpi, F. and Valente, S. (2002). Creep and fracture in concrete: A fractional order rate approach, *Engineering Fracture Mechanics* **70**, 611–623.
Barpi, F. and Valente, S. (2004). A fractional order rate approach for modeling concrete structures subjected to creep and fracture, *International Journal of Solids and Structures*, **41**, 2607–2621.
Barret, J.H. (1954). Differential equations of non-integer order, *Canad. J. Math.* **6**, 529–541.
Bartenev, G. M. and Zelenev, Yu. V. (1974). *Relaxation Phenomena in Polymers*, Wiley, New York.
Baeumer, B. and Meerschaert, M.M. (2001). Stochastic solutions for fractional Cauchy problems, *Fractional Calculus and Applied Analysis* **4**, 481–500.
Bazhlekova, E. (1996). Duhamel-type representations of the solutions of non-local boundary value problems for the fractional diffusion-wave

equation, in P. Rusev, I. Dimovski, V. Kiryakova (Editors), *Transform Methods & Special Functions, Varna '96*, (Inst. Maths & Informatics, Bulg. Acad. Sci, Sofia, 1998), pp. 32–40.

Bazhlekova, E. (1998). The abstract Cauchy problem for the fractional evolution equation, *Fractional Calculus and Applied Analysis* **1** No 1, 255–270.

Bazhlekova, E. (1999). Perturbation properties for abstract evolution equations of fractional order, *Fractional Calculus and Applied Analysis* **2** No 4, 359–366.

Bazhlekova, E. (2000). Subordination principle for fractional evolution equations, *Fractional Calculus and Applied Analysis* **3** No 3, 213–230.

Bazhlekova, E. (2001). *Fractional Evolution Equations in Banach Spaces*, PhD Thesis, University Press Facilities, Eindhoven University of Technology, pp. 107.

Bazzani, A. Bassi, G. and Turchetti, G. (2003). Diffusion and memory effects for stochastic processes and fractional Langevin equations *Physica A* **324**, 530–550.

Becker, R. (1926). Elastische Nachwirkung und Plastizität, *Zeit. Phys.* **33**, 185–213.

Becker, R. and Doring, W. (1939). *Ferromagetismus*, Springer-Verlag, Berlin.

Beda, T. and Chevalier, Y. (2004). New methods for identifying rheological parameter for fractional derivative modeling of viscoelastic behavior, *Mechanics of Time-Dependent Materials* **8** No 2, 105–118.

Beer, F.P. and Johnston, Jr. E.R. (1981). *Mechanics of Materials*, McGraw-Hill, New York.

Beghin, L. (2008). Pseudo-processes governed by higher-order fractional differential equations *Electronic Journal of Probability* **13**, 467–485 (Paper No 16).

Beghin, L. and Orsingher, E. (2003). The telegraph process stopped at stable-distributed times and its connection with the telegraph equation, *Fractional Calculus and Applied Analysis* **6**, 187–204.

Beghin, L. and Orsingher, E. (2005). The distribution of the local time for "pseudoprocess" and its connection with fractional diffusion equations, *Stochastic Processes and their Applications* **115**, 1017–1040.

Beghin, L. and Orsingher, E. (2009). Iterated elastic Brownian motions and fractional diffusion equations, *Stochastic Processes and their Applications* **119**, 1975–2003.

Bender, C.M. and Orszag, S.A. (1987). *Advanced Mathematical Methods for Scientists and Engineers*, McGraw-Hill, Singapore.

Ben-Menahem, A. (1965). Observed attenuation and Q values of seismic surface waves in the upper mantle, *J. Geophys. Res.* **70**, 4641–4651.

Ben-Menahem, A. and Jeffreys, H. (1971). The saddle approximation for damped surface waves, *Geophys. J. R. Astr. Soc.* **24**, 1-2.

Ben-Menahem, A., and Singh, S.J. (1981). *Seismic Waves and Sources*, Springer-Verlag, New York.

Bennewitz, K. and Rötger, H. (1936). Über die innere Reibung fester Körper; Absorptionsfrequenzen von Metallen in akustischen Gebeit, *Phys. Zeitschr.* **37**, 578-588.

Bennewitz, K. and Rötger, H. (1938). Thermische Dämpfung bei Biegeschwingungen, *Zeitschr. f. tech. Phys.* **19**, 521.

Benson, D.A., Wheatcraft, S.W. and Meerschaert, M.M. (2000). Application of a fractional advection-dispersion equation, *Water Resources Res.* **36**, 1403-1412.

Benvenuti, P. (1965). Sulla teoria ereditaria delle deformazioni lineari dinamiche, *Rend. Matem.* **24**, 255-290.

Berberan-Santos, M.N. (2005a). Analytic inversion of the Laplace transform without contour integration: application to luminescence decay laws and other relaxation functions, *J. Math. Chem.* **38**, 165-173.

Berberan-Santos, M.N. (2005b). Relation between the inverse Laplace transforms of $I(t^\beta)$ and $I(t)$: Application of the Mittag-Leffler and asymptotic power law relaxation functions, *J. Math. Chem.* **38**, 265-270.

Berberan-Santos, M.N. (2005c). Properties of the Mittag-Leffler relaxation function, *J. Math. Chem.* **38**, 629-635.

Berens, H. and Westphal, U. (1968). A Cauchy problem for a generalized wave equation, *Acta Sci. Math. (Szeged)* **29**, 93-106.

Berg, Ch. and Forst, G. (1975). *Potential Theory on Locally Compact Abelian Groups*, Springer, Berlin.

Beris, A. N. and Edwards, B. J. (1993). On the admissibility criteria for linear viscoelastic kernels, *Rheological Acta* **32**, 505-510.

Berry, D. S. (1958). A note on stress pulses in viscoelastic rods, *Phil. Mag* (Ser. VIII) **3**, 100-102.

Berry, D. S. and Hunter, S. C. (1956). The propagation of dynamic stresses in viscoelastic rods, *J. Mech. Phys. Solids* **4**, 72-95.

Betten, J. (2002). *Creep Mechanics*, Springer, Berlin.

Beyer, H. and Kempfle, S. (1994). Dämpfungsbeschreibung mittels gebrochener Ableitungen, *Z. angew. Math. Mech. (ZAMM)* **74**, T657-T660.

Beyer, H. and Kempfle, S. (1995). Definition of physically consistent damping laws with fractional derivatives, *Z. angew. Math. Mech. (ZAMM)* **75**, 623-635.

Bieberbach, L. (1931). *Lehrbuch der Funktionentheorie*, Teubner Verlag, Leipzig.

Bingham, N.H., Goldie, C.M. and Teugels, J.L. (1987). *Regular Variation*, Cambridge University Press, Cambridge.

Biot, M.A. (1952). The interaction of Rayleigh and Stonley waves in the ocean bottom, *Bull. Seism. Soc. Amer.* **42**, 81–93.

Biot, M.A. (1954). Anisotropic viscoelasticity and relaxation phenomena, *J. Appl. Phys.* **25**, 1385–1391.

Biot, M.A. (1956). Theory of propagation of elastic waves in a fluid-saturated porous solids, *J. Acoust. Soc. Amer.* **28**, 168–178.

Biot, M.A. (1958). Linear thermodynamics and the mechanics of solids, *Proc. Third. U.S. Nat. Congr. Appl. Mech.*, ASME, pp. 1–18.

Biot, M.A. (1961). Generalized theory of acoustic propagation in porous dissipative media, *J. Acoust. Soc. Amer.* **34**, 1254–1264.

Bird, R.B., Amstrong, R.C. and Hassager, O. (1987). *Dynamics of Polymeric Liquids*, Wiley, New York.

Bland, D.R. (1960). *The Theory of Linear Viscoelasticity*, Pergamon, Oxford.

Bland, D.R. (1988). *Wave Theory and Applications*, Clarendon, Oxford.

Bland, D.R. and Lee, E.H. (1956). On the determination of viscoelastic model for stress analysis of plastics, *J. Appl. Mech.* **23**, 416–420.

Blank, L. (1997). Numerical treatment of differential equations of fractional order, *Non-linear World* **4** No 4, 473–491.

Bleistein, N. (1984). *Mathematical Methods for Wave Phenomena*, Academic Press, Orlando, Florida.

Bleistein, N. and Handelsman, R.A. (1986). *Asymptotic Expansions of Integrals*, Dover, New York. [The Dover edition is a corrected version of the first edition published by Holt, Rinehart and Winston, New York 1975.]

Blizard, R.B. (1951). Viscoelasticity of rubber, *J. Appl. Phys.* **22**, 730–735.

Blumenfeld, R. and Mandelbrot, B.B. (1997). Lévy dusts, Mittag-Leffler statistics, mass fractal lacunarity and perceived dimensions, *Phys. Rev. E* **56**, 112–118.

Bode, H.W. (1945). *Network Analysis and Feedback Amplifier Design*, D. Van Nostrand Company, Inc., New York.

Boltzmann, L. (1876). Zur Theorie der elastischen Nachwirkung. *Ann. Phys. Chem.* **7**, 624–657.

Bonilla, B., Rivero, M. and Trujillo, J.J. (2007). Linear differential equations of fractional order, in [Sabatier *et al.* (2007)], pp. 77–92.

Born, W.T. (1944). The attenuaion constant of earth materials, *Geophysics* **6**, 132–148.

Bossemeyer, H.G. (2001). Evaluation technique for dynamic moduli, *Mechanics of Time-Dependent Materials* **5**, 273–291.

Brachman, M.K. and Macdonald, J.R. (1954). Relaxation-time distribution functions and the Kramers-Kronig relations, *Physica* **20**, 1266–1270.

Brachman, M.K. and Macdonald, J.R. (1956). Generalized admittance kernels and the Kronig-Kramers relations, *Physica* **22**, 141–148.
Brankov, J,G. (1989). Finite-size scaling for the mean spherical model with power law Interaction, *J. Stat. Phys.* **56** No 3/4, 309–339.
Brankov, J.G. and Tonchev, N.S. (1990). An investigation of finite-size scaling for systems with long-range interaction: the spherical model, *J. Stat. Phys.* **59** No 5/6, 1431–1450.
Brankov, J.G. and Tonchev, N.S. (1992). Finite-size scaling for systems with long-range interactions, *Physica A* **189**, 583–610.
Braun, H. (1990). Rheological rate type models including temperature, in Oliver, D. R. (Editor), Third European Rheology Conference, Elsevier Applied Science, Amsterdam, pp. 66–69.
Braun, H. (1991). A model for the thermorheological behavior of viscoelastic fluids, *Rheologica Acta* **30**, 523–529.
Braun, H. and Friedrich, Chr. (1990). Dissipative behaviour of viscoelastic fluids derived from rheological constitutive equations, *J. Non-Newtonian Fluid Mech.* **38**, 81–91.
Breuer, S. and Onat, E.T. (1962). On uniqueness in linear viscoelasticity, *Quarterly of Applied Mathematics* **19**, 355–359.
Breuer, S. and Onat, E.T. (1964). On recoverable work in linear viscoelasticity, *ZAMP* **15**, 13–21.
Breuer, S. and Onat, E.T. (1964). On the determination of free energy in viscoelastic solids, *ZAMP* **15**, 185–191.
Brilla, J. (1982). Generalized variational principles and methods in dynamic viscoelasticity, in [Mainardi (1982)], pp. 258–272.
Brillouin, L. (1960). *Wave Propagation and Group Velocity*, Academic Press, New York.
Bronskii, A.P. (1941). After-effect phenomena in solid bodies. *Prikl. Mat. i Meh. (PMM)* **5** No 1, 31–56 [In Russian].
Bruckshaw, McG. and Mahanta, F.C. (1954). The variation of the elastic constants of rocks with frequency, *Petroleum* **17**, 14–18.
Brun, L. (1974). L'onde simple viscoélastique linéaire, *J. Mécanique* **13**, 449–498.
Brun, L. and Molinari, A. (1982). Transient linear and weakly non-linear viscoelastic waves, in [Mainardi (1982)], pp. 65–94.
Brunner, H. and van der Houwen, P.J. (1986). *The Numerical Solution of Volterra Equations*, North-Holland, Amsterdam.
Buchen, P.W. (1971). Plane waves in linear viscoelastic media, *Geophys. J. R. Astr. Soc.* **23**, 531–542.
Buchen, P.W. (1974). Application of the ray-series method to linear viscoelastic wave propagation, *Pure Appl. Geophys.* **112**, 1011–1023.
Buchen, P.W. and Mainardi, F. (1975). Asymptotic expansions for transient viscoelastic waves, *Journal de Mécanique* **14**, 597–608.

Buckingham, M.J. (1997). Theory of acoustic attenuation, dispersion and pulse propagation in unconsolidated granulated materials including marine sediments, *J. Acoust. Soc. Am.* **102**, 2579–2596.

Buckingham, M.J. (1998). Theory of compressional and shear waves in fluid-like marine sediments, *J. Acoust. Soc. Am.* **103**, 288–299.

Buckwar, E. and Luchko, Yu. (1998). Invariance of a partial differential equation of fractional order under the Lie group of scaling transformations, *J. Math. Anal. Appl.* **227**, 81–97.

Budden, K. G. (1988). *Wave Propagation of Radio Waves*, Cambridge University Press, Cambridge.

Buhl, A. (1925). *Séries Analytiques. Sommabilité*, Mémorial des Sciences Mathématiques, Acad. Sci. Paris, Fasc. VII, Gauthier-Villars, Paris, Ch. 3.

Burgers, J.M. (1939). Mechanical Considerations - model systems - phenomenological theories of relaxation and viscosity. In: First Report on Viscosity and Plasticity, Second edition, Nordemann Publishing Company, Inc., New York. [Prepared by the committee of viscosity of the Academy of Sciences at Amsterdam]

Burridge, R., Mainardi, F. and Servizi, G. (1980). Soil amplification of plane seismic waves, *Phys. Earth Planet. Inter.* **22**, 122–136.

Burridge, R., De Hoop, M.V., Hsu, K., Le, L. and Norris, A. (1993). Waves in stratified viscoelastic media with microstructure, *J. Acoust. Soc. Am.* **94**, 2884–2894.

Butcher, E.A. and Segalman, D.J. (2000). Characterizing damping and restitution in compliant impact via modified K-V and higher order linear viscoelastic models, *ASME J. Appl. Mech.* **67**, 831–834.

Butzer, P.L., Kilbas, A.A., and Trujillo, J.J. (2002a). Compositions of Hadamardtype fractional integration operators and the semigroup property, *J. Math. Anal.Appl* **269** No 2, 387–400.

Butzer, P.L., Kilbas, A.A., and Trujillo, J.J. (2002b). Fractional calculus in the Mellin setting and Hadamard-type fractional integrals, *J. Math. Anal. Appl.* **269** No 1, 1–27.

Butzer, P.L., Kilbas, A.A., and Trujillo, J.J. (2002c). Mellin transform and integration by parts for Hadamard-type fractional integrals, *J. Math. Anal. Appl.*, **270** No 1, (2002d), 1–15.

Butzer, P.L., Kilbas, A.A., and Trujillo, J.J. (2003). Generalized Stirling functions of second kind and representations of fractional order differences via derivatives, *J. Diff. Equat. Appl.* **9** No 5, 503–533.

An access to fractional differentiation via fractional difference quotients, in [Ross (1975a)], pp. 116–145.

Butzer, P. and Westphal, U. (2000). Introduction to fractional calculus, in [Hilfer (2000a)], pp. 1–85.

Cafagna, D. (2007). Fractional calculus: a mathematical tool from the past for present engineers, *IEEE Industrial Electronics Magazine* **1**, 35–40.

Cafagna, D. and Grassi, G. (2008). Bifurcation and chaos in the fractional-order Chen system via a time-domain approach, *Int. Journal of Bifurcation and Chaos* **18** No 7, 1845–1863.

Cafagna, D. and Grassi, G. (2009). Fractional-order chaos: a novel four-wing attractor in coupled Lorenz systems, *Int. Journal of Bifurcation and Chaos*, to appear.

Caloi, P. (1948). Comportamento delle onde di Rayleigh in un mezzo firmo-elastico indefinito, *Annali di Geofisica* **1**, 550–567.

Capelas de Oliveira, E (2004). *Special Functions with Applications*, Editora Livraria da Fisica, Sao Paulo, Brazil. [in Portuguese]

Capelas de Oliveira, E. (2009). Fractional Langevin equation and the Caputo fractional derivative, *PRE-PRINT*, submitted for publication.

Caputo, M. (1966). Linear models of dissipation whose Q is almost frequency independent, *Annali di Geofisica* **19**, 383–393.

Caputo, M. (1967). Linear models of dissipation whose Q is almost frequency independent, Part II, *Geophys. J. R. Astr. Soc.* **13**, 529–539. [Reprinted in *Fract. Calc. Appl. Anal.* **11**, 4–14 (2008)]

Caputo, M. (1969). *Elasticità e Dissipazione*, Zanichelli, Bologna. [in Italian]

Caputo, M. (1974). Vibrations of an infinite viscoelastic layer with a dissipative memory, *J. Acoust. Soc. Am.* **56**, 897–904.

Caputo, M. (1976). Vibrations of an infinite plate with a frequency independent Q, *J. Acoust. Soc. Am.* **60**, 634–639.

Caputo, M. (1979). A model for the fatigue in elastic materials with frequency independent Q, *J. Acoust. Soc. Am.* **66**, 176–179.

Caputo, M. (1981). Elastic radiation from a source in a medium with an almost frequency independent Q, *J. Phys. Earth* **29**, 487–497.

Caputo, M. (1985). Generalized rheology and geophysical consequences, *Tectonophysics* **116**, 163–172.

Caputo, M. (1986). Linear and non-linear inverse rheologies of rocks, *Tectonophysics* **122**, 53–71.

Caputo, M. (1989). The rheology of an anelastic medium studied by means of the observation of the splitting of its eigenfrequencies, *J. Acoust. Soc. Am.* **86**, 1984–1989.

Caputo, M. (1993a). The Riemann sheets solutions of anelasticity, *Ann. Matematica Pura Appl.* (Ser. IV) **146**, 335–342.

Caputo, M. (1993b). The splitting of the seismic rays due to dispersion in the Earth's interior, *Rend. Fis. Acc. Lincei* (Ser. IX) **4**, 279–286.

Caputo, M. (1995). Mean fractional-order derivatives differential equations and filters, *Ann. Univ. Ferrara, Sez VII, Sc. Mat.* **41**, 73–84.

Caputo, M. (1996a). The Green function of the diffusion of fluids in porous media with memory, *Rend. Fis. Acc. Lincei* (Ser. 9) **7**, 243–250.

Caputo, M. (1996b). Modern rheology and dielectric induction: multivalued index of refraction, splitting of eigenvalues and fatigue, *Annali di Geofisica* **39**, 941–966.

Caputo, M. (1998). Three-dimensional physically consistent diffusion in anisotropic media with memory, *Rend. Mat. Acc. Lincei* (Ser. 9) **9**, 131–143.

Caputo, M. (1999a). *Lectures on Seismology and Rheological Tectonics*, Lecture Notes, Università "La Sapienza", Dipartimento di Fisica, Roma, pp. 319. [Enlarged and revised edition based on the first edition, 1992]

Caputo, M. (1999b). Diffusion of fluids in porous media with memory, *Geothermics* **28**, 113–130.

Caputo, M. (2000). Models of flux in porous media with memory, *Water Resources Research* **36**, 693–705.

Caputo, M. (2001). Distributed order differential equations modelling dielectric induction and diffusion, *Fractional Calculus and Applied Analysis* **4**, 421–442.

Caputo, M. and Mainardi, F. (1971a). A new dissipation model based on memory mechanism, *Pure and Appl. Geophys. (PAGEOPH)* **91**, 134–147. [Reprinted in *Fractional Calculus and Applied Analysis* **10** No 3, 309–324 (2007)]

Caputo, M. and Mainardi, F. (1971b). Linear models of dissipation in anelastic solids, *Riv. Nuovo Cimento* (Ser. II) **1**, 161–198.

Caputo, M. and Plastino, W. (1998). Rigorous time domain responses of polarizable media-II, *Annali di Geofisica* **41**, 399–407.

Caputo, M. and Plastino, W. (1999). Diffusion with space memory, in : F. Krumm and V.S. Schwarze (Editors), *Festschrift on Honor of Eric Grafarend*, Schriftenreihe der Institute des Fachbereichs Geodesie und Informatik, *Report Nr. 1999.6*, pp. 59–67.

Caputo, M. and Plastino W. (2004). Diffusion in porous layers with memory, *Geophysical Journal International* **158**, 385–396.

Carcione, J.M. (1990). Wave propagation in anisotropic linear viscoelastic media: theory and simulated wavefields, *Geophys. J. Int.* **101**, 739–750. Erratum: **111**, 191 (1992).

Carcione, J.M. (1995). Constitutive model and wave equations for linear, viscoelastic, anisotropic media, *Geophysics* **60**, 537–548.

Carcione, J.M. (1998). Viscoelastic effective rheologies for modelling wave propagation in porous media, *Geophys. Prosp.* **46**, 249–270.

Carcione, J.M. (2007). *Wave Fields in Real Media. Theory and Numerical Simulation of Wave Propagation in Anisotropic, Anelastic and Porous Media*, Elsevier, Amsterdam. (Second edition, extended and revised).

Carcione, J.M. (2009). Theory and modeling of constant-Q P- and S-waves using fractional time derivatives, *Geophysics* **74**, T1–T11.

Carcione, J.M., and Cavallini, F. (1993). Energy balance and fundamental relations in anisotropic-viscoelastic media, *Wave Motion* **18**, 11–20.

Carcione, J.M. and Cavallini, F. (1994a). A rheological model for anelastic anisotropic media with applications to seismic wave propagation, *Geophys. J. Int.* **119**, 338–348.

Carcione, J.M. and Cavallini, F. (1994b). Anisotropic-viscoelastic rheologies via eigenstrains, in Rionero, S. and Ruggeri, T. (Editors), *Waves and Stability in Continuous Media*, World Scientific, Singapore, pp. 40–46.

Carcione, J.M., and Cavallini, F. (1995a). On the acoustic-electromagnetic analogy, *Wave Motion* **21**, 149–162.

Carcione, J.M., and Cavallini, F. (1995b). Attenuation and quality factors surfaces in anisotropic-viscoelastic media, *Mech. Mater.* **19**, 311–327.

Carcione, J.M., Cavallini, F., Helbig, K. (1996). Anisotropic attenuation and material symmetry. *Acustica* **84**, 495–502.

Carcione, J.M., Cavallini, F., Mainardi, F. and Hanyga, A. (2002). Time-domain seismic modelling of constant-Q wave propagation using fractional derivatives, *Pure and Appl. Geophys. (PAGEOPH)* **159**, 1719–1736.

Carcione, J.M., Helle, H.B. and Gangi, F. (2006). Theory of borehole stability when drilling through salt formations, *Geophysics* **71**, F31–F47.

Carpinteri, A., Chiaia, B. and Cornetti, P. (2000). A fractional calculus approach to the mechanics of fractal media, *Rend. Sem. Mat. Univ. Pol. Torino* **58** No 1, 57–68.

Carpinteri, A. and Cornetti, P. (2002). A fractional calculus approach to the description of stress and strain localization in fractal media, *Chaos, Solitons and Fractals* **13**, 85–94.

Carpinteri, A., Cornetti, P., Sapora, A., Di Paola, M. and Zingales, M. (2009). Fractional calculus in solid mechanics: local versus non-local approach, *Physica Scripta* **T136**, 014003/1–7.

Carpinteri, A. and Mainardi, F. (Editors) (1997). *Fractals and Fractional Calculus in Continuum Mechanics*, Springer Verlag, Wien and New York. (Vol. no 378, series CISM Courses and Lecture Notes, ISBN 3-211-82913-X) [Lecture Notes of the Advanced School held at CISM, Udine, Italy, 23-27 September 1996]

Carrier, G.F., Krook, M. and Pearson, C.E. (1966). *Functions of a Complex Variable*, McGraw-Hill, New York, Chapter 6.

Casula, G. and Carcione, J.M. (1992). Generalized mechanical model analogies of linear viscoelastic behaviour, *Bull. Geofis. Teor. Appl.* **34**, 235–256.

Catania, G and Sorrentino, S. (2007). Analytical modelling and experimental identification of viscoelastic mechanical systems, in [Sabatier et al. (2007)], pp. 403–416.

Cavallini, F. and Carcione, J.M. (1994). Energy balance and inhomogeneous plane-wave analysis of a class of anisotropic viscoelastic constitutive laws, in Rionero, S. and Ruggeri, T. (Editors), *Waves and Stability in Continuous Media*, World Scientific, Singapore, pp. 47–53.

Caviglia G. and Morro, A. (1991). Wave propagation in inhomgeneous viscoelastic solids, *Quarterly Journal of Mechanics and Applied Mathematics* **44** No 1, 45–54.

Caviglia G. and Morro, A. (1992). *Inhomogenepus Waves in Solids and Fluids*, World Scientific, Singapore.

Cerveny, V. and Psencik, I. (2005a). Plane waves in viscoelastic anisotropic media. Part I: Theory, *Geophys. J. Int.* **161**, 197–212.

Cerveny, V. and Psencik, I. (2005b). Plane waves in viscoelastic anisotropic media. Part II: Numerical examples, *Geophys. J. Int.* **161**, 213–229.

Chamati, H., and Tonchev, N. S. (2006). Generalized Mittag-Leffler functions in the theory of finite-size scaling for systems with strong anisotropy and/or long-range interaction, *J. Phys. A* **39**, 469–478.

Chatterjee, A. (2005). Statistical origins of fractional derivatives in viscoelasticity, *Journal of Sound and Vibration* **284**, 1239–1245.

Chechkin, A.V., Gorenflo, R. and Sokolov, I.M. (2002). Retarding subdiffusion and accelerating superdiffusion governed by distributed-order fractional diffusion equations, *Physical Review E* **66**, 046129/1–6.

Chechkin, A.V., Gorenflo, R., Sokolov, I.M. and Gonchar, V.Yu. (2003a). Distributed order time fractional diffusion equation, *Fractional Calculus and Applied Analysis* **6**, 259–279.

Chechkin, A.V., Klafter, J. and Sokolov, I.M. (2003b). Fractional Fokker-Planck equation for ultraslow kinetics, *Europhys. Lett.* **63**, 326–332.

Chechkin, A.V., Gonchar, V.Yu, Gorenflo, R., Korabel, M and Sokolov, I.M. (2008). Generalized fractional diffusion equations for accelerating subdiffusion and truncated Lévy flights, *Phys. Rev. E* **78**, 021111/1–13.

Chen, Y., Vinagre, B.M. and Podlubny, I. (2004), Continued fractional expansion approaches to discretizing fractional order derivatives: An expository review, *Nonlinear Dyn.* **38**, 155–170.

Chen, W. and Holm, S. (2003). Modified Szabos wave equation models for lossy media obeying frequency power law, *J. Acoust. Soc. Am.* **114**, 2570–2574.

Chen, W. and Holm, S. (2009). A study on modified Szabo's wave equation modeling of frequency-dependent dissipation inultrasonic medical imaging, *Physica Scripta* **T136**, 014014/1–5.

Chin, R.C.Y. (1980). Wave propagation in viscoelastic media, in Dziewonski, A. and Boschi, E. (Editors), *Physics of the Earth's Interior*, North-Holland, Amsterdam, pp. 213–246. [E. Fermi Int. School, Course 78]

Chin, R.C.Y., Hedstrom, G.W. and G. Majda, G. (1986). Application of a correspondence principle to the free vibrations of some viscoelastic solids, *Geophysical J. Int.* **86**, 137–166.

Christensen, R.M. (1982). *Theory of Viscoelasticity*, Second edition, Academic Press, New York. [First edition (1972)]

Chu, B.-T. (1962). Stress waves in isotropic linear viscoelastic waves, *Journal de Mécanique* **1**, 439–462.

Chukbar, K.V. (1995). Stochastic transport and fractional derivatives, *Zh. Eksp. Teor. Fiz.* **108**, 1875–1884. [in Russian]

Cisotti, V. (1911a). L'ereditarietà lineare ed i fenomeni dispersivi, *Il Nuovo Cimento* (Ser. VI) **2**, 234–244.

Cisotti, V. (1911b). Sulla dispersività in relazione ad una assegnata frequenza, *Il Nuovo Cimento* (Ser. VI) **2**, 360–374.

Coffey, W.T. (2004). Dielectric relaxation: an overview, *J. Molecular Liquids* **114**, 5–25.

Coffey, W.T., Kalmykov, Yu.P. and Titov, S.V. (2002a). Inertial effects in anomalous dielectric relaxation, *Phys. Rev. E* **65**, 032102/1–4.

Coffey, W.T., Kalmykov, Yu.P. and Titov, S.V. (2002b). Inertial effects in anomalous dielectric relaxation of rotators in space, *Phys. Rev. E* **65**, 051105/1–9.

Coffey, W.T., Kalmykov, Yu.P. and Walderom, J.T. (2004). *The Langevin Equation with Applications to Stochastic Problems in Physics, Chemistry and Electrical Engineering*, Second edition, World Scientific, Singapore. [First edition (1996)]

Coffey, W.T., Kalmykov, Yu.P. and Titov, S.V. (2002a). Inertial effects in anomalous dielectric relaxation, *J. Molecular Liquids* **114**, 35–41.

Coffey, W.T., Crothers, D.S E., Holland, D. and Titov, S.V. (2004c). Green function for the diffusion limit of one-dimensional continuous time random walks, *J. Molecular Liquids* **114**, 165–171.

Cole, K.S. (1933). Electrical conductance of biological systems, Electrical excitation in nerves, in *Proceedings Symposium on Quantitative Biololgy*, Cold Spring Harbor, New York, Vol. 1, pp. 107–116.

Cole, K.S. and Cole, R.H. (1941). Dispersion and absorption in dielectrics, I. Alternating current characteristics, *J. Chemical Physics* **9**, 341–349.

Cole, K.S. and Cole, R.H. (1942). Dispersion and absorption in dielectrics, II. Direct current characteristics, *J. Chemical Physics* **10**, 98–105.

Cole, R.H. (1955). On the analysis of dielectric relaxation measurements, *J. Chemical Physics* **23**, 493–499.

Cole, R.H. and Davidson, D.W. (1952). High frequency dispersion in n-propanol, *J. Chemical Physics* **20**, 1389–1391.

Coleman, B. (1961). Thermodynamics of materials with memory, *Arch. Rat. Mech. Anal.* **17**, 1–46.

Coleman, B. and Gurtin, M.E. (1965). Waves in materials with memory, II, *Arch. Rat. Mech. Anal.* **19**, 239–265.

Coleman, B., Gurtin, M.E. and Herrera, I.R. (1965). Waves in materials with memory, I, *Arch. Rat. Mech. Anal.* **19**, 1–19.

Coleman, B. and Noll, W. (1961). Foundations of linear viscoelasticity, *Rev. Mod. Phys.* **33**, 239–249. [Errata **36** (1964), 1103.]

Collins, F. and Lee, C.C. (1956). Seismic wave attenuation characteristics from pulse experiments, *Geophysics* **1**, 16–40.

Compte, A. (1996). Stochastic foundations of fractional dynamics. *Phys. Rev. E* **53**, 4191–4193.

Compte, A., Jou D., Katayama, Y. (1997). Anomalous diffusion in linear shear flows. *J. Phys. A: Math. Gen.* **30**, 1023–1030.

Conlan, J. (1983). Hyperbolic differential equations of generalized order, *Appl. Anal.* **14** No 3, 167–177.

Corinaldesi, E. (1957). Causality and dispersion relations in wave mechanics, *Nuclear Physics* **2**, 420–440.

Courant, R. (1936). *Differential and Integral Calculus*, Vol. 2, Nordemann, New York, pp. 339–341.

Craiem, D., Rojo, F.J., Atienza, J.M., Armentano, R.L. and Guinea, G.V. (2008). Fractional-order viscoelasticity applied to describe uniaxial stress relaxation of human arteries, *Phys. Med. Biol.* **53**, 4543–4554.

Cristescu, N. (1986). *Rock Rheology*, Kluwer, Dordrecht.

Crothers, D.S.E., Holland, D., Kalmykov, Yu.P. and Coffey, W.T. (2004). The role of Mittag-Leffler functions in anomalous relaxation, *J. Molecular Liquids* **114**, 27–34.

Cruden, D.M. (1971a). The form of the creep law for rock under uniaxial compression, *Int. J. Rock Mech. Mining Sci. (& Geomechanics)* **8**, 105–126.

Cruden, D.M. (1971b). Single-increment creep experiments on rock under uniaxial compression, *Int. J. Rock Mech. Mining Sci. (& Geomechanics)* **8**, 127–141.

Crump, K.S. (1976). Numerical inversion of Laplace transforms using a Fourier series approximation, *Journal of the Association for Computing Machinery* **23**, 89–96.

Dattoli, G., Ricci, P.E., Cesarano, C. and Vázquez, L. (2002). Fractional operators, integral representations and special polynomials, *Int. J. Appl. Math.* **10**, 131–139.

Dattoli, G., Ricci, P.E., Cesarano, C. and Vázquez, L. (2003). Special polynomials and fractional calculus, *Math. Comput. Modelling* **37** No 7-8, 729–733.

Dattoli, G., Cesarano, C., Ricci, P.E. and Martinez, L.V. (2004). Fractional derivatives: Integral representations and generalized polynomials, *J. of Concrete and Applicable Mathematics* **2**, 59–66.

Davidson, D.W. and Cole, R.H. (1950). Dielectric relaxation in glycerine, *J. Chemical Physics* **18**, 1417. [Letter to the Editor]

Davidson, D.W. and Cole, R.H. (1951). Dielectric relaxation in glycerol, propylene glycol and n-propanol, *J. Chemical Physics* **19**, 1484–1490.

Davies, B. (2002). *Integral Transforms and Their Applications*, Third Edition, Springer Verlag, New York.

Davies, R.M. (Editor) (1960). *International Symposium Stress Wave Propagation in Materials, 1959*, Interscience, New York.

Davies, A.R. and Anderssen, R.S. (1997). Sampling localization in determining the relaxation spectrum, *J. Non-Newtonian Fluid Mech.* **73** (1997), 163–179.

Davis, H.T. (1936). *The Theory of Linear Operators*, The Principia Press, Bloomington, Indiana.

Davis, P.J. (1959). Leonard Euler's integral: a historical profile of the Gamma function, *The American Mathematical Monthly* **66**, 849–869.

Day, W. (1968). Thermodyanamics based on a work axiom, *Arch. Rat. Mech. Anal.* **31**, 1–34.

Day, W. (1970). On monotonicity of the relaxation functions of viscoelastic materials, *Proc. Cambridge Phil. Soc.* **67**, 503–508.

Day, W. (1970b). Restrictions on the relaxation functions in linear viscoelasticity, *Quart. J. Mech. and Appl. Math.* **24**, 487–497.

Day, S.M. and Minster, J.B. (1984). Numerical simulation of wave-fields using a Padé approximant method, *Geophys. J. R. Astron. Soc.* **78**, 105–118.

Debnath, L. (2003a). Fractional integral and fractional differential equations in fluid mechanics, *Fractional Calculus and Applied Analysis* **6**, 119–155.

Debnath, L. (2003b). Recent developments in fractional calculus and its applications to science and engineering, *Internat. Jour. Math. and Math. Sci.* 2003, 1–30.

Debnath, L., (2004). A brief historical introduction to fractional calculus, *Internat. J. Math. Edu. Sci. & Technol*, **35**, 487–501.

Debnath, L and Bhatta, D. (2007). *Integral Transforms and Their Applications*, Second Edition, Chapman & Hall/CRC, Boca Raton.

Debnath, L and Grum, W.J (1988). The fractional calculus and its role in the synthesis of special functions: Part I, *International Journal of Mathematical Education in Science and Technology* **19** No 2, 215–230.

Del Piero, G. (2004). The relaxed work in linear viscoelasticity, *Mathematics Mech. Solids* **9** No 2, 175–208.

Del Piero, G. and Deseri, L. (1995). Monotonic, completely monotonic and exponential relaxation in linear viscoelasticity, *Quart. Appl. Math.* **53**, 273–300.

Del Piero, G. and Deseri, L. (1997). On the concept of state in linear viscoelasticity, *Arch. Rat. Mech. Anal.* **138**, 1–35.

Del Piero, G. and Deseri, L. (1996). On the analytic expression of the free energy in linear viscoelasticity, *J. Elasticity* **47**, 247–278.

De Oliveira Castro, F.M. (1939). Zur Theorie der dielektrischen Nachwirkung, *Zeits. f. Physik* **114**, 116–126.

Desch, W. and Grimmer, R. (1989). Singular relaxation moduli and smoothing in three-dimensional viscoelasticity, *Trans. Amer. Math. Soc.* **314**, 381–404.

Deng, R., Davies, P. and Bajaj, A.K. (2004). A case study on the use of fractional derivatives: the low frequency viscoelastic uni-directional behavior of polyurethane foam, *Nonlinear Dynamics* **38** No 1-2, 247–265.

Derks, G. and Van Groesen, E. (1992). Energy propagation in dissipative systems, Part II: Centrovelocity for nonlinear systems, *Wave Motion* **15**, 159–172.

Desch, W. and Grimmer, R. (1986). Propagation of singularities for integro-differential equations. *J. Diff. Eqs.* **65**, 411–426.

Desch, W. and Grimmer, R. (1989a). Singular relaxation moduli and smoothing in three-dimensional viscoelasticity, *Trans. Am. Math. Soc.* **314**, 381–404.

Desch, W. and Grimmer, R. (1989b). Smoothing properties of linear Volterra integro-differential equations. *SIAM J. Math. Anal.* **20**, 116–132.

Deseri, L., Fabrizio, M. and Golden, M. (2006). The concept of a minimal state in viscoelasticity: new free energies and applications to PDEs, *Arch. Rational Mech. Anal.* **181**, 43–96.

Diethelm, K. (2007). Smoothness properties of solution of Caputo-type fractional differential equations, *Fractional Calculus and Applied Analysis* **10** No 2, 151–160.

Diethelm, K. (2008a). An investigation of some no-classical methods for the numerical approximation of Caputo-type fractional derivatives, *Numer. Algorithms* **47**, 361–390.

Diethelm, K. (2008b). Multi-term fractional differential equations, multi-order fractional differential systems and their numerical solution, *J. Europ. Syst. Autom.* **42**, 665–676.

Diethelm, K. (2009). *The Analysis of Fractional Differential Equations*, Lecture Notes in Mathematics, Springer, Berlin, forthcoming.

Diethelm, K. and Ford, N.J. (2001). Numerical solution methods for distributed order differential equations, *Fractional Calculus and Applied Analysis* **4** No 4, 531–542.

Diethelm, K. and Ford, N.J. (2002). Numerical solution of the Bagley-Torvik equation, *BIT (Numerical Mathematics)* **42** No 3, 490–507.

Diethelm, K. and Ford, N.J. (2002). Analysis of fractional differential equations, *J. Math. Anal. Appl.* **265**, 229–248.

Diethelm, K. and Ford, N.J. (2004). Multi-order fractional differential equations and their numerical solution, *Applied Mathematics and Computation* **154**, 621–640.

Diethelm, K.and Ford, N.J. (2009). Numerical analysis for distributed order differential equations, *J. Comput. Appl. Math.* **225**, 96–104.

Diethelm, K., Freed, A.D., Ford, N., 2002. A predictor - corrector approach to the numerical solution of fractional differential equations. *Nonlinear Dynamics* **22**, 3–22.

Diethelm, K., Ford, N. J., Freed, A. D. and Luchko, Yu. (2005). Algorithms for the fractional calculus: a selection of numerical methods, *Comput. Methods Appl. Mech. Engrg.* **194**, 743–773.

Dietrich, L., Lekszycki, T. and Turski, K. (1998). Problems of identification of mechanical characteristics of viscoelastic composites, *Acta Mechanica* **126**, 153–167.

Di Paola, M. and Zingales, M. (2008). Long-range cohesive interactions of non-local continuum faced by fractional calculus, *Int. J. of Solids and Structures* **45**, 5642–5659.

Dissado, L. A. and Hill, R.M. (1979). Non-exponential decay in dielectric and dynamics of correlated systems, *Nature* **279**, 685–689.

Djordjević, V.D., Jarić, J., Fabry, B., Fredberg, J.J. and Stamenović, D. (2003). Fractional derivatives embody essential features of cell rheological behavior, *Annals of Biomedical Engineering* **31** No 6, 692–699.

Doetsch, G. (1950)-(1955). *Handbuch der Laplace-Transformation*, I and II, Birkhäuser, Basel.

Doetsch, G. (1974). *Introduction to the Theory and Application of the Laplace Transformation*, Springer Verlag, Berlin.

Doi M. and Edwards, S.F. (1986). *The Theory of Polymer Dynamics*, Oxford University Press, Oxford.

Donato, R.J. and O'Brien, P.N.S. (1962). Absorption and dispersion of elastic energy in rocks, *Nature* **193**, 764–765.

Dostanić, M. (1996). A basic property of families of the Mittag-Leffler functions, *Mat. Vesn.* **48**, 77–82.

Dovstam, K. (2000). Augmented Hooke's law based on alternative stress relaxation, *Computational Mechanics* **26** No 1, 90–103.

Drozdov, A.D. (1997). Fractional differential models in finite viscoelasticity, *Acta Mech.* **124** No 1-4, 155–180.

Drozdov, A. (1998). *Viscoelastic Structures*, Academic Press, New York.
Douglas, J.F. (2000). Polymer science applications of path-integration, integral equations and fractional calculus, in [Hilfer (2000a)], pp. 241–330.
Dubkov, A.A., Spagnolo, B. and Uchaikin, V.V. (2008). Levy flight superdiffusion: an introduction, *International Journal of Bifurcation and Chaos* **18** No. 9, 2649–2672.
Durbin, F. (1974). Numerical inversion of Laplace transforms: an efficient improvement to Dubner and Abate's method, *The Computer Journal* **17**, 371–376.
Dzherbashyan, M.M. (1966). *Integral Transforms and Representations of Functions in the Complex Plane*, Nauka, Moscow. [in Russian]. There is also the transliteration as Djrbashyan.
Dzherbashyan, M.M. (1993). *Harmonic Analysis and Boundary Value Problems in the Complex Domain*, Birkhäuser Verlag, Basel.
Dzherbashyan, M.M. and Bagian, R.A. (1975). On integral representations and measures associated with Mittag-Leffler type functions, *Izv. Akad. Nauk Armjanskoy SSR, Matematika* **10**, 483–508. [in Russian]
Dzherbashyan, M.M. and Nersesyan, A.B. (1958). Some integro-differential operators, *Dokl. Acad. Nauk SSSR* **121** No 2, 210–213. [In Russian]
Dzherbashyan, M.M. and Nersesyan, A.B. (1968). Fractional derivatives and the Cauchy problem for differential equations of fractional order. *Izv. Acad. Nauk Armjanskvy SSR, Matematika* **3**, 3–29. [In Russian]
Eidelman, S.D., Ivasyshen, S.D. and Kochubei, A.N. (2004). *Analytic Methods in the Theory of Differential and Pseudo-Differential Equations of Parabolic Type*, Birkhäuser, Basel.
Eidelman, S.D. and Kochubei, A.N. (2004). Cauchy problem for fractional diffusion equations, *J. Diff. Eqs.* **199**, 211–255.
Eldred, L.B., Baker, W.P., and Palazzotto, A.N. (1995). Kelvin-Voigt vs fractional derivative model as constitutive relations for viscoelastic materials, *AIAA Journal* **33**, 547–550.
Eldred, L.B., Baker, W.P., and Palazzotto, A.N. (1996). Numerical application of fractional derivative model constitutive relations for viscoelastic materials, *Comput. Struct* **60** No 6, 975–882.
Eirich, F.R., Editor (1967). *Rheology, Theory and Applications*, Vol. 4, Academic Press, New York.
Elices, M. and García-Moliner, F. (1968), Wave packet propagation and frequency-dependent internal friction, in Mason, W.P. (Editor), *Physical Acoustics* Vol. 5, Academic Press, New York, pp. 163–219.
El-Sayed, A.M.A. (1995). Fractional-order evolution equations, *J. Fractional Calculus* **7**, 89–100.
El-Sayed, A.M.A. (1996). Fractional-order diffusion-wave equation, *Int. J. Theor. Phys.* **35**, 311–322.

Emmerich, M.and Korn, M. (1987). Incorporation of attenuation into time-domain computation of seismic wave-fields. *Geophysics* **52**, 1252–1264.

Enelund, M., Fenander, A. and Olsson, P.A. (1997). A fractional integral formulation of constitutive equations of viscoelasticity, *AIAA J.* **35**, 1356–1362.

Enelund, M., and Lesieutre, G.A. (1999). Time domain modeling of damping using anelastic displacement fields and fractional calculus, *Int. J. Solids Structures* **36**, 4447–4472.

Enelund, M., Mähler, L., Runesson, K. and Josefson, B.L. (1999a). Formulation and integration of the standard linear viscoelastic solid with fractional order rate laws, *Int. J. Solids Structures* **36**, 2417–2442.

Enelund, M. and Olsson, P. (1999). Damping described by fading memory - analysis and application to fractional derivative models, *Int. J. Solids Structures* **36**, 939–970.

Engler, H. (1997). Similarity solutions for a class of hyperbolic integro-differential equations, *Differential Integral Equations* **10**, 815–840.

Engler, H. (2005). Asymptotic self-similarity for solutions of partial integro-differential equations, E-print arXiv:math.AP/0510206v1, 10 Oct. 2005 pp. 21.

Erdélyi, A. (1956). *Asymptotic Expansions*, Dover, New York.

Erdélyi, A., Magnus, W., Oberhettinger, F and Tricomi, F. G. (1953-1954). *Tables of Integral Transforms*, 2 Volumes, McGraw-Hill, New York [Bateman Project].

Erdélyi, A., Magnus, W., Oberhettinger, F and Tricomi, F. G. (1953-1955). *Higher Transcendental Functions*, 3 Volumes, McGraw-Hill, New York [Bateman Project].

Escobedo-Torres, J., and Ricles, J. M. (1998). The fractional order elastic-viscoelastic equations of motion: formulation and solution methods, *J. Intell. Mater. Syst. Struct.* **9** No 7, 489–502.

de Espindola, J.J., da Silva Neto, J.M. and Lopes E.M.O. (2005). A generalized fractional derivative approach to viscoelastic material properties measurement, *Appl. Mathematics and Computation* **164**, 493–506.

Evgrafov, M.A. (1961), *Asymptotic Estimates and Entire Functions*, Gordon and Breach, New York.

Fabrizio, M. (1968). Sulla equazione di propagazione nei mezzi viscoelastici, *Rend. Sci. Istituto Lombardo* **A 102**, 437–441.

Fabrizio, M., Giorgi, C. and Morro, A. (1994). Free energies and dissipation properties for systems with memory, *Arch. Rat. Mech. Anal.* **125**, 341–373.

Fabrizio, M., Giorgi, C. and Morro, A. (1995). Internal dissipation, relaxation property and free energy in materials with fading memory, *J. Elasticity* **40**, 107–122.

Fabrizio, M., Giorgi, C. and Pata, V. (2009). A new approach to equations with memory, E-print arXiv:0901.4063v1 [math.DS], 26 Jan. 2009, pp. 39.

Fabrizio, M. and Golden, M.J. (2002). Maximum and minimum free energies for a linear viscoelastic material, *Quart. Appl. Math.* **60**, 341–381.

Fabrizio, M. and Golden, M.J. (2003). Minimum free energies for materials with finite memory, *J. Elasticity* **72**, 121–143.

Fabrizio, M. and Lazzari, B. (1991). On the existence and the asymptotic stability of solutions for linearly viscoelastic solids, *Arch. Rat. Mech. Anal.* **116**, 139–152.

Fabrizio, M. and Morro, A. (1985). Thermodynamic restrictions on relaxation functions in linear viscoelasticity, *Mech. Res. Comm.* **12**, 101–105.

Fabrizio, M. and Morro, A. (1988). Viscoelastic relaxation functions compatible with thermodynamics, *J. Elasticity* **19**, 63–75.

Fabrizio, M. and Morro, A. (1989). Fading memory spaces and approximate cycles in linear viscoelasticity, *Rend. Sem. Mat. Univ. Padova* **82**, 239–255.

Fabrizio, M. and Morro, A. (1991). Reversible processes in the thermodynamics of continuous media, *J. Nonequilbrium Thermodynamics* **16**, 1–12.

Fabrizio, M. and Morro, A. (1992). *Mathematical Problems in Linear Viscoelasticity*, SIAM, Philadelphia.

Farrell, O. J. and B. Ross, B. (1963). *Solved Problems: Gamma and Beta functions, Legendre Polynomials, Bessel Functions*, MacMillan, New York.

Farrell, O. J. and B. Ross, B. (1971). *Solved Problems in Analysis*, Dover Publications, New York.

Feller, W. (1952). On a generalization of Marcel Riesz' potentials and the semi-groups generated by them. *Meddelanden Lunds Universitets Matematiska Seminarium* (Comm. Sém. Mathém. Université de Lund), Tome suppl. dédié à M. Riesz, Lund, pp. 73–81.

Feller, W. (1971). *An Introduction to Probability Theory and its Applications*, Wiley, New York, Vol. II, Second Edition. [First edition (1966)]

Fernander, A. (1996). Modal syntesis when modelling damping by use of fractional derivatives, *AIAA J.* **34** No 5, 1051–1058.

Fernander, A. (1998). A fractional derivative railpad model included in a railway track model, *J. Sound Vibr.* **212**, 889–903.

Ferry, J. D. (1980). *Viscoelastic Properties of Polymers*, Second edition Wiley, New York.

Fichera, G. (1979a). Analytical problems of hereditary phenomena, in Graffi, D. (Dditor) *Materials with Memory*, Liguori, Napoli, pp. 111–169. [CIME Lectures, Bressanone, 2-11 June 1977]

Fichera, G. (1979b). Avere una memoria tenace crea gravi problemi, *Arch. Rat. Mech. Anal.* **70**, 101–112.

Fichera, G. (1982). Sul principio della memoria evanescente, *Rend. Sem. Mat. Univ. Padova* **68**, 245–259.

Fichera, G. (1985). On linear viscoelasticity, *Mech. Res. Comm.* **12**, 241–242.

Figueiredo Camargo, R. (2009). *Càlculo Fracionàrio e Aplicacoes*, PhD Thesis in Portuguese, Universidade Estadual de Campinas, Instituto de Matemàtica, Estatistica e Computacao Cientifica, March 2009. [Supervisor: Prof. Capelas de Oliveira, E.]

Figueiredo Camargo, R., Chiacchio, A.O. and Capelas de Oliveira, E. (2008). Differentiation to fractional orders and the fractional telegraph equation, *Journal of Mathematical Physics* **49**, 033505/1–11.

Figueiredo Camargo, R., Charnet, R. and Capelas de Oliveira, E. (2009a). On some fractional Greens functions, *Journal of Mathematical Physics* **50**, 043514/1–12.

Figueiredo Camargo, R., Chiacchio, A.O., Charnet, R. and Capelas de Oliveira, E. (2009b). Solution of the fractional Langevin equation and the Mittag-Leffler functions, *Journal of Mathematical Physics* **50**, 063507/1–8.

Findley, W.N., Lai, J.S. and Onaran, K. (1976). *Creep and Relaxation of Nonlinear Viscoelastic Materials with an Introduction to Linear Viscoelasticity*, North-Holland, Amsterdam.

Flügge, W. (1975). *Viscoelasticity*, Springer Verlag, Berlin, Second edition.

Fox, C. (1961). The G and H functions as symmetrical Fourier kernels, *Trans. Amer. Math. Soc.* **98**, 395–429.

Freed, A. and Diethelm, K. (2007a). Fractional calculus in Biomechanics: a 3D viscoelastic model using regularized fractional derivative kernels with application to the human calcaneal fat pad, *Biomechanics and Modeling in Mechanobiology* **5** No 4, 203–215.

Freed, A. and Diethelm, K. (2007b). Caputo derivatives in viscoelasticity: a nonlinear finite-deformation theory for tissue, *Fractional Calculus and Applied Analysis* **10** No 3, 219–248.

Freed, A., Diethelm, K. and Luchko, Yu. (2002). *Fractional-order Viscoelasticity (FOV): Constitutive Development using the Fractional Calculus*, First Annual Report, NASA/TM-2002-211914, Gleen Research Center, pp. XIV – 121.

Friedlander, F.G. and Keller, J.B. (1955). Asymptotic expansions of solutions of $(\Delta^2 + k^2)u = 0$. *Comm. Pure Appl. Math.* **8**, 387–394.

Friedrich, Chr. (1991a). Relaxation functions of rheological constitutive equations with fractional derivatives: thermodynamic constraints, in: Casaz-Vazques, J. and Jou, D. (Editors), *Rheological Modelling: Thermodynamical and Statistical Approaches*, Springer Verlag, Berlin, pp. 321–330. [Lecture Notes in Physics, Vol. 381]

Friedrich, Chr. (1991b). Relaxation and retardation functions of the Maxwell model with fractional derivatives, *Rheol. Acta* **30**, 151–158.

Friedrich, Chr. (1992). Rheological material functions for associating comb-shaped or H-shaped polymers. A fractional calculus approach, *Phil. Mag. Lett.* **66**, 287–292.

Friedrich, Chr. (1993). Mechanical stress relaxation in polymers: fractional integral model versus fractional differential model, *J. Non-Newtonian Fluid Mech.* **46**, 307–314.

Friedrich, Chr., Schiessel, H. and Blumen, A. (1999). Constitutive behavior modeling and fractional derivatives, in Siginer, D.A., Kee, D. and Chhabra, R.P. (Editors), *Advances in the Flow and Rheology of Non-Newtonian Fluids*, Elsevier, Amsterdam, pp. 429–466.

Friedrich, Chr. and Braun, H. (1992). Generalized Cole-Cole behavior and its rheological relevance, *Rheol. Acta* **31** No 4, 309–322.

Friedrich, Chr. and Braun, H. (1994). Linear viscoelastic behaviour of complex materials: a fractional mode representation, *Colloid Polym Sci.* **272**, 1536–1546.

Fröhlich, H. (1958). *Theory of Dielectrics*, Oxford University Press, Oxford, Second Edition.

Fujita, Y. (1990a). Integro-differential equation which interpolates the heat equation and the wave equation, I, II, *Osaka J. Math.* **27**, 309–321, 797–804.

Fujita, Y. (1990b). Cauchy problems of fractional order and stable processes, *Japan J. Appl. Math.* **7**, 459–476.

Fujita, Y. (1992). Energy inequalities for integro-partial differential with Riemann-Liouville integral, *SIAM J. Math. Anal* **23**, 1182–1188.

Fujiwara, S. and Yonezawa, F. (1992) Anomalous relaxation in fractal structures, *Phys. Rev. E* **51**, 2277–2285.

Fukunaga, M. (2002). On initial value problems in fractional differential equations, *Int. J. Appl. Mathematics* **9**, 219–236.

Fukunaga, M. and Shimizu, N. (2004). Role of prehistories in the initial value problems of fractional viscoelastic equations, *Nonlinear Dynamics* **38**, 207–220.

Fukunaga, M., Shimizu, N. and Nasuno, H. (2009). A nonlinear fractional derivative model of impulse motion for viscoelastic materials, *Physica Scripta* **T136**, 014010/1–6.

Fung, Y. C. (1965). *Foundation of Solid Mechanics*, Prentice-Hall, Englewood.

Fuoss, R.M. and Kirkwood, J.G. (1941). Anomalous dispersion and dielectric loss in polar polymers, *J. Chem. Phys.* **9**, 329–340.

Futterman, W.I. (1962). Dispersive body waves, *J. Geophys. Res.* **67**, 5279–5291.

Galucio, A.C., Deü,J.-F. and Ohayon, R. (2004). Finite element formulation of viscoelastic sandwich beams using fractional derivative operators, *Computational Mechanics* **33**, 282–291.

Galucio, A.C., Deü, J.-F. and Ohayon, R. (2005). A fractional derivative viscoelastic model for hybrid active-passive damping treatments in time domain-application to sandwich beams, *J. Intell. Mater. Syst. Struct.* **16** No 1, 33–45.

Gamby, D. (1975). Solution of one-dimensional impact problems, *Mech. Research Comm.* **2**, 131–135.

Garcia-Franco, C.A. and Mead, D.W. (1999). Rheological and molecular characterization of linear backbone flexible polymers with the Cole–Cole model relaxation spectrum, *Rheologica Acta*, **38**, 34–47.

Garret, C.G.B. and McCumber, D.E. (1970). Propagation of a Gaussian light pulse through an anomalous dispersion medium, *Phys. Rev. A* **1**, 305–313.

Gatteschi, L. (1973). *Funzioni Speciali*, UTET, Torino.

Gaul, L. (1999). The influence of damping on waves and vibrations, *Mech. Systems Signal Processing* **13**, 1–30.

Gaul, L., Bohlen, S. and Kempfle, S. (1985). Transient and forced oscillations of systems with constant hysteretic damping, *Mechanics Research Communications* **12**, 187–201.

Gaul, L., Kempfle, S. and Klein, P. (1990). Transientes Schwingungsverhalten bei der Dämpfungsbeschreibung mit nicht ganzzahligen Zeitableitungen, *Z. angew. Math. Mech. (ZAMM)* **70**, T139–T141.

Gaul, L., Klein, P. and Kempfle, S. (1989). Impulse response function of an oscillator with fractional derivative in damping description, *Mechanics Research Communications* **16**, 297–305.

Gaul, L., Klein, P. and Kempfle, S. (1991). Damping description involving fractional operators, *Mechanical Systems and Signal Processing* **5**, 81–88.

Gaul, L. and Schanz, M. (1994). Dynamics of viscoelastic solids treated by boundary element approaches in time domain, *European J. Mech. A/Solids* **13**, 43–59.

Gaul, L. and Schanz, M. (1997). Calculation of transient response of viscoelastic solids based on inverse transformation, *Meccanica* **32**, 171–178.

Gaunaurd, G.C., Madigosky, W., Überall, H. and Dragonette, L.R. (1982). Inverse scattering and the resonances of viscoelastic and electromagnetic waves, in [Mainardi (1982)], pp. 235–257.

Gaunaurd, G. C. and Überall, H. (1978). Theory of resonant scattering from spherical cavities in elastic and viscoelastic media, *J. Acoust. Soc. Am.* **63**, 1699–1712.

Gautschi, W. (1998). The incomplete Gamma functions since Tricomi, in *Atti Convegni Lincei 147, Tricomi's Ideas and Contemporary Applied Mathematics*, Proceedings of the International Workshop in the occasion of the centenary of the birth of Francesco G. Tricomi (Rome 28-29 November, Torino 1-2 December 1997). Accademia Naz. Lincei, Roma, pp. 203–237.

Gautschi, W. and Giordano, C. (2008). Luigi Gatteschis work on asymptotics of special functions and their zeros, *Numer. Algor.* (2008) **49**, 11–31.

Gawronski, W. (1984). On the bell-shape of stable distributions, *Annals of Probability* **12**, 230–242.

Gel'fand, I.M. and Shilov, G.E. (1964). *Generalized Functions*, Vol. 1, Academic Press, New York.

Geluk, J.L. and de Haan, L. (1980). *Regular Variation, Extensions and Tauberian Theorems*, CWI Tract **40**, Centrum voor Wiskunde en Informatica.

Gemant, A. (1936). A method of analyzing experimental results obtained from elastiviscous bodies, *Physics* **7**, 311–317.

Gemant, A. (1938). On fractional differentials, *Phil. Mag.* (Ser. 7) **25**, 540–549.

Gemant, A. (1942). Frictional phenomena: VIII, (1942) *J. Appl. Phys.* **13**, 210–221.

Gemant, A. (1950). *Frictional Phenomena*, Chemical Publ. Co, Brooklyn N.Y.

Gentili, G. (2002). Maximum recoverable work, minimum free energy and state space in linear viscoelasticity, *Quart. Appl. Math.* **60**, 153–182.

Gerasimov, A. (1948). A generalization of linear laws of deformation and its applications to problems of internal friction, *Prikl. Matem. i Mekh. (PMM)* **12** No 3, 251–260. [in Russian]

Gere, J.M. and Timoshenko, S.P. (1997). *Mechanics of Materials*, Fourth Edition, PWS Publishing Company, Boston.

Germano, G., Politi, M., Scalas, E. and Schilling, R.E. (2009). Stochastic calculus for uncoupled continuous-time random walks, *Phys. Rev. E* **79**, 066102/1–12.

Ghizzetti, A. and Ossicini, A. (1971). *Trasformate di Laplace e Calcolo Simbolico*, UTET, Torino. [in Italian]

Giannantoni, C. (2003). The problem of initial conditions and their physical meaning in linear differential equations of fractional order, *Appl. Math. Comp.* **141**, 87–102.

Giona, M. and Roman, H. E. (1992a). Fractional diffusion equation for transport phenomena in random media, *Physica A* **185**, 82–97.

Giona, M. and Roman, H. E. (1992b). A theory of transport phenomena in disordered systems, *The Chemical Engineering Journal* **49**, 1–10.

Giona, M., Cerbelli, S. and Roman, H.E. (1992). Fractional diffusion equation and relaxation in complex viscoelastic materials, *Physica A* **191**, 449–453.

Giorgi, G. (1912a). Sui problemi della elasticità ereditaria, *Atti Reale Accad. Lincei* (Ser. V) **21**, 412–418.

Giorgi, G. (1912b). Sugli operatori funzionali ereditari, *Atti Reale Accad. Lincei* (Ser. V) **21**, 683–687.

Giovagnoni, M. and Berti, G. (1992). A fractional derivative model for single-link mechanism vibration, *Meccanica* **27** No 2, 131–138.

Glöckle, W.G. and Nonnenmacher, T.F. (1991). Fractional integral operators and Fox functions in the theory of viscoelasticity, *Macromolecules* **24**, 6426–6434.

Glöckle, W.G. and Nonnenmacher, T.F. (1993). Fox function representation of non-Debye relaxation processes, *J. Stat. Phys.* **71**, 741–757.

Glöckle, W.G. and Nonnenmacher, T.F. (1994). Fractional relaxation and the time-temparature superposition principle, *Reological Acta*, **33**, 337–343.

Glöckle, W.G. and Nonnenmacher, T.F. (1995). A fractional calculus approach to self-similar protein dynamics, *Biophysical Journal*, **68**, 346–353.

Gnedenko, B.V. and Kolmogorov A.N. (1954). *Limit Distributions for Sums of Independent Random Variables*, Addison-Wesley, Cambridge, Mass. [Translated from the Russian edition, Moscow 1949, with notes by K.L. Chung, revised (1968)]

Gnedenko, B.V. and Kovalenko, I.N. (1968). *Introduction to Queueing Theory*, Israel Program for Scientific Translations, Jerusalem. [Translated from the 1966 Russian edition]

Golden, J.M. and Graham, G.A.C. (1988). *Boundary Value Problems in Linear Viscoelasticity*, Springer-Verlag, Berlin.

Goloviznin, V.M., and Korotkin, I.A. (2006). Numerical methods for some one-dimensional equations with fractional derivatives, *Differential Equations* **42** No 7, 967–973.

Gomes, M. I. and Pestana, D. D. (1978). The use of fractional calculus in probability, *Portugaliae Mathematica* **37** No 3-4, 259–271.

Gomi, S. and Yonezawa, F. (1995). Anomalous relaxation in the fractal time random walk model, *Phys. Rev. Lett.* **74**, 4125–4128; Erratum, *Phys. Rev. Lett.* **75**, 2454.

Gonsovski, V.L., Meshkov, S.M. and Rossikhin, Yu.A. (1972). Impact of a viscoelastic onto a rigid target, *Sov. Appl. Mekh.* **8** (10), 1109–1113.

Gonsovski, V.L. and Rossikhin, Yu.A. (1973). Stress waves in a viscoelastic medium with a singular hereditary kernel, *J. Appl. Mech. Tech. Physics* **14** No 4, 595–597.

González, M.O. (1992). *Complex Analysis. Selected Topics*, Marcel Dekker, New York.

Gordon, L. (1994). The stochastic approach to the gamma function, *Amer. Math. Monthly* **101**, 858–865.

Gordon, R.B. and Davis, L.A. (1968). Velocity and attenuation of seismic waves in imperfectly elastic rock, *J. Geophys. Res.* **73**, 3917–3935.

Gordon, R.B. and Nelson, C.W. (1966). Anelastic properties of the Earth, *Rev. Geophys.* **4**, 457–474.

Gorenflo, R. (1996). *Abel Integral Equations with special emphasis on Applications*, Lectures in Mathematical Sciences Vol. 13, The University of Tokyo, Graduate School of Mathematical Sciences.

Gorenflo, R. (1997). Fractional calculus: some numerical methods, in: Carpinteri, A. and Mainardi, F. (Editors) *Fractals and Fractional Calculus in Continuum Mechanics*, Springer Verlag, Wien, pp. 277–290. [Downloadable from http://www.fracalmo.org]

Gorenflo, R. and Abdel-Rehim, E.A. (2004). From power laws to fractional diffusion, *Vietnam J. of Mathematics* **32** SI, 65–75.

Gorenflo, R. and Abdel-Rehim, E.A. (2005). Discrete models of time-fractional diffusion in a potential well, *Fractional Calculus and Applied Analysis* **8** No 2, 173–200.

Gorenflo, R. and Abdel-Rehim, E.A. (2007). Convergence of the Grüwald–Letnikov scheme for time–fractional diffusion, *Journal of Computational and Applied Mathematics* **205**, 871–881.

Gorenflo, R., De Fabritiis, G. and Mainardi, F. (1999). Discrete random walk models for symmetric Lévy-Feller diffusion processes, *Physica A* **269** No 1, 79–89.

Gorenflo, R. Iskenderov, A. and Luchko, Yu. (2000). Mapping between solutions of fractional diffusion-wave equations, *Fractional Calculus and Applied Analysis* **3**, 75–86.

Gorenflo, R., Kilbas, A.A. and Rogosin, S.V. (1998). On the generalized Mittag-Leffler type function. *Integral Transforms Spec. Funct.* **7**, No. 3-4, 215–224.

Gorenflo, R., Loutchko, J. and Luchko, Yu. (2002). Computation of the Mittag-Leffler function $E_{\alpha,\beta}(z)$ and its derivatives, *Fractional Calculus and Applied Analysis* **5**, 491–518.

Gorenflo, R., Luchko, Yu. and Mainardi, F. (1999). Analytical properties and applications of the Wright function. *Fractional Calculus and Applied Analysis* **2**, 383–414.

Gorenflo, R., Luchko, Yu. and Mainardi, F. (2000). Wright functions as scale-invariant solutions of the diffusion-wave equation, *Journal of Computational and Applied Mathematics* **118**, 175–191.

Gorenflo, R., Luchko, Yu. and Rogosin, S.V. (1997). Mittag-Leffler type functions: notes on growth properties and distribution of zeros, Preprint No A-97-04, Fachbereich Mathematik und Informatik, Freie Universität Berlin, Serie Mathematik, pp. 23. [E-print: http://www.math.fu-berlin.de/publ/index.html]

Gorenflo, R. and Mainardi, F. (1996). Fractional oscillations and Mittag-Leffler functions, Preprint No A-96-14, Fachbereich Mathematik und Informatik, Freie Universität Berlin, Serie Mathematik, pp. 22. [E-print: http://www.math.fu-berlin.de/publ/index.html]

Gorenflo, R. and Mainardi, F. (1997). Fractional calculus: integral and differential equations of fractional order, in: Carpinteri, A. and Mainardi, F. (Editors) *Fractals and Fractional Calculus in Continuum Mechanics*, Springer Verlag, Wien, pp. 223–276. [E-print: http://arxiv.org/abs/0805.3823]

Gorenflo, R. and Mainardi, F. (1998a). Fractional calculus and stable probability distributions, *Archives of Mechanics* **50**, 377–388.

Gorenflo, R. and Mainardi, F. (1998b). Random walk models for space-fractional diffusion processes. *Fractional Calculus and Applied Analysis* **1**, 167–191.

Gorenflo, R. and Mainardi, F. (1998c). Signalling problem and Dirichlet-Neumann map for time-fractional diffusion-wave equations, *Matimyás Matematika* **21**, 109–118. Pre-print No. A-07/98, Freie Universität Berlin, Serie A Mathematik, pp. 10. [E-print: http://www.math.fu-berlin.de/publ/index.html]

Gorenflo, R. and Mainardi, F. (1999). Approximation of Lévy-Feller diffusion by random walk, *J. Analysis and its Applications (ZAA)* **18** No 2, p 231–246.

Gorenflo, R. and Mainardi, F. Random walk models approximating symmetric space-fractional diffusion processes, in J. Elschner, I. Gohberg and B. Silbermann (Editors), *Problems in Mathematical Physics* (Siegfried Prössdorf Memorial Volume), Birkhäuser Verlag, Boston-Basel-Berlin, pp. 120–145.

Gorenflo, R. and Mainardi, F. (2003). Fractional diffusion processes: probability distributions and continuous time random walk, in G. Rangarajan and M. Ding (Editors), *Processes with Long Range Correlations*, Springer-Verlag, Berlin, pp. 148–166. [Lecture Notes in Physics, No. 621]
E-print http://arxiv.org/abs/0709.3990

Gorenflo, R. and Mainardi, F. (2008). Continuous time random walk, Mittag-Leffler waiting time and fractional diffusion: mathematical aspects, Chap. 4, In: Klages, R., Radons, G. and Sokolov, I.M. (Editors): *Anomalous Transport: Foundations and Applications*, Wiley-VCH, Weinheim, Germany, 2008, pp. 93–127. [E-print http://arxiv.org/abs/0705.0797]

Gorenflo, R. and Mainardi, F. (2009). Some recent advances in theory and simulation of fractional diffusion processes, *Journal of Computational and Applied Mathematics* **229** No 2, 400–415. [E-print: http://arxiv.org/abs/0801.0146]

Gorenflo, R., Mainardi, F., Moretti, D., Pagnini, G. and Paradisi, P. (2002a). Discrete random walk models for space-time fractional diffusion, *Chem. Phys.* **284**, 521–541.

Gorenflo, R., Mainardi, F., Moretti, D. and Paradisi, P. (2002a) Time-fractional diffusion: a discrete random walk approach, *Nonlinear Dynamics* **29**, 129–143.

Gorenflo, R. and Rutman, R. (1994). On ultraslow and intermediate processes, in P. Rusev, I. Dimovski, V. Kiryakova (Editors) *Transform Methods and Special Functions, Sofia 1994*, Science Culture Technology, Singapore, pp. 171–183.

Gorenflo, R. and Vessella, S. (1991). *Abel Integral Equations: Analysis and Applications*, Springer-Verlag, Berlin. [Lecture Notes in Mathematics, Vol. 1461]

Gorenflo, R. and Vivoli, A. (2003). Fully discrete random walks for space-time fractional diffusion equations, *Signal Process.* **83**, 2411–2420.

Gorenflo, R., Vivoli, A. and Mainardi, F. (2004). Discrete and continuous random walk models for space-time fractional diffusion, *Nonlinear Dynamics* **38**, 101–116.

Gottenberg, W.G. and Christensen, R.M.C. (1964). An experiment for determination of the mechanical property in shear for a linear, isotropic viscoelastic solids, *Int. J. Engineering Science* **2**, 45–57.

Graffi, D. (1928a). Sui problemi dell'ereditarietà lineare, *Il Nuovo Cimento* **5**, 53–71.

Graffi, D. (1928b). Sulla teoria delle oscillazioni elastiche con ereditarietà, *Il Nuovo Cimento* **5**, 310–317.

Graffi, D. (1952). Sulla teoria dei materiali elastico-viscosi, *Atti Accad. Ligure Sci. Lett.* **9**, 1–6.

Graffi, D. (1963). Sulla propagazione in mezzi dispersivi, *Ann. Mat. Pura ed Appl.* (Ser. IV) **60**, 173–194.

Graffi, D. (1974). Sull'espressione dell'energia libera nei materiali viscoelastici lineari, *Annali di Matematica Pura ed Applicata* (Ser. IV) **98**, 273–279.

Graffi, D. (1982a). Sull'espressione di alcune grandezze termodinamiche nei materiali con memoria, *Rend. Sem. Mat. Univ. Padova* **68**, 17–29.

Graffi, D. (1982). Mathematical models and waves in linear viscoelasticity, in [Mainardi (1982)], pp. 1–27.
Graffi, D. (1983). On the fading memory, *Appl. Anal.* **15**, 295–311.
Graffi, D. and Fabrizio, M. (1989). Non unicità dell'energia libera per i materiali viscoelastici, *Atti Acc. Naz. Lincei* **83**, 209–214.
Graham, G.A.C. and Walton, J.R., Editors (1995). *Crack and Contact Problems for Viscoelastic Bodies*, Springer-Verlag, Vienna.
Grassi, E. and Mainardi, F. (1993). On attenuation of pressure waves in large arteries, in Barbieri, I., Grassi, E., Pallotti, G. and Pettazzoni, P. (Editors), *Topics on Biomathematics*, World Scientific, Singapore, pp. 265–271.
Griggs, D. (1939). Creep of rocks, *Journal of Geology* **47**, 225–251.
Grigolini, P. (1982). A generalized Langevin equation for dealing with nonadditive fluctuations, *Journal of Statistical Physics* **27** No 2, 283–316.
Grigolini, P., Rocco, A. and West, B.J. (1999). Fractional calculus as a macroscopic manifestation of randomness, *Phys. Rev. E* **59**, 2603–2613.
Grigolini, P., Allegrini, P. and West, B.J. (2007). In search of a theory of complexity: An overview on the Denton workshop, *Chaos, Solitons and Fractals* **34**, 3–10.
Gripenberg, G., Londen, S.-O. and Staffans, O.J. (1990). *Volterra Integral and Functional Equations* Cambridge University Press, Cambridge.
Gromov, V.G. (1967). On the solution of boundary value problems in linear viscoelasticity, *Mekhanika of Polimerov* **3** No 6, 999–1011. [in Russian]
Gromov, V.G. (1971a). Representation of the resolvents of viscoelastic operators by relaxation distribution functions, *Prikl. Mat. i Mekh. (PMM)* **35** No 4, 750–759. [in Russian]
Gromov, V.G. (1971b). On the mathematical contents of Volterra's principle in the boundary value problem of viscoelasticity, *Prikl. Matem. i Mech. (PMM)* **35** No 5, 869–878. [in Russian]
Gross, B. (1941). Theory of dielectric loss, *Phys. Rev.* **59**, 748–750.
Gross, B. (1947a). On creep and relaxation, *J. Appl. Phys.* **18**, 212–221.
Gross, B. (1948). On creep and relaxation, II. *J. Appl. Phys.* **19**, 257–264.
Gross, B. (1951). Relations between creep function and relaxation function, *Journal of Polymer Science* **6**, 123–124.
Gross, B. (1953). *Mathematical Structure of the Theories of Viscoelasticity*, Hermann & C., Paris.
Gross, B. (1956a). Lineare Systeme, *Il Nuovo Cimento* (Ser. X) **3** Suppl. 2, 235–296.
Gross, B. (1956a). Ladder structures for representation of viscoelastic systems. II. *Journal of Polymer Science* **20**, 123–131.

Gross, B. (1956b). Electrical analogs for viscoelastic systems. *Journal of Polymer Science* **20**, 371–380.
Gross, B. (1975). Applications of the Kronig-Kramers relations. *J. Phys. C: Solid State Phys.* **8**, L226–L227.
Gross, B. and Braga, E.P. (1961). *Singularities of Linear System Functions*, Elsevier, Amsterdam.
Gross, B. and Fuoss, R.M. (1956). Ladder structures for viscoelastic systems, *J. Polymer Sci.* **19**, 39–50.
Gross, B. and Pelzer, H. (1951). On creep and relaxation, III. *J. Appl. Phys.* **22**, 1035–1039.
Gross, B. and Pelzer, H. (1952). Relations between delta functions, *Proc. Roy. Soc. (London) A* **210**, 434–437.
Grum, W.J. and Debnath, L (1988). The fractional calculus and its role in the synthesis of special functions: Part II, *International Journal of Mathematical Education in Science and Technology* **19** No 3, 347–362.
Gubreev, G.M. (1987). Basic properties of Mittag-Leffler's type functions, Djarbashyan's transforms and the Mackenhoupt condition. *Functs. Anal. Ego Prilozhen* **21** No 4, 71–72.
Gupta, I.S. and Debnath, L. (2007). Some properties of the Mittag-Leffler functions. *Integral Transforms and Special Functions* **18** No 5, 329–336.
Guillemin, E.A. (1957). *Synthesis of Passive Networks*, Wiley, New York.
Guran, A., Boström, A., Leroy, O. and Maze, G., Editors (1994). *Acoustic Interactions with Submerged Elastic Structures, Part IV: Non-destructive Testing, Acoustic Wave Propagation and Scattering*, World Scientific, Singapore. [Vol. 5 on the Series B on Stability, Vibration and Control of Systems (Series Founder and Editor: A. Guran)]
Gurevich, B. and Lopatnikov, S.L. (1997). Velocity and attenuation of elastic waves in finely layered porous rocks, *Geophys. J. Int.* **121**, 933–947.
Gurtin, M.E. (1963). Variational principles in the linear theory of viscoelasticity, *Arch. Rational Mech. Anal.* **13**, 179–191.
Gurtin, M.E. (1964). Variational principles for linear initial-value problems, *Quart. Appl. Math.* **22**, 252–256.
Gurtin, M.E. (1981). *An Introduction to Continuum Mechanics*, Academic Press, New York.
Gurtin, M.E. and Herrera, I. (1965). On dissipation inequalities and linear viscoelasticity, *Quart. Appl. Math.* **23**, 235–245.
Gurtin, M.E. and Hrusa, W.J. (1988). On energies for nonlinear viscoelastic materials of single-integral type, *Quart. Appl. Math.* **46**, 381–392.
Gurtin, M.E. and Hrusa, W.J. (1991). On the thermodynamics of viscoelastic materials of single-integral type, *Quart. Appl. Math.* **49**, 67–85.

Gurtin, M.E. and Sternberg, E. (1962). On the linear theory of viscoelasticity, *Arch. Rat. Mech. Anal.* **11**, 291–356.
Hamilton, E. L. (1972). Compressional wave attenuation in marine sediments, *Geophysics* **37**, 620–646.
Hanneken, J.W., Achar, B.N.N., Puzio, R., and Vaught, D.M. (2009). Properties of the Mittag-Leffler function for negative α, *Physica Scripta* **T136**, 014037/1–5.
Hanneken, J.W., Vaught, D.M. and Achar, B.N.N. (2005). The number of real zeros of the single parameter Mittag-Leffler function for parameter values between 1 and 2, *Proceedings of IDETC/CIE 2005*: DETC2005-84172, pp. 1–5.
Hanneken, J.W., Vaught, D.M. and Achar, B.N.N. (2007). Enumeration of the real zeros of the Mittag-Leffler function $E_\alpha(z)$, $1 < \alpha < 2$, in [Sabatier *et al.* (2007)], pp. 15–26.
Hanyga, A. (2001a). Multi-dimensional solutions of space-fractional diffusion-wave equation, *Proc. R. Soc. London* **457**, 2993–3005.
Hanyga, A. (2001a). Wave propagation in media with singular memory, *Mathematical and Computer Modelling* **34**, 1399–1421.
Hanyga, A. (2002a). Multi-dimensional solutions of space-time-fractional diffusion-wave equation, *Proc. R. Soc. London* **458**, 429–450.
Hanyga, A. (2002b). Multi-dimensional solutions of time-fractional diffusion-wave equation, *Proc. R. Soc. London* **458**, 933–957.
Hanyga, A. (2002c). Propagation of pulses in viscoelastic media, *Pure Appl. Geophys.* **159**, 1749–1767.
Hanyga, A. (2003a). Anisotropic viscoelastic models with singular memory, *J. Appl. Geophys.* **54**, 411–425.
Hanyga, A. (2003b). Well-posedness and regularity for a class of linear thermoviscoelastic, materials, *Proc. R. Soc. London* **459**, 2281–2296.
Hanyga, A. (2005a). Realizable constitutive equations in linear viscoelasticity, in [Le Méhauté *et al.* (2005)], pp. 353–364.
Hanyga, A. (2005b). Physically acceptable viscoelastic models, in Hutter, K. and Y. Wang, Y. (Editors), *Trends in Applications of Mathematics to Mechanics*, Shaker Verlag GmbH, Aachen. 12 pp. [www.geo.uib.no/hjemmesider/andrzej/index.html]
Hanyga, A. (2005c). Viscous dissipation and completely monotonic relaxation moduli, *Rheologica Acta* **44**, 614–621.
Hanyga, A. (2007a). Fractional-order relaxation laws in non-linear viscoelasticity, *Continuum Mechanics and Thermodynamics* **19**, 25–36.
Hanyga, A. (2007b). Long-range asymptotics of a step signal propagating in a hereditary viscoelastic medium, *Q. Jl Mech. Appl. Math.* **60**, 85–98.
Hanyga, A. (2007c). Anomalous diffusion without scale invariance, *J. Phys. A: Math. Theor.* **40**, 5551–5563.

Hanyga, A. and Seredyńska, M. (2002). Asymptotic and exact fundamental solutions in hereditary media with singular memory, *Quart. Appl. Math.* **60**, 213–244.

Hanyga, A. and Seredyńska, M. (2003). Power law attenuation in acoustic and isotropic anelastic media, *Geophys. J. Int.* **155**, 830–838.

Hanyga, A. and Seredyńska, M. (2007a). Multiple-integral viscoelastic constitutive equations, *Int. J. Non-Linear Mech.* **42**, 722–732.

Hanyga, A. and Seredyńska (2007b). Relations between relaxation modulus and creep compliance in anisotropic linear viscoelasticity, *J. of Elasticity* **88**, 41–71.

Hanyga, A. and Seredyńska (2008a). On a mathematical framework for the constitutive equations of anisotropic dielectric relaxation, *J. Stat. Phys.* **131**, 269–303.

Hanyga, A. and Seredyńska (2008b). A rigorous construction of maximum recoverable energy, *Z. angew. Math. Phys* **59** No 5, 780–789.

Hartmann, B., Lee, G.F. and Lee, J.D. (1994). Loss factor height and width limits for polymer relaxations, *J. Acoust. Soc. Amer.* **95** No 1, 226–233.

Haubold, H.J. and Mathai, A.M. (2000). The fractional kinetic equation and thermonuclear functions, *Astrophysics and Space Science* **273**, 53–63.

Haubold, H.J., Mathai, A.M. and Saxena, R.K. (2007). Solution of fractional reaction-diffusion equations in terms of the H-function, *Bull. Astro. Soc., India* **35**, 681–689.

Haubold, H.J, Mathai, A.M. and Saxena, R.K. (2009). Mittag-Leffler functions and their applications, *E-Print arXiv: 0909.0230*, pp. 49.

Haupt, P. and Lion, A. (2002). On finite linear viscoelasticity of incompressible isotropic materials. *Acta Mechanica* **159** No 1-4, 87–124.

Haupt, P., Lion, A. and Backhaus, E. (2000). On the dynamic behaviour of polymers under finite strains: constitutive modelling and identification of parameters, *Int. J. Solids and Structures* **37**, 3633–3646.

Havriliak, S.Jr. and Havriliak, S.J. (1994). Results from an unbiased analysis of nearly 1000 sets of relaxation data, *J. Non-Crystalline Solids* **172–174**, 297–310.

Havriliak, S.Jr. and Havriliak, S.J. (1995). Time to frequency domain transforms, *Polymer* **36** No 14, 2675–2690.

Havriliak, S.Jr. and Havriliak, S.J. (1996). Comparison of the Havriliak-Negami and stretched exponential time, *Polymer* **37** No 18, 4107–4110.

Havriliak, S.Jr. and Havriliak, S.J. (1997). *Dielectrical and Mechanical Relaxation in Materials*, Hanser Publ., Munich.

Havriliak, S.Jr. and Negami, S. (1966). Analysis of α dispersion in some polymer systems by the complex variables method, *Journal of Polymer Science C* **14**, 99–117.

Havriliak, S.Jr. and Negami, S. (1967). A complex plane representation of dielectric and mechanical relaxation processes in some polymers, *Polymer* **8**, 161–210.

Havriliak, S.Jr. and Negami, S. (1969). On the equivalence of dielectric and mechanical dispersions in some polymers; e.g. Poly(n-Octyl Methacrylate), *Polymer* **10**, 859–872.

Hayes, M. (1982). Viscoelastic plane waves, in [Mainardi (1982)], pp. 28–40.

Hayes, M.A. and Rivlin, R.S. (1974). Plane waves in linear viscoelastic materials, *Quart. Appl. Math.* **32**, 113–121.

Helbig, K. (1994). *Foundations of Anisotropy for Exploration Seismics*, Pergamon, London.

Hellinckx, S., Heymans, N. and Bauwens, J.-C. (1994). Analytical and fractal descriptions of non-linear mechanical behaviour of polymers. *J. Non-Crystalline Solids* **172-174**, 1058–1061.

Henrici, P. (1977). *Applied and Computational Complex Analysis*, Vol. 2: *Special Functions, Integral Transforms, Asymptotics, Continued Fractions*, Wiley, New York.

Herrera, I. and Gurtin, M.E. (1965). A correspondence principle for viscoelastic wave propagation, *Quart. Appl. Math.* **22**, 360–364.

Heymans, N. (1996). Hierarchical models for viscoelasticity: dynamic behaviour in the linear range, *Rheol. Acta* **35**, 508–519.

Heymans, N. (2000). A novel look at models for polymer entanglement, *Macromlecules* **33**, 4226–4234.

Heymans, N. (2003). Constitutive equations for polymer viscoelasticity derived from hierarchical models in cases of failure of time-temperature superposition, *Signal Processing* **83**, 2345–2357.

Heymans, N. (2004). Fractional calculus description of non-linear viscoelastic behaviour of polymers, *Nonlinear Dynamics* **38** No 1-2, 221–234.

Heymans, N. (2008). Dynamic measurements in long-memory materials: fractional calculus evaluation of approach to steady state, *Journal of Vibration and Control* **14**, 1587–1596.

Heymans, H. and Bauwens, J.-C. (1994). Fractal rheological models and fractional differential equations for viscoelastic behaviour. *Rheol. Acta* **33** No 3, 210–219.

Heymans, N. and Kitagawa, M. (2004). Modelling "unusual" behaviour after strain reversal with hierarchical fractional models, *Rheol. Acta* **43** No 4, 383–389.

Heymans, N. and Podlubny, I. (2006). Physical interpretation of initial conditions for fractional differential equations with Riemann-Liouville fractional derivatives, *Rheol. Acta* **45** No 5, 765–771.

Hibbeler, R.C. (1997). *Mechanics of Materials*, Third Edition, Princeton-Hall, Upper Saddle River, NJ.
Hilfer, R. (1995a). Exact solutions for a class of fractal time random walks, *Fractals* **3**, 211–216.
Hilfer, R. (1995b). Foundations of fractional dynamics, *Fractals* **3**, 549–555.
Hilfer, R. (1995c). An extension of the dynamical foundation of the statistical equilibrium concept, *Physica A* **221**, 89–96.
Hilfer, R. (1997). Fractional derivatives in static and dynamic scaling, in B. Dubrulle, F. Graner and D. Sornette (Editors), *Scale Invariance and Beyond*, Springer Verlag, Berlin and EDP Science, France, pp. 53–62.
Hilfer, R., Editor (2000a). *Fractional Calculus, Applications in Physics*, World Scientific, Singapore.
Hilfer, R. (2000b). Fractional time evolution, in [Hilfer (2000a)], pp. 87–130.
Hilfer, R. (2000c). Fractional calculus and regular variation in thermodynamics, in [Hilfer (2000a)], pp. 429–463.
Hilfer, R. (2002a). Experimental evidence for fractional time evolution in glass forming materials, *J. Chem. Phys.* **284**, 399–408.
Hilfer, R. (2002b). Analytical representations for relaxation functions of glasses, *J. Non-Crystalline Solids* **305**, 122–126.
Hilfer, R. (2002c). H-function representations of stretched exponential relaxation and non-Debye susceptibilities in glass systems, *Phys. Rev. E* **65**, 061510/1-5.
Hilfer, R. (2003). On fractional diffusion and continuous time random walks, *Physica A* **329**, 35–40.
Hilfer, R. (2008). Threefold introduction to fractional derivatives, Chap. 2, in Klages, R., Radons, G. and Sokolov, I.M., Editors: *Anomalous Transport: Foundations and Applications*, Wiley-VCH, Weinheim, Germany, pp. 17–73.
Hilfer, R. and Anton, L. (1995). Fractional master equations and fractal time random walks, *Phys. Rev. E* **51**, R848–R851.
Hilfer, R. and Seybold, H.J. (2006). Computation of the generalized Mittag-Leffler function and its inverse in the complex plane, *Integral Transforms and Special Functions* **17** No 9, pp. 637–652.
Hille, E. and Tamarkin, J. D. (1930). On the theory of linear integral equations, *Ann. Math.* **31**, 479–528.
Hillier, K.W. (1949) Measurements of the dynamic elasticity of rubber, *Proc. Phys. Soc B* **62**, 701–707.
Horr, A.M. and Schmidt, L.C. (1995). Dynamic response of a damped large space structure: a new fractional-spectral approach, *International Journal of Space Structures* **2**, 113–120.
Horr, A.M., and Schmidt, L.C. (1996a). Frequency domain dynamic analysis of large space structures with added elastomeric dampers, *International Journal of Space Structures* **11**, 279–289.

Horr, A.M., and Schmidt, L.C. (1996b). Fractional-spectral method for vibration of damped space structures, *Engineering Structures* **18**, 947–956.
Hrusa, W., Nohel, J.A. and Renardy, M. (1988). Initial value problems in viscoelasticity, *Quart. Appl. Math.* **43**, 237–253.
Hrusa, W. and Renardy, M. (1985). On wave propagation in linear viscoelasticity, *Quart. Appl. Math.* **43**, 237–253.
Humbert, P. (1945). Nouvelles correspondances symboliques, *Bull. Sci. Mathém.* (Paris, II ser.) **69**, 121–129.
Humbert, P. (1953). Quelques résultats relatifs à la fonction de Mittag-Leffler, *C.R. Acad. Sci. Paris* **236**, 1467–1468.
Humbert, P. and Agarwal, R.P. (1953). Sur la fonction de Mittag-Leffler et quelques-unes de ses généralisations, *Bull. Sci. Math* (Ser. II) **77**, 180–185.
Humbert, P. and Delrue, P. (1953). Sur une extension à deux variables de la fonction de Mittag-Leffler, *C.R. Acad. Sci. Paris* **237**, 1059–1060.
Hunter, S.C. (1960). Viscoelastic Waves, in Sneddon, I. and Hill, R., Editors, *Progress in Solid Mechanics*, North-Holland, Amsterdam, Vol 1, pp. 3–60.
Hunter, S.C. (1983). *Mechanics of Continuous Media*, Wiley, New York.
Husain, S.A. and Anderssen, R.S. (2005). Modelling the relaxation modulus of linear viscoelasticity using Kohlrausch functions, *J. Non-Newtonian Fluid Mech.* **125**, 159–170.
Il'yushin, A.A. and Pobedrya, B.E. (1970). *Foundations of the Mathematical Theory of Viscoelasticity*, Nauka, Moscow. [In Russian]
Ingman, D. and Suzdalnitsky, J. (2001). Iteration method for equation of viscoelastic motion with fractional differential operator of damping, *Computer Methods in Applied Mechanics and Engineering* **190**, 5027–5036.
Ingman, D. and Suzdalnisky, J. (2002). Application of dynamic fractional differentiation to the study of oscillating viscoelastic medium with cylindrical cavity, *J. of Vibration and Acoustics* (Transaction of ASME) **124**, 642–659.
Ingman, D. and Suzdalnitsky, J. (2004). Control of damping oscillations by fractional differential operator with time-dependent order, *Computer Methods in Applied Mechanics and Engineering* **193**, 5585–5595.
Ishlinskii, A.Yu. (1940). Longitudinal vibrations of a bar with a linear law of aftereffect and relaxation, *Prikl. Matem. i Mech. (PMM)* **4** No 1, 79–92. [In Russian]
Jackson, D.D. (1969). Elastic relaxation model for seismic wave attenuation, *Phys. Earth. Planet. Interiors* **2**, 30–34.
Jackson, D.D. and Anderson, D.L. (1970). Physical mechanisms for seismic wave attenuation, *Rev. Geophys.* **8**, 1–63.

Jahnke, E. and Emde, F. (1943). *Tables of Functions with Formulas and Curves*, Dover, New York.
Jayakamur, K. (2003). Mittag-Leffler process, *Mathematical and Computer Modelling* **37**, 1427–1434.
Jaeger, J.C. (1956). *Elasticity, Fracture and Flow*, Methuen, London.
Jeffrey, A. and Engelbrecht, J (1982). Waves in non-linear relaxing media, in [Mainardi (1982)], pp. 95–123.
Jeffreys, H. (1931). Damping in bodily seismic waves, *Mon. Not. Roy. Astr. Soc. Geophys. Suppl.* **2**, 318–323.
Jeffreys, H. (1932). On plasticity and creep in solids, *Proc. Roy. Soc. London A* **138**, 283–297.
Jeffreys, H. (1958). A modification of Lomnitz's law of creep in rocks, *Geophys. J. Roy. Astr. Soc.* **1**, 92–95.
Jeffreys, H. (1976). *The Earth*, Cambridge University Press, Cambridge [Sixth Edition].
Jeffreys, H. and Crampin, S. (1960). Rock creep: a correction, *J. Roy. Astr. Soc., Monthly Notices* **121**, 571–577.
Jeffreys, H. and Crampin, S. (1970). On the modified Lomnitz law of damping, *Mon. Not. Roy. Astr. Soc. (London)* **147**, 295–301.
Jeffreys, H. and Swirles Jeffreys, B. (1972). *Methods of Mathematical Physics*, Cambridge University Press, Cambridge [3-rd Edition].
Jia, J-H., Shen, X-Y. and Hua, H-X. (2007). Viscoelastic behavior and application of the fractional derivative Maxwell model, *J. Vibration and Control* **13**, 385–401.
Jiménez, A.H., Vinagre Jara, B. and Santiago, J.H. (2002). Relaxation modulus in the fitting of polycarbonate and poly(viinyl chloride) viscoelastic polymers by a fractional Maxwell model, *Colloid & Polymer Science* **280** No 5, 485–489.
Jiménez, A.H., Santiago, J.H., Garcia, M.A. and Gonzáles, S.J. (2002). Relaxation modulus in PMMA and PTFE fitting by fractional Maxwell model, *Polymer Testing* **21**, 325–331.
Johnson, D.L. (2001). Theory of frequency dependent acoustics in patchy-saturated porous media. *J. Acoust. Soc. Am.* **110**, 682–694.
Jones,T.D. (1986). Pore fluids and frequency-dependent wave propagation in rocks. *Geophysics* **51**, 1939–1953.
Jonscher, A.H. (1983). *Dielectric Relaxation in Solids*, Chelsea Dielectric Press, London.
Jonscher, A.H. (1996). *Universal Relaxation Law*, Chelsea Dielectric Press, London.
Joseph, D.D. (1976). *Stability of Fluid Motion*, Springer Verlag, Berlin.
Kalia, R.N., Editor (1993). *Recent Advances on Fractional Calculus*, Global Publishing Company, Sauk Rapids, Minnesota/.

Kalla, S.L. (1987). Fractional relations by means of Riemann-Liouville operator, *Serdica* **13**, 170–173.

Kalmykov, Yu. P., Coffey, W. T., Crothers, D. S. F. and Titov, S. V. (2004). Microscopic models for dielectric relaxation in disordered systems, *Phys. Rev. E* **70**, 041103/1–11.

Kaminskii, A.A. and Selivanov, M.F. (2005a). A method for determining the viscoelastic characteristics of composites, *International Applied Mechanics* **41** No 5, 469–480.

Kaminskii, A.A. and Selivanov, M.F. (2005b). An approach to the determination of the deformation characteristics of viscoelastic materials, *International Applied Mechanics* **41** No 8, 867–875.

Kanamori, H. and. Anderson, D.L. (1977). Importance of physical dispersion in surface wave and free oscillation problems: a review, *Rev. Geophys. and Space Phys.* **15**, 105–112.

Karczewska, A. and Lizama, C. (2009). Solutions to stochastic fractional relaxation equations, *Physica Scripta* **T136**, 014030/1–7.

Kegel, G. (1954). Electro-mechaninische Analogien, *Kolloid. Z.* **135**, 125–133.

Kang, I.B. and McMechan, G.A. (1993). Effects of viscoelasticity on wave propagation in fault zones, near-surfaces sediments and inclusions, *Bull. Seism. Soc. Am.* **83**, 890–906.

Karlsson, A. and Rikte, S. (1998). The time-domain theory of forerunners, *J. Opt. Soc. Am.* **15**, 487–502.

Kempfle, S. (1998b). Causality criteria for solutions of linear fractional differential equations, *Fractional Calculus and Applied Analysis* **1**, 351–364.

Kempfle, S. and Beyer, H. (2003). The scope of a functional calculus approach to fractional differential equations, In Begehr, H.G.W. et al. (Editors), *Progress in Analysis* [Proceedings of the 3-rd Int. ISAAC Congress, Berlin (Germany), July 2001], World Scientific, Singapore (2003), pp. 69–81.

Kempfle, S. and Gaul, L. (1996). Global solutions of fractional linear differential equations, *Z. angew. Math. Mech. (ZAMM)* **76**, Supl.2, 571–572. [Proceedings ICIAM95]

Kempfle, S. Krüger, K. and Schäfer, I. (2009). Fractional modeling and applications, in Begehr, H.G.W and Nicolosi, F. (Editors), *More Progresses in Analysis* [Proceedings of the Fifth Int. ISAAC Congress, Catania (Italy), July 2005], World Scientific, Singapore, pp. 311–324.

Kempfle, S. and Schäfer, I. (1999). Functional calculus method versus Riemann-Liouville approach, *Fractional Calculus and Applied Analysis* **2** No 4, 415–427.

Kempfle, S. and Schäfer, I. (2000). Fractional differential equations and initial conditions, *Fractional Calculus and Applied Analysis* **3** No 4, 387–400.
Kempfle, S., Schäfer, I. and Beyer, H. (2002a). Functional calculus and a link to fractional calculus, *Fractional Calculus and Applied Analysis* **5**, 411–426.
Kempfle, S., Schäfer, I. and Beyer, H. (2002b). Fractional calculus via functional calculus, *Nonlinear Dynamics* **29**, 99–127.
Kennedy, A.J. (1963). *Processes of Creep and Fatigue in Metals*, Wiley, New York.
Kilbas, A.A. and Saigo, M. (1996). On Mittag-Leffler type functions, fractional calculus operators and solution of integral equations, *Integral Transforms and Special Functions* **4**, 355–370.
Kilbas, A.A. and Saigo, M. (1999). On the H functions, *Journal of Applied Mathematics and Stochastic Analysis* **12**, 191–204.
Kilbas, A.A. and Saigo, M. (2004). *H-transforms. Theory and Applications*, Chapman and Hall/CRC, Boca Raton, FL. [Series on Analytic Methods and Special Functions, Vol. 9.]
Kilbas, A.A., Saigo, M. and Trujillo, J.J. (2002). On the generalized Wright function, *Fractional Calculus and Applied Analysis* **5** No. 4, 437–460.
Kilbas, A.A., Srivastava, H.M. and Trujillo, J.J. (2006). *Theory and Applications of Fractional Differential Equations*, Elsevier, Amsterdam. [North-Holland Series on Mathematics Studies No 204]
Kilbas, A.A. and Trujillo, J.J. (2001). Differential equation of fractional order: methods, results and problems, I, *Appl. Anal.* **78**, 153–192.
Kilbas, A.A. and Trujillo, J.J. (2002). Differential equation of fractional order: methods, results and problems, II, *Appl. Anal.* **81**, 435–493.
Kirane, M. (2003). Wave equation with fractional damping, *Zeit. Anal. Anw.* **22**, 609–617.
Kiryakova, V. (1994). *Generalized Fractional Calculus and Applications*, Longman, Harlow. [Pitman Research Notes in Mathematics, Vol. 301]
Kiryakova, V. (1997). All the special functions are fractional differintegrals of elementary functions, *J. Phys. A: Math. Gen.* **30**, 5085–5103.
Kiryakova, V. (1999). Multi-index Mittag-Leffler functions, related Gelfond-Leontiev operators and Laplace type integral transforms, *Fractional Calculus and Applied Analysis* **2** No 4, 445–462.
Kiryakova, V. (2000). Multiple (multi-index) Mittag-Leffler functions and relations to generalized fractional calculus, *J. Comput. Appl. Math.* **118** No 1-2, 241–259.
Kiryakova, V. (2008). Some special functions related to fractional calculus and fractional (non-integer) order control systems and equations, *"Facta Universitatis" (Sci. J. of University of Nis), Series:*

Automatic Control and Robotics, **7** No 1, 79–98. [available on line http://facta.junis.ni.ac.rs/acar/acar200801/acar2008-07.pdf]

Kiryakova, V. (2009a). The multi-index Mittag-Leffler functions as important class of special functions of fractional calculus, *Computers and Mathematics with Applications*, in press. [doi:10.1016/j.camwa.2009.08.025]

Kiryakova, V. (2009b). The special functions of fractional calculus as generalized fractional calculus operators of some basic functions, *Computers and Mathematics with Applications*, in press. [doi:10.1016/j.camwa.2009.05.014]

Kjartansson, E. (1979). Constant-Q wave propagation and attenuation, *J. Geophys. Res.* **94**, 4737–4748.

Klafter, J., Rubin, R.J. and Schlesinger, M.F. (Editors) (1986). *Transport and Relaxation in Random Materials*, World-Scientific, Singapore.

Klafter, J., Blumen, A., Zumofen, G. and M Shlesinger, M.F. (1990). Lévy walks approach to anomalous diffusion, *Physica A* **168**, 637–645.

Klafter, J., Shlesinger, M.F., Zumofen, G. and Blumen, A. (1992). Scale-invariance in anomalous diffusion, *Phis. Mag. B* **65**, 755–765.

Klafter, J., Shlesinger, M.F. and Zumofen, G. (1996). Beyond Brownian motion, *Physics Today* **49**, 33–39.

Klages, R., Radons, G. and Sokolov, I.M., Editors (2008), *Anomalous Transport: Foundations and Applications*, Wiley-VCH, Weinheim, Germany.

Klasztorny, M. (2004a). Constitutive modelling of resins in the compliance domain, *Mechanics of Composite Materials* **40** No 4, 349–358.

Klasztorny, M. (2004b). Constitutive modelling of resins in the stiffness domain, *Mechanics of Composite Materials* **40** No 5, 443–452.

Klausner, Y.(1991). *Fundamentals of Continuum Mechanics of Soils*, Springer Verlag, Berlin.

Knopoff, L. (1954). On the dissipative constants of higher order, *J. Acous. Soc. Amer.* **26**, 183–186.

Knopoff, L. (1956). The seismic pulse in materials possessing solid friction, *Bull. Seismol. Soc. Amer.* **46**, 175–184.

Knopoff, L. (1964). Q, *Rev. Geophys.* **2**, 625–660.

Knopoff, L. and MacDonald, G.J.F. (1958). Attenuation of small amplitude stress waves in solids, *Rev. Mod. Phys.* **30**, 1178–1192.

Knopoff, L. and MacDonald, G.J.F. (1960). Models for acoustic loss in solids, *J. Geophys. Res.* **65**, 2191–2197.

Knopoff, L. and Porter, L.D. (1963). Attenuation of surface waves in a granular material, *J. Geophys. Res.* **68**, 6317–6321.

Kobelev, V.V. (2007). Linear non-conservative systems with fractional damping and the derivatives of critical load parameter, *GAMM-Mitt.* **30** No 2, 287–299.

Kochubei, A.N. (1989). A Cauchy problem for evolution equations of fractional order, *Differential Equations* **25**, 967–974. [English translation from the Russian Journal *Differentsial'nye Uravneniya*]

Kochubei, A.N. (1990). Fractional order diffusion, *Differential Equations* **26**, 485–492. [English translation from the Russian Journal *Differentsial'nye Uravneniya*]

Koeller, R.C. (1984). Applications of fractional calculus to the theory of viscoelasticity, *ASME J. Appl. Mech.* **51**, 299–307.

Koeller, R.C. (1986). Polynomial operators, Stieltjes convolution and fractional calculus in hereditary mechanics, *Acta Mechanica* **58**, 251–264.

Koeller, R.C. (2007). Towards an equation of state for solid materials with memory by use of the half-order derivative, *Acta Mechanica* **191**, 125–133.

Koeller, R.C. (2010). A theory relating creep and relaxation for linear materials with memory, *ASME J. Appl. Mech.* **77**, forthcoming in March 2010, pp. 1–7.

Koh, C.G. and Kelly, J.M. (1990). Application of fractional derivatives to seismic analysis of base isolated models, *Earthquake Engineering and Structural Dynamics.* **19** No 8, 229–241.

Kohlrausch, R. (1847). Ueber das Dellmann'sche Elektrometer, *Annal. Phys.* **58** No 11, 353–405.

Kohlrausch, R. (1854). Theorie des Elektrischen Rückstandes in der Leidener Flasche, *Prog. Ann. Phys. Chem.* **91**, 179–214.

Kolsky, H. (1953). *Stress Waves in Solids*, Clarendon Press, Oxford. [reprinted by Dover, New York (1963)]

Kolsky, H. (1956). The propagation of stress pulses in viscoelastic solids, *Phil. Mag.* (Ser. 8) **2**, 693–710.

Kolsky, H. (1960). Viscoelastic waves, in Davis, N. (Editor), *Proc Int. Symposium Stress Wave Propagation in Materials*, Interscience Publ. Inc., New York, pp. 59–88.

Kolsky, H. (1965). Experimental studies in stress wave propagation, *Proc. Fifth Congress Applied Mechanics* **5**, 21–36.

König, H. and Meixner, J. (1958). Lineare Systeme und lineare Transformationen, *Math. Nachr.* **19**, 265–322.

Körnig, M. and Müller, G.M. (1989). Rheological models and interpretation of postglacial uplift, *Geophys. J. Int.* **98**, 243-2-53.

Kovach, R.L. and Anderson, D.L. (1964). Attenuation of shear waves in the upper mantle and lower mantle, *Bull. Seism. Soc. Amer.* **54**, 1855–1864.

Kreis, A. and Pipkin, A.C. (1986). Viscoelastic pulse propagation and stable probability distributions, *Quart. Appl. Math.* **44**, 353–360.

Kuhn, W., Kunzle, O and Preissmann, A. (1947). Relaxationszeitspektrum Elastizität und Viskosität von Kautschuk, *Helv. Chim. Acta* **30**, 307–328, 464–486.

Kulish, V.V. and Lage, J.L. (2000). Fractional-diffusion solutions for transient local temperature and heat flux *Journal of Heat Transfer* **122**, 372–376.

Kulish, V.V. and Lage, J.L. (2002). Application of fractional calculus to fluid mechanics, *J. Fluids Eng.* **124** No 3, 803–806.

Lakes, R. S. (1998). *Viscoelastic Solids*, CRC Press, Boca Raton.

Lavoie, J.L., Osler, T.J. and Tremblay, R. (1976). Fractional derivatives and special functions, *SIAM Rev.* **18**, 240–268.

Non-local continuum mechanics and fractional calculus Lazopoulos, K.A. (2006). Non-local continuum mechanics and fractional calculus, *Mechanics Research Communications* **33**, 753–757.

Leaderman, H. (1957). Proposed nomenclature for linear viscoelastic behavior, *Trans. Soc. Rheology* **1**, 213–222.

Lebedev, N.N. (1972). *Special Functions and Their Applications*, Dover, New York.

LeBlond, P.H. and Mainardi, F. (1987). The viscous damping of capillary - gravity waves, *Acta Mechanica* **68**, 203–222.

Lee, E.H. (1961). Stress analysis for linear viscoelastic materials, *Rheologica Acta* **1**, 426–430.

Lee, E.H. (1962). *Viscoelasticity* in Flügge, W. (Editor), *Handbook of Engineering Mechanics*, McGraw-Hill, New York, Chapter 53.

Lee, E.H. and Kanter, T. (1956). Wave propagation in finite rods of viscoelastic materials, *J. Appl. Phys.* **24**, 1115–1122.

Lee, E.H. and Morrison, J.A. (1956). A comparison of the propagation of longitudinal waves in rods of viscoelastic materials, *J. Polym. Sci.* **19**, 93–110.

Leigh, d.C. (1968). Asymptotic constitutive approximations for rapid deformations of viscoelastic materials, *Acta Mechanica* **5**, 274–288.

Leith, J.R. (2003). Fractal scaling of fractional diffusion processes, *Signal Process.* **83** No 11, 2397–2409.

Leitman, M.J. (1966). Variational principles in the linear dynamic theory of viscoelasticity, *Quart. Appl. Math.* **24**, 37–46.

Leitman, M. J. and Fisher, G.M. G (1973). *The linear theory of viscoelasticity* in Truesdell, C. (Editor), *Encyclopedia of Physics*, Vol. VIa/3, Springer Verlag, Berlin, pp. 1–123.

Le Méhauté, A., Tenreiro Machado, J.A., Trigeassou, J.C and Sabatier, J (Editors), *Fractional Differentiation and its Applications*, U Books, Germany.

Lenzi, E.K., Mendes, E.S., Fa, K.S., da Silva, L.R. and Lucena, L.S. (2004). Solutions for a fractional nonlinear diffusion equation: Spatial time dependent diffusion coefficient and external forces, *J. Math. Phys.* **45** No 9, 3444–3452.

Lenzi, E.K., Mendes, E.S., Goncalves, G., Lenzi, M.K. and da Silva, L.R. (2006). Fractional diffusion equation and Green function approach: Exact solutions, *Physica A* **360**, 215–226.

Lévy, P. (1923a). Sur une application de la la dérivée d'ordre non entier au calcul des probabilités, *Compt. Rendus Acad, Sci. Paris* **176**, 1118–1120. [Reprinted in *Oeuvres de Paul Lévy* edited by D. Dugué, Vol 3 (Eléments aléatoires) No 115 pp. 339–341, Gauthier-Villars, Paris (1976)]

Lévy, P. (1923b). Sur les lois stables en calcul des probabilités, *Compt. Rendus Acad. Sci. Paris* **176**, 1284–1286. [Reprinted in *Oeuvres de Paul Lévy* edited by D. Dugué, Vol 3 (Eléments aléatoires) No 116 pp. 342–344, Gauthier-Villars, Paris (1976)]

Lévy, P. (1923c). Sur une opération fonctionnelle généralisant la dérivation d'ordre non entier, *Compt. Rendus Acad. Sci. Paris* **176** (1923), 1441–1444. [Reprinted in *Oeuvres de Paul Lévy* edited by D. Dugué, Vol 3 (Eléments aléatoires) No 114 pp. 336–3338, Gauthier-Villars, Paris (1976)]

Lévy, P. (1923d). Sur la dérivation et l'intégration généralisées *Bull. Sci. Math.* (Ser. 2) **47**, 307–320, 343–352.

Li, Y. and Xu, M. (2006). Hysteresis and precondition of viscoelastic solid models, *Mech. Time-Depend. Mater.* **10**, 113–123.

Li, Y., Chen, Y. and Podlubny, I. (2009). Mittag-Leffler stability of fractional order nonlinear dynamic systems, *Automatica* **45** No 8, 1965–1969.

Lighthill, M. J. (1965). Group velocity, *J. Inst. Maths. Appl.* **1**, 1–28.

Lin, G.D. (1998). On the Mittag-Leffler distributions, *J. Statistical Planning and Inference* **74**, 1–9.

Lion, A. (1997). On the thermodynamics of fractional damping elements, *Continuum Mechanics and Thermodynamics* **9** No 2, 83–96.

Lion, A. (1998). Thixotropic behaviour of rubber under dynamic loading histories: experiments and theory, *J. Mech. Phys. Solids* **46**, 895–930.

Liouville, J. (1832). Mémoire sur quelques questions de geometrie et de mecanique et sur un nouveau genre de calcul pour resudre ces questions, *J. École Polytech.* **13** (21), 1–69.

Liouville, J. (1835). Mémoire sur le changemant de variables dans calcul des differentielles d'indices quelconques, *J. École Polytech.* **15** (24), 17–54.

Liu, H.P., Anderson, D.L. and Kanamori, H. (1976). Velocity dispersion due to anelasticity: implications for seismology and mantle composition, *Geophys. J. R. Astr. Soc.* **47**, 41–58.

Liu, H.P. and Archambeau, C. (1975). The effect of anelasticity on periods of the Earth's free oscillation (toroidal modes), *Geophys. J. R. Astr. Soc.* **43**, 795–814.

Liu, H.P. and Archambeau, C. (1976). Correction to the effect of anelasticity on periods of the Earth's free oscillation (toroidal modes), *Geophys. J. R. Astr. Soc.* **47**, 1–7.

Liu, H.P., Anderson, D.L. and Kanamori, H. (1975). Velocity dispersion due to anelasticity implications for seismology and mantle composition, *Geophys. J. R. Astr. Soc.* **47**, 41–58.

Liu, J.G. and Xu, M.Y. (2006). Higher-order fractional constitutive equations of viscoelastic materials involving three different parameters and their relaxation and creep functions, *Mechanics of Time-Dependent Materials* **10** No 4, 263–279.

Lockett, F. J. (1972). *Nonlinear Viscoelastic Solids*, Academic Press, London, New York.

Lokshin, A.A. and Rok, V.E. (1978a). Automodel solutions of wave equations with time lag, *Russ. Math. Surv.* **33**, 243–244.

Lokshin, A.A. and Rok, V.E. (1978b). Fundamental solutions of the wave equation with delayed time, *Doklady AN SSSR* **239**, 1305–1308.

Lokshin, A. A. and Suvorova, Yu. V. (1982). *Mathematical Theory of Wave Propagation in Media with Memory*, Moscow University Press, Moscow. [in Russian]

Lomnitz, C. (1956). Creep measurements in igneous rocks, *J. Geol.* **64**, 473–479.

Lomnitz, C. (1957). Linear dissipation in solids, *J. Appl. Phys.* **28**, 201–205.

Lomnitz, C. (1962). Application of the logarithmic creep law to stress wave attenuation in the solid earth, *J. Geophys. Res.* **67**, 365–368.

Lorenzo, C. and Hartley, T. (2000). Initialized fractional calculus, *Int. J. Appl. Mathematics* **3**, 249–265.

Lorenzo, C. and Hartley, T. (2002). Variable order and distributed order fractional operators, *Nonlinear Dynamics* **29**, 57–98.

Lovell, R. (1974). Applications of the Kronig-Kramers relations to the interpretation of dielectric data, *J. Phys. C: Solid State Phys.* **7**, 4378–4384.

Loudon, R. (1970). The propagation of electromagnetic energy through an absorbing medium, *J. Physics A* **3**, 233–245.

Lu, Y.C. (2006). Fractional derivative viscoelastic model for frequency-dependent complex moduli of automative elastomers, *Int. J. Mech. Mater. Des.* **3**, 329–336.

Lubich, C. (1986). Discretized fractional calculus, *SIAM J. Math. Anal.* **17**, 704–719.

Lubich, C (1988a). Convolution quadrature and discretized fractional calculus: I, *Numer. Math.* **52**, 129–145.

Lubich, C. (1988b). Convolution quadrature and discretized fractional calculus: II, *Numer. Math.* **52**, 413–425.

Lubliner, J. and Panoskaltsis, V.P. (1992). The modified Kuhn model of linear viscoelasticity, *Int. J. Solids Struct.* **29**, 3099–3112.

Luchko, Yu. (1999). Operational method in fractional calculus, *Fractional Calculus and Applied Analysis* **2**, 463–488.

Luchko, Yu. (2000). Asymptotics of zeros of the Wright function, *Zeit. Anal. Anwendungen* **19**, 583–595.

Luchko, Yu. (2001). On the distribution of zeros of the Wright function, *Integral Transforms and Special Functions* **11**, 195–200.

Luchko, Yu. (2008). Algorithms for evaluation of the Wright function for the real arguments' values, *Fractional Calculus and Applied Analysis* **11**, 57–75.

Luchko, Yu. and Gorenflo, R. (1998). Scale-invariant solutions of a partial differential equation of fractional order, *Fractional Calculus and Applied Analysis* **1**, 63–78.

Luchko, Yu. and Gorenflo, R. (1999). An operational method for solving fractional differential equations with the derivatives, *Acta Math. Vietnam.* **24**, 207–233.

Luchko, Yu., Rivero, M., Trujillo, J.J. and Pilar Velasco, M. (2010). Fractional models, non-locality, and complex systems, *Computers and Mathematics with Applications*, in press.

Luchko, Yu. and Srivastava, H.M. (1995). The exact solution of certain differential equations of fractional order by using operational calculus, *Comput. Math. Appl.* **29**, 73–85.

Lv, L-J., Xiao, J-B., Ren F.Y and Gao L. (2008). Solutions for multidimensional fractional anomalous diffusion equations, *J. Math. Phys.* **49**, 073302.

Lv, L-J., Xiao, J-B., Zhang, L. and Gao L. (2008). Solutions for a generalized fractional anomalous diffusion equation, *J. Comp. Appl. Math.* **225**, 301–308.

Macdonald, J.R. (1959). Raileigh-wave dissipation function in low-loss media, *Geophys. J. R. Astr. Soc.* **2**, 132–135.

Macdonald, J.R. (1961). Theory and application of superposition model of internal friction, *J. Appl. Phys.* **32**, 2385–2398.

Macdonald, J.R. and Brachman, M.K. (1956). Linear-system integral transform relations, *Rev. Modern Physics* **28** No 4, 393–422.

Magin, R.L. (2004). Fractional Calculus in Bioengineering: Part1, Part 2 and Part 3, *Critical Reviews in Biomedical Engineering* **32**, 1–104, 105–193, 195–377.

Magin, R.L. (2006). *Fractional Calculus in Bioengineering*, Begell Houuse Publishers, Connecticut.

Magin, R.L. (2008). Anomalous diffusion expressed through fractional order differential operators in the Bloch-Torrey equation, *J. Magn. Reson.* **190**, 255–270.

Magin, R.L. and Ovadia, M. (2008). Modeling the cardiac tissue electrode interface using fractional calculus, *Journal of Vibration and Control* **14**, 1431–1442.

Mainardi, F. (1972). On the seismic pulse in a standard linear solid, *Pure and Appl. Geophys. (Pageoph)* **99**, 72–84.

Mainardi, F. (1973). On energy velocity of viscoelastic waves, *Lett. Il Nuovo Cimento* **6**, 443–449.

Mainardi, F., Editor (1982). *Wave Propagation in Viscoelastic Media*, Pitman, London. [Res. Notes in Maths, Vol. 52]

Mainardi, F. (1983a). On signal velocity of anomalous dispersive waves, *Il Nuovo Cimento B* **74**, 52–58.

Mainardi, F. (1983b). Signal velocity for transient waves in linear dissipative media, *Wave Motion* **5**, 33–41.

Mainardi, F. (1984). Linear dispersive waves with dissipation, in Rogers, C. and Moodie, T. B. (Editors), *Wave Phenomena: Modern Theory and Applications*, North-Holland, Amsterdam, pp. 307–317. [Mathematics Studies Vol. 97]

Mainardi, F. (1987). Energy velocity for hyperbolic dispersive waves, *Wave Motion* **9**, 201–208.

Mainardi, F. (1993). Energy propagation for dispersive waves in dissipative media, *Radiophysics and Quantum Electronics* **36** No 7, 650–664. [English translation of *Radiofisika* from the Russian]

Mainardi, F. (1994a). On the initial value problem for the fractional diffusion-wave equation, in: Rionero, S. and Ruggeri, T. (Editors), *Waves and Stability in Continuous Media*, World Scientific, Singapore, pp. 246–251. [Proc. VII-th WASCOM, Int. Conf. "Waves and Stability in Continuous Media", Bologna, Italy, 4-7 October 1993]

Mainardi, F. (1994b). Fractional relaxation in anelastic solids, *Journal of Alloys and Compounds* **211/212**, 534–538.

Mainardi, F. (1995a). Fractional diffusive waves in viscoelastic solids, in: Wegner, J.L. and Norwood, F.R. (Editors), *Nonlinear Waves in Solids*, Fairfield NJ, pp. 93–97, ASME book No AMR 137, Proc. IUTAM Symposium, University of Victoria, Canada, 15–20 August 1993. [Abstract in *Appl. Mech. Rev.* **46** No 12 (1993) p. 549]

Mainardi, F. (1995b). The time fractional diffusion-wave equation, *Radiophysics and Quantum Electronics* **38** No 1-2, 20–36. [English translation from the Russian of *Radiofisika*]

Mainardi, F. (1996a). The fundamental solutions for the fractional diffusion-wave equation, *Applied Mathematics Letters* **9** No 6, 23–28.

Mainardi, F. (1996b). Fractional relaxation-oscillation and fractional diffusion-wave phenomena, *Chaos, Solitons & Fractals* **7**, 1461–1477.

Mainardi, F. (1997). Fractional calculus: some basic problems in continuum and statistical mechanics, in: Carpinteri, A. and Mainardi, F. (Editors) *Fractals and Fractional Calculus in Continuum Mechanics*, Springer Verlag, Wien, pp. 291–348. [Down-loadable from http://www.fracalmo.org]

Mainardi, F. (2002a). Linear viscoelasticity, Chapter 4 in [Guran *et al.* (2002)], pp. 97–126.

Mainardi, F. (2002b). Transient waves in linear viscoelastic media, Chapter 5 in [Guran *et al.* (2002)], pp. 127–161.

Mainardi, F. and Bonetti, E. (1988). The application of real-order derivatives in linear visco-elasticity, *Rheologica Acta* **26** Suppl., 64–67.

Mainardi, F. and Buggisch, H. (1983). On non-linear waves in liquid-filled elastic tubes, in Nigul, U. and Engelbrecht, J. (Editors), *Nonlinear Deformation Waves*, Springer Verlag, Berlin, pp. 87–100.

Mainardi, F. and Gorenflo, R. (2000). On Mittag-Leffler-type functions in fractional evolution processes, *J. Computational and Appl. Mathematics* **118**, 283–299.

Mainardi, F. and Gorenflo, R. (2007). Time-fractional derivatives in relaxation processes: a tutorial survey, *Fractional Calculus and Applied Analysis* **10**, 269–308. [E-print: http://arxiv.org/abs/0801.4914]

Mainardi, F., Gorenflo, R. and Scalas, E. (2004). A fractional generalization of the Poisson processes, *Vietnam Journal of Mathematics* **32** SI, 53–64. [E-print http://arxiv.org/abs/math/0701454]

Mainardi, F., Gorenflo, R. and Vivoli, A. (2005). Renewal processes of Mittag-Leffler and Wright type, *Fractional Calculus and Applied Analysis* **8**, 7–38. [E-print http://arxiv.org/abs/math/0701455]

Mainardi, F., Luchko, Yu. and Pagnini, G. (2001). The fundamental solution of the space-time fractional diffusion equation, *Fractional Calculus and Applied Analysis* **4**, 153–192. [E-print http://arxiv.org/abs/cond-mat/0702419]

Mainardi, F., Mura, A., Gorenflo, R. and Stojanovic, M. (2007). The two forms of fractional relaxation of distributed order, *Journal of Vibration and Control* **13** No 9-10, 1249–1268. [E-print http://arxiv.org/abs/cond-mat/0701131]

Mainardi, F., Mura, A., Pagnini, G. and Gorenflo, R. (2008). Time-fractional diffusion of distributed order, *Journal of Vibration and Control* **14** No 9-10, 1267–1290. [arxiv.org/abs/cond-mat/0701132]

Mainardi, F., Mura, A. and Pagnini, G. (2009). The M-Wright function in time-fractional diffusion processes: a tutorial survey, *Int. Journal of Differential Equations*, in press.

Mainardi, F. and Nervosi, R. (1980). Transient waves in finite viscoelastic rods, *Lett. Nuovo Cimento* **29**, 443–447.

Mainardi, F. and Pagnini, G. (2001). The fundamental solutions of the time-fractional diffusion equation, in Fabrizio, M., B. Lazzari, B. and Morro, A. (Editors), *Mathematical Models and Methods for Smart Materials*, World Scientific, Singapore, pp. 207–224.

Mainardi, F. and Pagnini, G. (2003). The Wright functions as solutions of the time-fractional diffusion equations, *Appl. Math. Comp.* **141**, 51–62.

Mainardi, F., Pagnini, G. and Gorenflo, R. (2003). Mellin transform and subordination laws in fractional diffusion processes, *Fractional Calculus and Applied Analysis* **6** No 4, 441–459. [E-print: http://arxiv.org/abs/math/0702133]

Mainardi, F., Pagnini, G. and Saxena, R.K. (2005). Fox H functions in fractional diffusion, *J. Comp. Appl. Math.* **178**, 321–331.

Mainardi, F. and Paradisi, P. (2001). Fractional diffusive waves, *J. Computational Acoustics* **9**, 1417–1436.

Mainardi, F., Raberto M., Gorenflo, R. and Scalas, E. (2000). Fractional calculus and continuous-time finance II: the waiting-time distribution, *Physica A* **287**, No 3-4, 468–481.

Mainardi, F., Servizi, G. and Turchetti, G. (1977). On the propagation of seismic pulses in a porous elastic solid, *J. Geophys.* **43**, 83–94.

Mainardi, F., Tampieri, F. and Vitali, G. (1991). Dissipative effects on internal gravity waves in geophysical fluids, *Il Nuovo Cimento C* **14**, 391–399.

Mainardi, F., Tocci, D. and Tampieri, F. (1993). On energy propagation for internal waves in dissipative fluids, *Il Nuovo Cimento B* **107**, 1337–1342.

Mainardi, F. and Tocci, D. (1993). Energy propagation in linear hyperbolic systems in the presence of dissipation, in Donato, A. and Oliveri, F. (Editors), *Nonlinear Hyperbolic Problems: Theoretical, Applied, and Numerical Aspects*, Vieweg, Braunschweig, pp. 409–415.

Mainardi, F. and Tomirotti, M. (1995). On a special function arising in the time fractional diffusion-wave equation, in: Rusev, P., Dimovski, I. and Kiryakova, V., Editors, *Transform Methods and Special Functions, Sofia 1994*. Science Culture Technology Publ., Singapore, pp. 171–183.

Mainardi, F. and Tomirotti, M. (1997). Seismic pulse propagation with constant Q and stable probability distributions, *Annali di Geofisica* **40**, 1311–1328.

Mainardi, F. and Turchetti, G. (1975). Wave front expansion for transient viscoelastic waves, *Mech. Research Comm.* **2**, 107–112.

Mainardi, F. and Turchetti, G. (1979). Positivity constraints and approximation methods in linear viscoelasticity, *Lett. Il Nuovo Cimento* (Ser. II) **26** No 2, 38–40.

Mainardi, F., Turchetti, G. and Vitali, G. (1976). Rational approximations and relations among the linear viscoelastic functions, Proc. Third National Congress of Italian Association of Theoretical and Applied Mechanics (AIMETA), Vol. I, pp. 18.1–18.10.

Mainardi, F. and Van Groesen, E. (1989). Energy propagation in linear hyperbolic systems, *Il Nuovo Cimento B* **104**, 487–496.

Mainardi, F. and Vitali, G. (1990). Applications of the method of steepest descents in wave propagation problems, in [Wong (1990)], pp. 639–651.

Makosko, C. W. (1994). *Rheology: Principles, Measurements and Applications*, Wiley, New York.

Makris, N. (1997). Three-dimensional constitutive viscoelastic laws with fractional order time derivatives, *Journal of Rheology* **41** No 5, 1007–1020.

Makris, N. and Constantinou, M.C. (1991). Fractional derivative Maxwell model for viscous dampers, *J. Structural Engineering ASCE* **117** No 9, 2708–2724.

Makris, N. and Constantinou, M.C. (1992). Spring-viscous damper systems for combined seismic and vibration isolation, *Earthquake Engineering and Structural Dynamics* **21** No 8, 649–664.

Makris, N. and Constantinou, M.C. (1993). Models of viscoelasticity with complex-order derivatives, *J. Engineering Mechanics ASCE* **119** No 7, 1453–1464.

Makris, N., Dargush, G.F. and Constantinou, M.C. (1993). Dynamic analysis of generalized viscoelastic fluids, *J. Engineering Mechanics ASCE* **119**, 1663–1679.

Makris, N., Dargush, G.F. and Constantinou, M.C. (1995). Dynamic analysis of viscoelastic-fluid dampers, *J. Engineering Mechanics ASCE* **121**, 1114–1121.

Marden, M. (1949). *The Geometry of Zeros of a Polynomial in a Complex Variable*, American Mathematical Society, New York.

Marichev, O.I. (1983). *Handbook of Integral Transforms of Higher Transcendental Functions, Theory and Algorithmic Tables*, Chichester, Ellis Horwood.

Markowitz, H. (1963). Free vibration experiment in the theory of linear viscoelsticity, *Journal of Applied Physics* **14**, 21–25.

Marshall, P.D. and Carpenter, E.W. (1966). Estimates of Q for Raileigh waves, *Geophys. J. R. astr. Soc.* **10**, 549–550.

Marvin, R.S. (1960). The linear viscoelastic behaviour of rubber-like polymers and its molecular interpretation, in Bergen, T. (Editor), *Viscoelasticity: Phenomenological Aspects*, Academic Press, New York. pp. 27–54.

Marvin, R.S. (1962). Derivation of the relaxation spectrum representation of the mechanical response, *J. Res. Nat. Bureau Standards* **66A**, 349–350.

Mathai, A.M. (1993). *Handbook of Generalized Special Functions in Statistics and Physical Sciences*, Clarendon Press, Oxford.

Mathai, A.M. and Haubold, H.J. (2008). *Special Functions for Applied Scientists*, Springer Science, New York.

Mathai, A.M. and Saxena, R.K. (1973). *Generalized Hypergeometric Function with Applications in Statistics and Physical Sciences*, Springer Verlag, Berlin. [Lecture Notes in Mathematics, No 348]

Mathai, A.M. and Saxena, R.K. (1978). *The H-function with Applications in Statistics and Other Disciplines*, Wiley Eastern Ltd, New Delhi.

Mathai, A.M., Saxena, R.K. and Haubold, H.J. (2006). A certain class of Laplace transforms with applications to reaction and reaction-diffusion equations, *Astrophysics and Space Science* **305**, 283–288.

Mathai, A.M., Saxena, R.K. and Haubold, H.J. (2010). *The H-function: Theory and Applications*, Springer verlag, New York.

Matignon, D. (1994). *Représentations en variables détat de modèles de guides dondes avec dérivation fractionnaire*, Ph-D. Thesis, Université Paris XI.

Matignon, D. and Montseny, G., Editors (1998). *Fractional Differential Systems: Models, Methods and Applications*, Proceedings of the Colloquium FDS '98, ESAIM (European Ser. Appl. & Ind. Maths), Proceedings **5**. [http://www.emath.fr/Maths/Proc/Vol.5/index.htm]

Matkowsky, B.J. and Reiss, E.I. (1971). On the asymptotic theory of dissipative wave motion, *Arch. Rat. Mech. Anal.* **42**, 194–212.

McBride, A.C. and Roach, G.F., Editors (1985). *Fractional Calculus*, Pitman Research Notes in Mathematics Vol. 138, Pitman, London. [Proc. Int. Workshop. held at Univ. of Strathclyde (UK), 1984]

McCrum, N. G., Read, B. E. amd Williams, G. (1991). *Anelastic and Dielectric Effects in Polymeric Solids*, Dover, New York.

McDonal, F.J., Angona, F.A., Mills, R.L., Sengbush, R.L., Van Nostrand, R.G. and White, J.E. (1958). Attenuation of shear and compressional waves in Pierre shale, *Geophysics* **23**, 421–438.

Meidav, T. (1960). Viscoelastic properties of the standard linear solid, *Geophys. Prospecting* **12**, 80–99.
Meixner, J. (1964). On the theory of linear passive systems, *Arch. Rat. Mech. Anal.* **17** No 4, 278–296.
Meixner, J. and Koenig, H. (1958). Zur Theorie der Linearen Dissipativen Systeme, *Rheol. Acta* **1**, 190–193.
Meral, F.C., Royston, T.J. and Magin, R., Fractional calculus in viscoelasticity: An experimental study, *Commun. Nonlinear Sci. Numer. Simulat.* **15**, 939–945.
Meshkov, S.I. (1967). Description of internal friction in the memory theory of elasticity using kernels with a weak singularity, *J. Appl. Mech. Tech. Physics* **8** No 4, 100–102.
Meshkov, S.I. (1969). On the steady regime of an hereditary elastic oscillator, *J. Appl. Mech. Tech. Physics* **10** No 5, 780–785.
Meshkov, S.I. (1970). The integral representation of fractionally exponential functions and their application to dynamic problems of linear viscoelasticity, *J. Appl. Mech. Tech. Physics* **11** No 1, 100–107.
Meshkov, S.I., Pachevkaya, G.N. and Shermergor, T.D. (1966). Internal friction described with the aid of fractionally–exponential kernels, *J. Appl. Mech. Tech. Physics* **7** No 3, 63–65.
Meshkov, S.I., Pachevkaya, G.N., Postnikov, V.S. and Rossikhin, Yu.A. (1971). Integral representation of $\mathcal{E}_\gamma$ functions and their application to problems in linear viscoelasticity, *Int. J. Engng. Sci.* **9**, 387–398.
Meshkov, S.I. and Rossikhin, Yu.A. (1968). Propagation of acoustic waves in a hereditary elastic medium, *J. Appl. Mech. Tech. Physics* **9** No 5, 589–592.
Meshkov, S.I. and Rossikhin, Yu.A. (1970). Sound wave propagation in a viscoelastic medium whose hereditary properties are determined by weakly singular kernels, in Rabotnov, Yu.N. (Editor), *Waves in Inelastic Media*, (Kishniev), pp. 162–172. [in Russian]
Metzler, R., Barkai, E., Klafter, J. (1999). Anomalous diffusion and relaxation close to equilibrium: a fractional Fokker – Planck equation approach, *Phys. Rev. Lett.* **82**, 3563–3567.
Metzler, R., Glöckle, W.G. and Nonnenmacher, T.F. (1994). Fractional model equation for anomalous diffusion, *Physica A* **211**, 13–24.
Metzler, R. and Klafter, J. (2000a) Boundary value problems for fractional diffusion equations, *Physica A* **278**, 107–125.
Metzler, R. and Klafter, J. (2000b). The random walk's guide to anomalous diffusion: a fractional dynamics approach, *Phys. Reports* **339**, 1–77.
Metzler, R. and Klafter, J. (2001). Lévy meets Boltzmann: strange initial conditions for Brownian and fractional Fokker-Planck equations, *Physica A* **302**, 290–296.

Metzler, R. and Klafter, J. (2002). From stretched exponential to inverse power-law: fractional dynamics, Cole-Cole relaxation processes, and beyond, *J. Non-Crystalline Solids* **305**, 81–87.

Metzler, R. and Klafter, J. (2004). The restaurant at the end of the random walk: Recent developments in the description of anomalous transport by fractional dynamics, *J. Phys. A. Math. Gen.* **37**, R161–R208.

Metzler, R. and Nonnenmacher, T. F. (2003). Fractional relaxation processes and fractional rheological models for the description of a class of viscoelastic materials *Int. J. Plasticity* **19**, 941–959.

Metzler, R., Schick, W., Kilian, H.-G., Nonnenmacher, T. F. (1995). Relaxation in filled polymers: A fractional calculus approach, *J. Chem. Phys.* **103**, 7180–7186.

Mikusiński, J. (1959a). On the function whose Laplace transform is $\exp(-s^\alpha)$, *Studia Math.* **18**, 191–198.

Mikusiński, J. (1959b). *Operational Calculus*, Pergamon Press, New York [translated from the Second Polish Edition]

Miller, K.S. (1975). The Weyl fractional calculus, in [Ross (1975a)], pp. 80–89.

Miller, K.S. (1993). The Mittag-Leffler and related functions, *Integral Transforms and Special Functions* **1**, 41–49.

Miller, K.S. (2001). Some simple representations of the generalized Mittag-Leffler functions, *Integral Transforms and Special Functions* **11** No 1, 13–24.

Miller, K.S. and Ross, B. (1993). *An Introduction to the Fractional Calculus and Fractional Differential Equations*, Wiley, New York.

Miller, K.S. and Samko, S.G. (1997). A note on the complete monotonicity of the generalized Mittag-Leffler function, *Real Anal. Exchange* **23** No 2, 753–755.

Miller, K.S. and Samko, S.G. (2001). Completely monotonic functions, *Integral Transforms and Special Functions* **12**, 389–402.

Misra, A.K., and Murrell, A.F. (1965). An experimental study of the effect of temperature and stress on the creep of rocks, *Geophys. J., Roy. Astr. Soc.* **9**, 509–535.

Mitchell, B.J. (1995). Anelastic structure and evolution of the continental crust and upper mantle from seismic surface wave attenuation, *Rev. Geophys.* **33**, 441–462.

Mittag-Leffler, G.M. (1903a). Une généralisation de l'intégrale de Laplace-Abel, *C.R. Acad. Sci. Paris* (Ser. II) **137**, 537–539.

Mittag-Leffler, G.M. (1903b). Sur la nouvelle fonction $E_\alpha(x)$, *C.R. Acad. Sci. Paris* (Ser. II) **137**, 554–558.

Mittag-Leffler, G.M. (1904). Sopra la funzione $E_\alpha(x)$, *Rendiconti R. Accademia Lincei* (Ser. V) **13**, 3–5.

Mittag-Leffler, G.M. (1905). Sur la représentation analytique d'une branche uniforme d'une fonction monogène, *Acta Math.* **29**, 101–181.

Molinari, A. (1975). Viscoélasticité linéaire et function complètement monotones, *Journal de Mécanique* **12**, 541–553.

Moshrefi-Torbati, M. and Hammond, J.K. (1998). Physical and geometrical interpretation of fractional operators, *J. Franklin Inst.* **335B** No 6, 1077–1086.

Moodie, T.B., Mainardi, F. and Tait, R.J. (1985). Pressure pulses in fluid-filled distensible tubes, *Meccanica* **20**, 33–37.

Moodie, T.B., Tait, J. and Haddow, J.B. (1982). Waves in compliant tubes, in [Mainardi (1982)], pp. 124–168.

Morrison, J.A. (1956). Wave propagation in rods of Voigt materials and viscoelastic materials with three parameter models, *Quart. Appl. Math.* **14**, 153–169.

Morro, A. (1985). Negative semi-definiteness of viscoelastic attenuation tensor via the Clausius-Duhem inequality, *Arch. Mech.* **37**, 255–259.

Morro, A. and Vianello, M. (1990). Minimal and maximal free energy for materials with memory, *Boll. Unione Matematica Italiana (UMI)* **A4**, 45–55.

Müller, G. (1983). Rheological properties and velocity dispersion of a medium with power law dependence of Q on frequency, *J. Geophys.* **54**, 20–29.

Müller, G. (1986). Generalized Maxwell bodies and estimates of mantle viscosity, *Geophys. J. R. Astr. Soc.* **87**, 1113–1145. Erratum, *Geophys. J. R. Astr. Soc.* **91** (1987), 1135.

Mura, A. and Mainardi, F. (2009). A class of self-similar stochastic processes with stationary increments to model anomalous diffusion in physics, *Integral Transforms and Special Functions* **20** No 3-4, 185–198. [E-print: http://arxiv.org/abs/0711.0665]

Mura, A. and Pagnini, G. (2008). Characterizations and simulations of a class of stochastic processes to model anomalous diffusion, *Journal of Physics A: Math. Theor.* **41**, 285003/1–22. [E-print http://arxiv.org/abs/0801.4879]

Mura, A., Taqqu, M.S. and Mainardi, F. (2008). Non-Markovian diffusion equations and processes: analysis and simulation, *Physica A* **387**, 5033–5064. [E-print: http://arxiv.org/abs/0712.0240]

Murphy, W.F. (1982). Effect of partial water saturation on attenuation in sandstones, *J. Acoust. Soc. Am.* **71**, 1458–1468.

Nakhushev, A.N. (2003). *Fractional Calculus and its Applications*, Fizmatlit, Moscow. [in Russian]

Narain, A. and Joseph, D.D. (1982). Linearized dynamics for step jumps of velocity and displacement of shearing flows of a simple fluid, *Rheol. Acta* **21**, 228–250.

Narain, A. and Joseph, D.D. (1983). Remarks about the interpretation of impulse experiments in shear flows of viscoelastic liquids, *Rheol. Acta* **22**, 528–538.

Nielsen, L.E. (1962). *Mechanical Properties of Polymers*, Reinhold Publishing, New York. Dielectric relaxation phenomenon based on the fractional kinetics: theory and its experimental confirmation R R Nigmatullin.

Nigmatullin, R.R. (1984a) To the theoretical explanation of the "universal" response, *Phys. Status Solidi B* **123**, 739–745.

Nigmatullin, R.R. (1984b). On the theory of relaxation with "remnant" memory, *Phys. Status Solidi B* **124**, 389–393.

Nigmatullin, R.R. (1986). The realization of the generalized transfer equation in a medium with fractal geometry, *Phys. Status Solidi B* **133**, 425–430.

Nigmatullin, R.R. (1992a). *The Physics of Fractional Calculus and its Realization on the Fractal Structures*, Doctorate Thesis, Kazan University. [in Russian]

Nigmatullin, R.R. (1992b). A fractional integral ant its physical interpretation, *J. Theoret. Math. Phys.* **90** No 3, 242–251.

Nigmatullin, R.R. (2009). Dielectric relaxation phenomenon based on the fractional kinetics: theory and its experimental confirmation, *Physica Scripta* **T136**, 014001/1–6.

Nigmatullin, R.R. and Le Mehauté, A. (2005). Is there a physical/geometrical meaning of the fractional integral with complex exponent? *J. Non-Cryst. Solids* **351**, 2888–2899.

Nigmatullin, R.R. and Osokin, S.I. (2003). Signal processing and recognition of true kinetic equations containing non-integer derivatives from raw dielectric data, *J. Signal Process.* **83**, 2433–2453.

Nigmatullin, R.R. and Ryabov, Y. E. (1987). Cole-Davidson dielectric relaxation as a self-similar relaxation process, *Phys. Solid State* **39**, 87–99.

Nishihara, M. (1952). Creep of shale and sandy shale, *J. Geological Soc. Japan* **43**, 373–382.

Nishimoto, K. (Editor) (1990). *Fractional Calculus and its Applications*, Nihon University, Tokyo. [Proc. Int. Conf. held at Nihon Univ., Tokyo (Japan), 1989]

Nishimoto, K. (1991). *An Essence of Nishimoto's Fractional Calculus*, Descartes Press, Koriyama.

Nolte, B., Kempfle, S. and Schäfer, I (2003). Does a real material behave fractionally? Applications of fractional differential operators to the damped structure borne sound in viscoelastic solids, *Journal of Computational Acoustics* **11** No 3, 451–489.

Nonnenmacher, T.F. (1990a). On Lévy distributions and some applications to biophysical systems, in Albeverio, S., Casati, G., Cattaneo, U., Merlini, D. and Moresi, R. (Editors), *Stochastic Processes, Physics and Geometry*, World Scientific, Singapore, pp. 627–638. [Proc. 2-nd Ascona-Locarno, Ticino (Switzerland), Intern. Conf., 4-9 July 1988].

Nonnenmacher, T.F. (1990b). Fractional integral and differential equations for a class of Lévy-type probability densities, *J. Phys A: Math. Gen.* **23**, L697–L700.

Nonnenmacher, T.F. (1991). Fractional relaxation equations for viscoelasticity and related phenomena, in Casaz-Vazques, J. and Jou, D. (Editors), *Rheological Modelling: Thermodynamical and Statistical Approaches*, Springer Verlag, Berlin, pp. 309–320. [Lectures Notes in Physics, Vol. 381]

Nonnenmacher, T.F. and Glöckle, W.G. (1991). A fractional model for mechanical stress relaxation, *Phil. Mag. Lett.* **64**, 89–93.

Nonnenmacher, T.F. and Metzler, R. (1995). On the Riemann-Liouville fractional calculus and some recent applications, *Fractals* **3**, 557–566.

Nonnenmacher, T.F. and Nonnenmacher, D.J.F. (1989). Towards the formulation of a nonlinear fractional extended thermodynamics, *Acta Physica Hungarica* **66**, 145–154.

Norris, A.N. (1986). On the viscodynamic operator in Biots theory, *J. Wave-Material Interact.* **1**, 365–380.

Novikov, V.V., Wojciechowski, K.W., Komikova, O.A. and Thiel, T. (2005). Anomalous relaxation in dielectrics: Equations with fractional derivatives. *Materials Science-Poland* **23** No. 4, 977–984.

Nowick, A.S. and Berry, B.S. (1972). *Anelastic Relaxation in Crystalline Solids*, Academic Press, New York.

Nutting, P.G. (1921). A new general law of deformation, *J. Frankline Inst.* **191**, 679–685.

Nutting, P. G. (1943). A general stress-strain-time formula, *J. Frankline Inst.* **235**, 513–524.

Nutting, P. G. (1946). Deformation in relation to time, pressure and temperature, *J. Franklin Inst.* **242**, 449–458.

Ochmann, M. and Makarov, S. (1993). Representation of the absorption of nonlinear waves by fractional derivatives, *J. Acoust. Soc. Am.* **94**, 3392–3399.

O'Connell, R.J. and Budiansky, B. (1978). Measures of dissipation in viscoelastic media, *Geophys. Res. Lett.* **5**, 5–8.

Oldham, K.B. and J. Spanier, J. (1974). *The Fractional Calculus*, Academic Press, New York.

Oldroyd, J.G. (1950). On the formulation of theological equations of state, *Proc. Roy. Soc. London A* **200**, 523–591.

Olver, F.W.J. (1974). *Asymptotics and Special Functions*, Academic Press, New York. [Reprinted by AK Peters, Wellesley, 1997]

Orowan, E. (1967). Seismic damping and creep in the mantle, *Geophys. J. R. Astr. Soc.* **14**, 191–218.

Orsingher, E. and Beghin, L. (2004). Time-fractional telegraph equations and telegraph processes with Brownian time, *Probab. Theory Related Fields* **128**, 141–160.

Orsingher, E. and Beghin, L. (2009). Fractional diffusion equations and processes with randomly varying time, *Ann. Probab.* **37**, 206–249.

Ortigueira, M.D. (2006). A coherent approach to non integer order derivatives, *Signal Processing* **86** No 10, 2505–2515.

Ortigueira, M.D. (2008). Fractional central differences and derivatives, *Journal of Vibration and Control* **14** No 9/10, 1255–1266.

Ortigueira, M.D. and Coito, F. (2004). From differences to differintegrations, *Fractional Calculus and Applied Analysis* **7** No 4, 459–471.

Oughstun, K. E. and Sherman, G. C. (1994). *Electromagnetic Pulse Propagation in Causal Dielectrics*, Springer Verlag, Berlin.

Oustaloup, A. (1995). *La Dérivation Non Entière: Théorie, Synthèse et Applications*, Hermes, Paris.

Padovan, J. (1987). Computational algorithm for FE formulation involving fractional operators, *Computational Mechanics* **2**, 271–287.

Padovan, J. and Guo, Y.H. (1988). General response of viscoelastic systems modeled by fractional operators, *J. Franklin Inst.* **325** No 2, 247–275.

Padovan, J. Chung, S. V. and Guo, Y.H. (1987). Asymptotic steady state behavior of fractionally damped systems, *J. Franklin Inst.* **324** No 3, 491–511.

Padovan, J. and Sawicki, J.T. (1997). Diophantine type fractional derivative representation of structural hysteresis - Part I: Formulation, *Computational Mechanics* **19**, 335–340.

Padovan, J. and Sawicki, J.T. (1998). Diophantinized fractional representations for nonlinear elastomeric media. *Computers & Structures* **66**, 613–626.

Palade, L.-I. and DeSanto, J. A. (2001). Dispersion equations which model high-frequency linear viscoelastic behavior as described by fractional derivative models, *Int. J. of Non-Linear Mechanics* **36**, 13–24.

Palade, L.-I., Attané, P., Huilgol, R.R. and Mena, B. (1999). Anomalous stability behavior of a properly invariant constitutive equation which generalises fractional derivative models, *Int. J. Engng. Sci.* **37**, 315–329.

Palade, L.-I., Veney, V. and Attané, P. (1996). A modified fractional model to describe the entire viscoelastic behavior of polybutadienes from flow to glassy regime, *Reol. Acta* **35**, 265–273.

Paley, R.E.A.C. and Wiener, N. (1934). *Fourier Transforms in the Complex Domain*, American Mathematical Society, New York.

Papoulis, A. (1962). *The Fourier Integral and its Applications*, McGraw-Hill, New York.

Paradisi, P., Cesari, R., Mainardi, F. and Tampieri, F. (2001a). The fractional Fick's law for non-local transport processes, *Physica A* **293**, 130–142.

Paradisi, P., Cesari, R., Mainardi, F., Maurizi, A. and Tampieri, F. (2001a). A generalized Fick's law to describe non-local transport processes, *Phys. Chem. Earth (B)* **26**, 275–279.

Paris, R.B. (2002). Exponential asymptotics of the Mittag-Leffler function, *Proc R. Soc. London A* **458**, 3041–3052.

Paris, R.B. and Kaminski, D. (2001). *Asymptotic and Mellin-Barnes Integrals*, Cambridge University Press, Cambridge.

Pathak R.S. (1972). A general differential equation satisfied by a special function, *Progress of Math.* **6**, 46–50.

Pelzer, H. (1955). Realizability, Kramers-Kronig relations and Fuoss-Kirkwood dielectrics, *Physica* **22**, 103–108.

Pelzer, H. (1957). Models of material with loss per cycle independent of frequency, *J. Polymer Science* **25**, 51–60.

Petrovic, Lj.M., Spasic, D.T. and Atanackovic T.M. (2005). On a mathematical model of a human root dentin, *Dental Materials* **21**, 125–128.

Pfitzenreiter, T. (2004). A physical basis for fractional derivatives in constitutive equations *Z. angew. Math. Mech. (ZAMM)* **84**, 284–287.

Phragmén, E. (1904). Sur une extension d'un théoreme classique de la théorie des fonctions, *Acta Math.* **28**, 351–368.

Picozzi, S. and West, B.J. (2002). Fractional Langevin model of memory in financial markets, *Phys. Rev. E* **66**, 046118/1–12.

Pillai, R. N. (1990). On Mittag-Leffler functions and related distributions, *Ann. Inst. Statist. Math.* **42**, 157–161.

Pipkin, A.C. (1986). *Lectures on Viscoelastic Theory*, Springer Verlag, New York, Second Edition. [First Edition 1972]

Pipkin, A.C. (1988). Asymptotic behaviour of viscoelastic waves. *Quart. J. Mech. Appl. Math.* **41**, 51–64.

Podlubny, I. (1994). *The Laplace Transform Method for Linear Differential Equations of Fractional Order*, Report no UEF-02-94, Inst. Exp. Phys. Slovak Acad. Sci, Kosice.

Podlubny, I. (1999). *Fractional Differential Equations*, Academic Press, San Diego. [Mathematics in Science and Engineering, Vol. 198]

Podlubny, I. (2002). Geometric and physical interpretation of fractional integration and fractional differentiation, *Fractional Calculus and Applied Analysis* **5**, 367–386.

Podlubny, I. (2006). *Mittag-Leffler function*. WEB Site of MATLAB Central: http://www.mathworks.com/matlabcentral/fileexchange

Podlubny, I., Petras, I, O'Leary, P., Dorcak, P. and Vinagre, B. (2002). Analogue realization of fractional order controllers, *Nonlinear Dynamics* **29** No 1-4, 281–296.

Pollard, H. (1946). The representation of $\exp(-x^\lambda)$ as a Laplace integral, *Bull. Amer. Math. Soc.* **52**, 908–910.

Pollard, H. (1948). The completely monotonic character of the Mittag-Leffler function $E_\alpha(-x)$, *Bull. Amer. Math. Soc.* **54**, 1115–1116.

Post, E.L. (1930). Generalized differentiation, *Trans. Amer. Math. Soc.* **32**, 723–781.

Postvenko, Y.Z. (2007). Two-dimensional axisymmentric stresses exerted by instantaneous pulses and sources of diffusion in an infinite space in a case of time-fractional diffusion equation, *Int. J. Solids Struct.* **44**, 2324–2348.

Postvenko, Y.Z. (2008). Signalling problem for time-fractional diffusion-wave equation in a half-plane, *Fractional Calculus nd Applied Analysis* **11** No 3, 329–352.

Postvenko, Y.Z. (2009). Theory of thermoelasticity based on the space-time-fractional heat conduction equation, *Physica Scripta* **T136**, 014017/1–6.

Prabhakar, T.R. (1971). A singular integral equation with a generalized Mittag-Leffler function in the kernel, *Yokohama Math. J.* **19**, 7–15.

Pritz, T. (1996). Analysis of four-parameter fractional derivative model of real solid materials, *J. Sound and Vibration* **195**, 103–115.

Pritz, T. (1998). Frequency dependences of complex moduli and complex Poisson's ratio of real solid materials, *J. Sound and Vibration* **214**, 83–104.

Pritz, T. (1999). Verification of local Kramers-Kronig relations for complex modulus by means of fractional derivative model, *J. Sound and Vibration* **228**, 1145–1165.

Pritz, T. (2001). Loss factor peak of viscoelastic materials: magnitude to width relations, *J. Sound and Vibration* **246**, 265–280.

Pritz, T. (2003). Five-parameter fractional derivative model for polymeric damping materials, *J. Sound and Vibration* **265**, 935–952.

Pritz, T. (2004). Frequency power law of material damping, *Applied Acoustics* **65**, 1027–1036.

Pritz, T. (2005). Unbounded complex modulus of viscoelastic materials and the l Kramers-Kronig relations, *J. Sound and Vibration* **279**, 687–697.

Pritz, T. (2007). The Poisson's loss factor of solid viscoelastic materials, *J. Sound and Vibration* **306**, 790–802.

Prudnikov, A. P., Brychkov, Y. A. and Marichev, O. I. (1986). *Integrals and Series*, Vol I, II, III, Gordon and Breach, New York.

Prüsse, J. (1993). *Evolutionary Integral Equations and Applications*, Birkhauser Verlag, Basel.

Pskhu, A.V. (2003). Solution of boundary value problems for the fractional diffusion equation by the Green function method, *Differential Equations* **39** No 10, 1509–1513 [English translation from the Russian Journal *Differentsial'nye Uravneniya*]

Pskhu, A.V. (2005). *Partial Differential Equations of Fractional Order*, Nauka, Moscow. [in Russian]

Pskhu, A.V. (2009). The fundamental solution of a diffusion-wave equation of fractional order, *Izv. Math.* **73** No 2, 351–392.

Qaisar, M. (1989). Attenuation properties of viscoelastic materials, *Pure Appl. Geophys. (PAGEOPH)* **131**, 703–713.

Rabotnov, Yu.N. (1948). Equilibrium of an elastic medium with after effect, *Prikl. Matem. i Mekh. (PMM)* **12** No 1, 81–91. [in Russian]

Rabotnov, Yu.N. (1969). *Creep Problems in Structural Members*, North-Holland, Amsterdam. [English translation of the 1966 Russian edition]

Rabotnov, Yu.N. (1973). *On the Use of Singular Operators in the Theory of Viscoelasticity*, Moscow, 1973, pp. 50. Unpublished Lecture Notes for the CISM course on Rheology held in Udine, October 1973. [http://www.cism.it]

Rabotnov, Yu.N. (1974). *Experimental Evidence of the Principle of Hereditary in Mechanics of Solids*, Moscow, 1974, pp. 80. Unpublished Lecture Notes for the CISM course on *Experimental Methods in Mechanics, A) Rheology* held in Udine, 24-29 October 1974. [http://www.cism.it]

Rabotnov, Yu.N. (1980). *Elements of Hereditary Solid Mechanics*, MIR, Moscow. [English translation, revised from the 1977 Russian edition]

Rabotnov, Yu.N., Papernik, L.Kh., and Zvonov, E.N. (1969). *Tables of a Fractional Exponential Function of Negative Parameters and its Integral*, Nauka, Moscow. [In Russian]

Rainville, E.D. (1960). *Special Functions*, Macmillan, New York.

Rajagopal, J.R. (2009). A note on a reappraisal and generalization of the Kelvin-Voigt model, *Mechanics Research Communications* **36** No 2, 232–235.

Rajagopal, J.R. and Wineman A.S. (1980). A useful correspondence principle in the theory of linear materials, *Journal of Elasticity* **10**, 429–434.

Ramirez, L.E.S and C.F.M. Coimbra, C.F.M. (2007). A variable order constitutive relation for viscoelasticity, *Ann. Physik* **16** No 7/8, 543–552.

Ramirez, L.E.S and C.F.M. Coimbra, C.F.M. (2010). On the selection and meaning of variable order operators for dynamic modeling, *Intern. Journal of Differential Equations*, Article ID 846107, 16 pages.

Ranalli, G. (1987). *Rheology of the Earth*, Allen & Unwin, London.
Rangarajan, G. and Ding, M.Z. (2000a). Anomalous diffusion and the first passage time problem, *Phys. Rev. E* **62**, 120–133.
Rangarajan, G. and Ding, M.Z. (2000b). First passage time distribution for anomalous diffusion, *Phys. Lett. A* **273**, 322–330.
Rao, M.A. and Steffe, J.F. (1992). *Viscoelasticity of Foods*, Elsevier Applied Science, New York.
Reiner, M. (1949). *Twelve Lectures on Rheology*, North-Holland, Amsterdam.
Reiss, E.I. (1969). The impact problem for the Klein-Gordon equation, *SIAM J. Appl. Math.* **17**, 526–542.
Renardy, M. (1982). Some remarks on the propagation and non-propagation of discontinuities in linearly viscoelastic liquids, *Rheol. Acta* **21**, 251–254.
Renardy, M., Hrusa, W.J., and Nohel, J.A. (1987). *Mathematical Problems in Viscoelasticity*, Longman, Essex. [Pitman Monographs and Surveys in Pure and Applied Mathematics, Vol. 35]
Renno, P. (1983). On the Cauchy problem in linear viscoelasticity, *Atti Acc. Lincei, Rend. Fis.* (Ser. VIII) **75**, 195–204.
Richards, P.G. (1984). On wave fronts and interfaces in anelastic media. *Bull. Seism. Soc. Am.* **74**, 2157–2165.
Ricker, N.H. (1977). *Transient Waves in Visco-Elastic Media*, Elsevier, Amsterdam.
Rigby, B.J. (1960). Power law for creep, *Br. J. Appl. Phys.* **11**, 281–283.
Riemann, B. (1892). Versuch einer allgemeinen Auffassung der Integration und Differentiation, *Bernard Riemann: Gesammelte Mathematische Werke*, Teubner Verlag, Leipzig, XIX, pp. 353–366. [Reprinted in *Bernard Riemann: Collected Papers*, Springer Verlag, Berlin (1990) XIX, pp. 385–398.]
Riesz, M. (1949). L'integral de Riemann-Liouville et le probleme de Cauchy, *Acta Math.* **81**, 1–223.
Rivero, M., Rodríguez-Germá, Trujillo, J.J and Pilar Velasco, M. (2010). Fractional operators and some special functions, *Computers and Mathematics with Applications*, in press.
Rocco, A. and West B.J. (1999). Fractional calculus and the evolution of fractal phenomena *Physica A* **265**, 535–546.
Rogers, L. (1981). On modeling viscoelastic behavior, *Shock and Vibration Bulletin* **51**, 55–69.
Rogers, L. (1983). Operators and fractional derivatives for viscoelastic constitutive equations, *J. Rheology* **27**, 351–372.
Ross, B. (Editor) (1975a). *Fractional Calculus and its Applications*, Springer Verlag, Berlin. [Lecture Notes in Mathematics, Vol. 457]

Ross, B. (1975b). A brief history and exposition of the fundamental theory of fractional calculus, in [Ross (1975a)], pp. 1–36.

Ross, B. (1977). The development of fractional calculus 1695-1900, *Historia Mathematica* **4**, 75–89.

Rossikhin, Yu.A., *Dynamic problems of linear viscoelasticity connected with the investigation of retardation and relaxation spectra*, PhD Dissertation, Voronezh Polytechnic Institute, Voronezh. [in Russian]

Rossikhin, Yu.A., Reflections on two parallel ways in the progress of fractional calculus in mechanics of solids, *Appl. Mech. Review* **63**, 010701/1–12.

Rossikhin, Yu.A. and Shitikova, M.V. (1995). Ray method for solving dynamic problems connected with propagation of wave surfaces of strong and weak discontinuity, *Appl. Mech. Review* **48**, 1–39.

Rossikhin, Yu.A. and Shitikova, M.V. (1997a). Applications of fractional calculus to dynamic problems of linear and nonlinear fractional mechanics of solids, *Appl. Mech. Review* **50**, 15–67.

Rossikhin, Yu.A. and Shitikova, M.V. (1997b). Applications of fractional derivatives to the analysis of damped vibrations of viscoelastic single mass systems, *Acta Mechanica* **120**, 109–125.

Rossikhin, Yu.A. and Shitikova, M.V. (1997c). Applications of fractional operators to the analysis of damped vibrations of viscoelastic single-mass systems, *J. Sound and Vibration* **199**, 567–586.

Rossikhin, Yu.A. and Shitikova, M.V. (1998). Application of fractional calculus for analysis of nonlinear damped vibrations of suspension bridges, *J. Eng. Mech.* **124** No 9, 1029–1036.

Rossikhin, Yu.A., and Shitikova, M.V. (1999). Vibrations of a hereditarily elastic 2dof mechanical system whose hereditary properties are described by fractional derivatives, *Applied Mechanics in the Americas* **8**, 1409–1412.

Rossikhin, Yu.A. and Shitikova, M.V. (2000a). Application of weakly anisotropic models of a continuous medium for solving the problems of wave dynamics, *Appl. Mech. Review* **53**, 37–86.

Rossikhin, Yu.A. and Shitikova, M.V. (2000b). Analysis of nonlinear vibrations of a two-degree-of-freedom mechanical system with damping modelled by a fractional derivative, *J. Eng. Math.* **37**, 343–362.

Rossikhin, Yu.A. and Shitikova, M.V. (2001a) Analysis of dynamic behaviour of viscoelastic rods whose rheological models contain fractional derivatives of two different orders, *Z. angew. Math. Mech. (ZAMM)* **81**, 363–376.

Rossikhin, Yu.A. and Shitikova, M.V. (2001b). A new method for solving dynamic problems of fractional derivative viscoelasticity, *Int. J. Engineering Science* **39**, 149–176.

Rossikhin, Yu.A., and Shitikova, M.V. (2001c). Analysis of rheological equations involving more than one fractional parameter by the use of the simplest mechanical systems based on these equations, *Mechanics of Time-Dependent Materials* **5**, 131–175.

Rossikhin, Yu.A. and Shitikova, M.V. (2004). Analysis of the viscoelastic rod dynamics via models involving fractional derivatives or operators of two different orders, *The Shock and Vibration Digest* **36**, 3–26.

Rossikhin, Yu.A. and Shitikova, M.V. (2007). Comparative analysis of viscoelastic models involving fractional derivatives of different orders, *Fractional Calculus and Applied Analysis* **10** No 2, 111–121.

Rossikhin, Yu.A. and Shitikova, M.V. (2007b). Transient response of thin bodies subjected to impact wave approach, *Shock Vib. Dig.* **39** No 4, 273–309.

Rossikhin, Yu.A. and Shitikova, M.V. (2010). Applications of fractional calculus to dynamic problems of solid mechanics: novel trends and recent results, *Appl. Mech. Review* **63**, 010801/1–52.

Rossikhin, Yu.A., Shitikova, M.V. and Shcheglova, T.A. (2010). Analysis of free vibrations of a viscoelastic oscillator via the models involving several fractional parameters and relaxation/retardation times, *Computers and Mathematics with Applications*, in press.

Rouse, P.E. (1953). A theory of the linear viscoelastic properties of dilute solutions of coiling polymers, *J. Chem. Phys.* **21** No 7, 1272–1280.

Rozovskii, M.I. (1959). Certain properties of special operators used in creep theory, *Prikl. Matem. i Meckh. (PMM)* **23** No 5, 1388–1462. [English Edition]

Rubin, M.J. (1954). Propagation of longitudinal deformation waves in a prestressed rod of material exhibiting a strain-rate effect, *J. Appl. Phys.* **25**, 528–536.

Rubin, B. (1996). *Fractional Integrals and Potentials*, Addison-Wesley & Longman, Harlow. [Pitman Monographs and Surveys in Pure and Applied Mathematics No 82]

Rumpker, G. and Wolff, D. (1996). Viscoelastic relaxation of a Burgers half-space: implications for the interpretation of the Fennoscandian uplift, *Geophysical J. Int.* **124**, 541–555.

Sabadini, R., Yuen, D.A. and Gasperini, P. (1985). The effects of transient rheology on the interpretation of lower mantle viscosity, *Geophys. Res. Lett.* **12**, 361–364.

Sabadini, R., Smith, B. K. and Yuen, D. A. (1987). Consequences of experimental transient rheology, *Geophys. Res. Lett.* **14**, 816–819.

Sabatier, J., Agrawal, O.P., Tenreiro Machado, J.A., Editors (2007). *Advances in Fractional Calculus: Theoretical Developments and Applications in Physics and Engineering*, Springer Verlag, New York.

Sabatier, J., Merveillaut, M. Malti, R. and Oustaloup, A. How to impose physically coherent initial conditions to a fractional system? *Commun. Nonlinear Sci. Numer. Simulat.* **15**, 1318–1326.

Sackman, J. L. (1982). Prediction and identification in viscoelastic wave propagation, in [Mainardi (1982)], pp. 218–234.

Sackman, J. L. and I. Kaya, I. (1968). On the propagation of transient pulses in linearly viscoelastic media. *J. Mech. Phys. Solids* **16**, 349–356.

Saichev, A. and Zaslavsky, G. (1997). Fractional kinetic equations: solutions and applications, *Chaos* **7**, 753–764.

Saigo, M. and Kilbas, A.A. (1998). On Mittag-Leffler type function and applications. *Integral Transforms Special Functions* **7** No 1-2, 97–112.

Saigo, M. and Kilbas, A.A. (2000). Solution of a class of linear differential equations in terms of functions of Mittag-Leffler type, *Differential Equations* **36** No 2, 193–200.

Sakakibara, S. (1997). Properties of vibration with fractional derivative damping of order 1/2, *JSME Int. Jour. C* **40**, 393–399.

Sakakibara, S. (2001). Relaxation properties of fractional derivative viscoelasticity models, *Nonlinear Analysis* **47**, 5449–5454.

Sakakibara, S. (2004). Continued fractions and fractional derivative viscoelasticity, Departmental Bulletin Paper, Tokyo Denki University, No 1381, 21–41. [E-print http://hdl.handle.net/2433/25674]

Samko, S.G., Kilbas, A.A. and Marichev, O.I. (1993). *Fractional Integrals and Derivatives, Theory and Applications*, Gordon and Breach, Amsterdam. [English translation from the Russian, Nauka i Tekhnika, Minsk, 1987]

Sansone, G. and Gerretsen, J. (1960). *Lectures on the Theory of Functions of a Complex Variable*, Vol. I. *Holomorphic Functions*, Nordhoff, Groningen.

Sanz-Serna, J.M. (1988). A numerical method for a partial integro-differential equation, *SIAM J. Numer. Anal.* **25**, 319–327.

Sato, K.-I. (1999). *Lévy Processes and Infinitely Divisible Distributions*, Cambridge University Press, Cambridge.

Savage, J.C. and O'Neill, M.E. (1975). The relation between the Lomnitz and the Futterman theories of internal friction, *J. Geophys. Res.* **80**, 249–251.

Sawicki, J.T. and Padovan J. (1997). Diophantine type fractional derivative representation of structural hysteresis - Part II: Fitting, *Computational Mechanics* **19**, 341–355.

Saxena, R.K. and Kalla, S.L. (2008). On the solution of certain kinetic equations, *Appl. Math. Comput.* **199**, 504–511.

Saxena, R.K., Kalla, S.L. and Kiryakova, V. (2003). Relations connecting multiindex Mittag-Leffler functions and Riemann-Liouville fractional calculus, *Algebras, Groups and Geometries* **20**, 363–385.

Saxena, R.K., Mathai, A.M. and Haubold, H.J. (2004a). On generalized fractional kinetic equations, *Physica A* **344**, 657–664.

Saxena, R.K., Mathai, A.M. and Haubold, H.J. (2004b). Unified fractional kinetic equations and a fractional diffusion, *Astrophysics and Space Science* **290**, 299–310.

Saxena, R.K., Mathai, A.M. and Haubold, H.J. (2006a). Fractional reaction-diffusion equations, *Astrophysics and Space Science* **305**, 289–296.

Saxena, R.K., Mathai, A.M. and Haubold, H.J. (2006b). Reaction-diffusion systems and nonlinear waves, *Astrophysics and Space Science* **305**, 297–303.

Saxena, R.K., Mathai, A.M. and Haubold, H.J. (2006c). Solution of generalized fractional reaction-diffusion equations, *Astrophysics and Space Science* **305**, 305–313.

Saxena, R.K., Mathai, A.M. and Haubold, H.J. (2006d). Solution of fractional reaction-diffusion equation in terms of Mittag-Leffler functions, *Int. J. Sci. Res.* **15**, 1–17.

Saxena, R.K., Mathai, A.M. and Haubold, H.J. (2008). Solutions of certain fractional kinetic equations a fractional diffusion equation, *Int. J. Sci. Res.* **17**, 1–8.

Scaife, B.K.P. (1998). *Principles of Dielectrics*, Oxford University Press, London.

Scalas, E., Gorenflo, R. and Mainardi, F. (2000). Fractional calculus and continuous-time finance, *Physica A* **284** No 1-4, 376–384.

Scalas, E., Gorenflo, R., Mainardi, F. and Raberto, M. (2003). Revisiting the derivation of the fractional diffusion equation, *Fractals* **11** Suppl. S, 281–289.

Scalas, E., Gorenflo, R. and Mainardi, F. (2004). Uncoupled continuous-time random walks: Solution and limiting behavior of the master equation, *Physical Review E* **69**, 011107/1–8.

Scalas, E., Gorenflo, R., Mainardi, F. and Meerschaert, M.M. (2005). Speculative option valuation and the fractional diffusion equation, in [Le Méhauté et al. (2005)], pp. 265–274.

Scarpi, G.B. (1972). Sulla possibilità di un modello reologico intermedio di tipo evolutivo, *Rend. Acc. Naz. Lincei*, Cl. Sci. Fis., Mat. Natur. (Ser. VIII), **52**, 913–917.

Scarpi, G.B. (1973). Sui modelli reologici intermedi per liquidi viscoelatici, *Atti Accad. Sci. Torino*, Cl. Sci. Fis. Mat. Natur. **107**, 239–243.

Schäfer, I. (2000). Beschreibung der Dämpfung in Staeben mittels fraktionaler Zeitableutungen, *Z. angew. Math. Mech. (ZAMM)* **80**, 1–5.

Schäfer, I. (2001). *Fraktionale Zeitableutungen, zur Beschreibung viskoelastischen Materialverhaltens* PhD Dissertation Thesis, Hamburg.

Schäfer, I and Kempfle, S. (2004). Impulse responses of fractional damped systems, *Nonlinear Dynamics* **38** No 1-2, 61–68.

Schapery, R.A. (1962). Approximate methods of transform inversion for viscoelastic stress analysis, *Proc. 4th U.S. National Congress Applied Mechanics*, Berkeley, California, USA, Vol. 2, ASME Publ., pp. 1075–1085.

Schapery, R.A. (1967). Stress analysis of viscoelastic composite materials, *J. Composite Materials* **1** No 3, 228–267.

Schelkunoff, S.A. (1944). Proposed symbols for the modified cosine and integral exponential integral, *Quart. Appl. Math.* **2**, p. 90.

Schiessel, H. and Blumen, A. (1993). Hierarchical analogues to fractional relaxation equations, *J. Phys. A: Math. Gen.* **26**, 5057–5069.

Schiessel, H., Alemany, P. and Blumen, A. (1994). Dynamics in disordered systems, *Progr. Colloid Polymer Sci.* **96**, 16–21.

Schiessel, H. and Blumen, A. (1995). Mesoscopic pictures of the sol–gel transition: ladder models and fractal networks, *Macromolecules* **28**, 4013–4019.

Schiessel, H., Metzler, R., Blumen, A. and Nonnenmacher, T.F. (1995). Generalized viscoelastic models: their fractional equations with solutions, *J. Physics A: Math. Gen.* **28**, 6567–6584.

Schiessel, H., Friedrich, Chr. and Blumen, A. (2000). Applications to problems in polymer physics and rheology, [Hilfer (2000a)], pp. 331–376.

Schmidt, A. and Gaul, L. (2002). Finite element formulation,of viscoelastic constitutive equations using fractional time derivative, *Nonlinear Dynamics* **29** No 1-4, 37–55.

Schmidt, A. and Gaul, L. (2003). Implementation von Stoffgesetzen mit fraktionalen Ableitungen in die Finite Elemente Methode, *Zeitschrift angew. Math. Mech. (ZAMM)* **83**, 26–37.

Schmidt, A. and Gaul, L. (2006). On the numerical evaluation of fractional derivatives in multi-degree-of-freedom systems, *Signal Processing* **86**, 2592–2601.

Schneider, W.R. (1986). Stable distributions: Fox function representation and generalization, in Albeverio, S., Casati, G. and Merlini, D. (Editors), *Stochastic Processes in Classical and Quantum Systems*, Springer Verlag, Berlin, pp. 497–511. [Lecture Notes in Physics, Vol. 262]

Schneider, W.R. (1990). Fractional diffusion, in: Lima, R., Streit, L. and Vilela Mendes, D (Editors), *Dynamics and Stochastic Processes,*

Theory and Applications, Springer Verlag, Heidelberg, pp. 276–286. [Lecture Notes in Physics, Vol. 355]

Schneider, W.R. (1996). Completely monotone generalized Mittag-Leffler functions, *Expositiones Mathematicae* **14**, 3–16.

Schneider, W.R. and Wyss, W. (1989). Fractional diffusion and wave equations, *J. Math. Phys.* **30**, 134–144.

Schwartz, L. (1957). *Théorie des distributions*, Second Edition, Hermannn, Paris.

Scott-Blair, G.W. (1944). Analytical and integrative aspects of the stress-strain-time problem, *J. Scientific Instruments* **21**, 80–84.

Scott-Blair, G.W. (1947). The role of psychophysics in rheology, *J. Colloid Sci.* **2**, 21–32.

Scott-Blair, G.W. (1949). *Survey of General and Applied Rheology*, Pitman, London.

Scott-Blair, G.W. (1974). *An Introduction to Biorheology*, Elsevier, New York.

Scott-Blair, G.W. and Caffyn, J.E. (1949). An application of the theory of quasi-properties to the treatment of anomalous stress-strain relations, *Phil. Mag.* (Ser. 7) **40**, 80–94.

Scott-Blair, G.W., Veinoglou, B.C. and Caffyn, J.E. (1947). Limitations of the Newtonian time scale in relation to non-equilibrium rheological states and a theory of quasi-properties, *Proc. R. Soc. London A* **189**, 69–87.

Sedletski, A.M. (2000). On zeros of a function of Mittag-Leffler type, *Math. Notes.* **68** No. 5, 117–132. [translation from Russian *Matematicheskie Zametki*]

Sedletski, A.M. (2004). Non asymptotic properties of roots of a Mittag-Leffler type function, *Math. Notes.* **75** No. 3, 372–386. [translation from Russian *Matematicheskie Zametki*]

Seneta, E. (1976). *Regularly Varying Functions*, Springer Verlag, Berlin. [Lecture Notes in Mathematics, No 508]

Seredyńska, M. and Hanyga, A. (2000). Nonlinear Hamiltonian equations with fractional damping, *J. Math. Phys.* **41**, 2135–2155.

Seybold, H.J. and Hilfer, R. (2005). Numerical results for the generalized Mittag-Leffler function, *Fractional Calculus and Applied Analysis* **8**, 127–139.

Shermergor, T.D. (1966). On the use of fractional differentiation operators for the description of elastic–after effect properties of materials, *J. Appl. Mech. Tech. Phys.* **7** No 6, 85–87.

Shimuzu, N. and Zhang, W. (1999). Fractional calculus approach to dynamic problems of viscoelastic materials, *JSME Int. Jour. C* **42**, 825–837.

Shlesingher, M.F., Zaslavsky, G.M. and Klafter, J. (1993). Strange kinetics, *Nature* **363**, 31–37.
Shlesinger, M.F., Zaslavsky, G.M., and Frisch, U., Editors (1995). *Lévy Flights and Related Topics in Physics*, Springer, Berlin.
Shokooh, A. and Suarez, L. (1999). A comparison of numerical methods applied to a fractional model of damping, *Journal of Vibration and Control* **5**, 331–354.
Shukla, A.K. and Prajapati, J.C. (2007). On a generalization of Mittag-Leffler function and its properties, *J. Math. Anal. Appl.* **336**, 797–811.
Simo, J.C., Hughes, T.J.R. (1998). *Computational Inelasticity*, Springer, New York.
Sing, S.J. and Chatterjee, A. (2007). Fractional damping: stochastic origin and finite approximations, in [Sabatier *et al.* (2007)], pp. 389–402.
Slonimsky, G.L. (1961). On the laws of deformation of visco-elastic polymeric bodies, *Dokl. Akad. Nauk BSSR* **140** No 2, 343–346. [in Russian]
Slonimsky, G.L. (1967). Laws of mechanical relaxation processes in polymers, *J. Polymer Science, Part C* **16**, 1667–1672.
Smit, W. and de Vries, H. (1970). Rheological models containing fractional derivatives, *Rheol. Acta* **9**, 525–534.
Smith, R.L. (1970). The velocity of light, *Amer. J. Physics* **38**, 978–984.
Smith, S.W. (1972). The anelasticity of the mantle, *Tectonophysics* **13**, 601–622.
Sneddon, I.N. (1956). *Special Functions of Mathematical Physics and Chemistry*, Oliver and Boyd, London.
Soczkiewicz, E. (2002). Application of fractional calculus in the theory of viscoelasticity, *Molecular and Quantum Acoustics* **23**, 397–404.
Sokolov, I.M. and Klafter, J. (2005). From diffusion to anomalous diffusion: A century after Einsteins Brownian motion, *Chaos* **15**, 026103/1–7.
Sokolov, I.M., Klafter, J. and Blumen, A. (2002). Fractional kinetics, *Physics Today* **55**, 48–54.
Sokolov, I.M., Chechkin, A.V. and Klafter, J. (2004). Distributed-order fractional kinetics, *Acta Phys. Polonica B* **35**, 1323–1341.
Song, D.Y. and Jiang, T.Q. (1998). Study on the constitutive equation with fractional derivative for the viscoelastic fluids - Modified Jeffreys model and its application, *Rheol. Acta* **37** No 5, 512–517.
Sorrentino, S., Marchesiello, S and Piombo, B.A.D. (2003). A new analytical technique for vibration analysis of non-proportionally damped beams, *Journal of Sound and Vibration* **265**, 765–782.
Soula, M. and Chevalier,Y. (1998). La dérivee fractionnaire en rhéologie des polymères: Application aux comportements élastiques et viscoélastiques linéaireset non-linéaires des élastomères. *ESAIM: Proceedings Fractional Differential Systems: Models, Methods and Applications*, Vol. 5, pp. 193–204.

Soula, M., Vinh, T., Chevalier, Y., Beda, T. and Esteoule, C. (1997a). Measurements of isothermal complex moduli of viscoelastic materials over a large range of frequencies, *J. Sound and Vibration* **205**, 167–184.

Soula, M., Vinh, T. and Chevalier, Y. (1997b). Transient responses of polymers and elastomers deduced from harmonic responses, *J. Sound and Vibration* **205**, 185–203.

Spada, G. (2009). Generalized Maxwell Love numbers, E-print http://arxiv.org/abs/0911.0834 [math-ph], 4 Nov 20079, pp. 17.

Spada, G., Yuen, D.A., Sabadini, R. and Boschi, E. (1991). Lower-mantle viscosity constrained by seismicity around deglaciated regions, *Nature* **351**, 53–55.

Spada, G., Sabadini, R., Yuen, D.A. and Ricard, Y. (1992). Effects on post-glacial rebound from the hard rheology in the transition zone, *Geophys. J. Int.* **109**, 683–700.

Spada, G. and Boschi, L. (2006). Using the Post-Widder formula to compute the Earth's viscoelastic Love numbers, *Geophys. J. Int.* **166**, 309–321.

Spencer, J.W. (1981). Stress relaxation at low frequencies in fluid saturated rocks; attenuation and modulus dispersion, *J. Geophys. Res.* **86**, 1803–1812.

Sperry, W.C. (1964). Rheological model concept, *J. Acoust. Soc. Amer.* **36**, 376–385.

Srivastava, H.M. (1968). On an extension of the Mittag-Leffler function, *Yokohama Math. J.* **16**, 77–88.

Srivastava, H.M., Gupta, K.C. and Goyal, S.P. (1982). *The H-Functions of One and Two Variables with Applications*, South Asian Publishers, New Delhi and Madras.

Srivastava, H.M. and Owa, S., Editors (1989). *Univalent Functions, Fractional Calculus, and Their Applications*, Ellis Horwood and Wiley, New York.

Srivastava, H.M., Owa, S. and Nishimoto, K. (1984). A note on a certain class of fractional differintegral equations, *J. College Engrg. Nihon Univ. Ser. B* **25**, 69–73.

Srivastava, H.M., Owa, S. and Nishimoto, K. (1985). Some fractional differintegral equations, *J. Math. Anal. Appl.* **106** No 2, 360–366.

Srivastava, H.M. and Saxena, R.K. (2001). Operators of fractional integration and their applications, *Appl. Math. Comput.* **118**, 1–52.

Stanislavski, A.A. (2000). Memory effects and macroscopic manifestation of randomness, *Phys. Rev. E* **61**, 4752–4759.

Stanislavski, A.A. (2003). Subordinated Brownian motion and its fractional Fokker-Planck equation, *Phys. Scripta* **67** No 4, 265–268.

Stankoviĉ, B. (1970). On the function of E.M. Wright, *Publ. de l'Institut Mathèmatique, Beograd, Nouvelle Sèr.* **10**, 113–124.

Stankovic, B. (2002). Differential equations with fractional derivatives and nonconstant coefficients, *Integral Transforms & Special Functions* **6**, 489–496.

Stankovic, B. and Atanackovic, T.M. (2001). On a model of a viscoelastic rod, *Fractional Calculus and Applied Analysis* **4**, 501–522.

Stankovic, B. and Atanackovic, T.M. (2002). Dynamics of a rod made of generalized Kelvin-Voigt viscoelastic material, *J. Math. Anal. Appl.* **268**, 550–563.

Stankovic, B. and Atanackovic, T.M. (2004a). On an inequality arising in fractional oscillator theory, *Fractional Calculus and Applied Analysis* **7**, 11–20.

Stankovic, B. and Atanackovic, T.M. (2004b). On a viscoelastic rod with constitutive equations containing fractional derivatives of two different order, *Mathematics and Mechanics of Solids* **9**, 629–656.

Stastna, J., DeKee, D. and Powley, M.B. (1985). Complex viscosity as a generalized response function *J. Rheol.* **29**, 457–469.

Stastna, J., Zanzotto, L. and Ho, K. (1994). Fractional complex modulus manifested in asphalts, *Rheol. Acta* **33**, 344–354.

Stastna, J. and Zanzotto, L. (1999). Linear response of regular asphalts to external harmonic fields, *J. Rheol.* **43**, 719–734.

Staverman, A. J. and Schwarzl, F. (1952). Thermodynamics of viscoelastic behavior, *Proc. Konnink. Nederlands Akad. van Wetenschchapen* **B55**, 474–485.

Steffe, J.F. (1992). *Rheological Methods in Food Process Engineering*, Freeman Press, East Lansing, Michigan.

Stiassnie, M. (1979). On the application of fractional calculus on the formulation of viscoelastic models, *Appl. Math. Modelling* **3**, 300–302.

Stojanović, M. (2009). Solving the n-term time fractional diffusion-wave problem via a method of approximation of tempered convolution, *Physica Scripta* **T136**, 014018/1–6.

Stratton, J.A. (1941). *Electromagnetic Theory*, McGraw-Hill, New York.

Strick, E. (1967). The determination of Q, dynamic viscosity, and transient creep from wave propagation measurements, *Geophysical J. R. Astr. Soc.* **13**, 197–219.

Strick, E. (1970). A predicted pedestal effect for pulse propagation in constant-Q solids, *Geophysics* **35**, 387–403.

Strick, E. (1971). An explanation of observed time discrepancies between continuous and conventional well velocity surveys, *Geophysics* **36**, 285–295.

Strick, E. (1976). *The Mechanical Behavior of Complex Anelastic Solids*, Lecture Notes, University of Pittsburg, July 1976, pp. 220.

Strick, E. (1982a). Application of linear viscoelasticity to seismic wave propagation, in [Mainardi (1982)], pp. 169–173.

Strick, E. (1982b). Application of general complex compliance model to direct-detection problem, *Geophys. Prospecting* **30**, 401–412.

Strick, E. (1982c). A uniformly convergent alternative to the Davidson-Cole distribution function, *Adv. Molecular Relaxation & Interactions Processes* **24**, 37–60.

Strick, E. (1984a). Implications of Jeffreys-Lomnitz transient creep, *J. Geophys. Res.* **89**, 437–452.

Strick, E. (1984b). Anelasticity and P-waves distortion, *Adavances in Geophysical Data Processing* (JAI Press, Inc.) **1**, 39–128.

Strick, E. and Mainardi, F. (1982). On a general class of constant Q solid, *Geophys. J. Roy. Astr. Soc.* **69**, 415–429.

Suarez, L. and Shokooh, A. (1995). Response of systems with damping materials modeled using fractional calculus, *ASME Appl. Mech. Rev.* **48**, S118–S126.

Suarez, L. and Shokooh, A. (1997). An eigenvector expansion method for the solution of motion containing fractional derivatives, *ASME J. Appl. Mech.* **64**, 629–635.

Subramanian, R. and Gunasekaran, S. (1997a). Small amplitude oscillatory shear studies on mozzarella cheese. Part I. Region of linear viscoelasticity, *J. Texture Studies* **28**, 633–642.

Subramanian, R. and Gunasekaran, S. (1997b). Small amplitude oscillatory shear studies on mozzarella cheese. Part II. Relaxation spectrum, *J. Texture Studies* **28**, 643–656.

Suki, B., Barabasi, A.L. and Lutchen, K.R. (1994). Lung tissue viscoelasticity: a mathematical framework and its molecular basis, *J. Appl. Physiol.* **76**, 2749–2759.

Sugimoto, N. (1989). Generalized' Burgers equations and fractional calculus, in Jeffrey, A. (Editor), *Non Linear Wave Motion*, Longman, London, pp. 172–179.

Sugimoto, N. (1991). Burgers equation with a fractional derivative; hereditary effects of nonlinear acoustic waves, *J. Fluid Mech.* **225**, 631–653.

Sugimoto, N. and Kakutani, T. (1985). Generalized Burgers' equation for nonlinear viscoelastic waves, *Wave Motion* **7**, 447–458.

Suki, B., Barabási, A-L., Lutchen, K.R. (1994). Lung tissue viscoelasticity: a mathematical framework and its molecular basis, *J. Appl. Physiol.* **76**, 2749-2759.

Sun, C.T. (1970). Transient wave propagation in viscoelastic rods, *J. Appl. Mech.* **37**, 1141–1144.

Sun, H.G. Chen, W. and Chen, Y.Q (2009) Variable-order fractional differential operators in anomalous diffusion modeling, *Physica A* **388** No. 21, 4586–4592.

Surguladze, T.A. (2002). On certain applications of fractional calculus to viscoelasticity, *J. Mathematical Sciences* **112** No 5, 4517–4557.

Szabo, T.L. (1994). Time domain wave equations for lossy media obeying a frequency power law, *J. Acoust. Soc. Amer.* **96**, 491–500.

Szabo, T.L. (1995). Causal theories and data for acoustic attenuation obeying a frequency power law, *J. Acoust. Soc. Amer.* **97**, 14–24.

Szabo, T.L. and Wu, J. (2000). A model for longitudinal and shear wave propagation in viscoelastic media. *J. Acoust. Soc. Amer.* **107**, 2437–2446.

Talbot, A. (1979). The accurate numerical inversion of Laplace transform, *Journal of the Institute of Mathematics and its Applications* **23**, 97–120.

Tan, W. and Xu, M. (2002). Plane surface suddenly set in motion in a viscoelastic fluid with fractional Maxwell model, *Acta Mechanica Sinica* **18** No 4, 342–349.

Tarasov, V.E. and Zaslavsky, G.M. (2006). Dynamics with low-level fractionality, and temporal memory, *Physica A* **368**, 399–415.

Tarasov, V.E. and Zaslavsky, G.M. (2007). Fractional dynamics of systems with long range space interaction and temporal memory, *Physica A* **383**, 291–308.

Tatar, N.E. (2004). The decay rate for a fractional differential equation, *J. Math. Anal. Appl.* **295** No 2, 303–314.

Tatar, N.E. and Kirane, M. (2003). Exponential growth for a fractionally damped wave equation, *Zeit. Anal. Anw.*, **22** No 1, 167–177.

Testa, R.B. (1966). Longitudinal impact of a semi-infinite circular viscoelastic rod, *J. Appl. Mech. (Trans. ASME, Ser. E)* **33** No. 3, 687–689.

Thau, S.A. (1974). Linear dispersive waves, in Leibovich, S. and Seebass, A.R. (Editors), *Non Linear Waves*, Cornell Univ. Press, Ithaca, pp. 44–81.

Thomson, W. (Lord Kelvin) (1865). On the elasticity and viscosity of metals. *Proc. Roy. Soc. London A* **14**, 289–297.

Titchmarsh, E.C. (1986). *Introduction to the Theory of Fourier Integrals*, Chelsea, New York. [First Edition, Oxford University Press, Oxford 1937]

Temme, N.M. (1996). *Special Functions: An Introduction to the Classical Functions of Mathematical Physics*, Wiley, New York.

Tobolsky, A.V. (1960). *Properties and Structure of Polymers*, Wiley, New York.

Tobolsky, A.V. and Catsiff, E. (1964). Elastoviscous properties of polysobutylene (and other amorphous polymers) from stress-relaxation studies. IX. A summary of results, *International J. Engineering Science* **2**, 111–121.

Tong, D. and Wang, R. (2004). Analysis of the low of non-Newtonian viscoelastic fluids in fractal reservoir with the fractional derivative, *Science in China, Ser. G* **47** No 4, 424–441.

Tong, D., Wang, R. and yang, H. (2005). Exact solutions for the flow of non-Newtonian fluid with fractional derivative in an anular pipe, *Science in China, Ser. G* **48** No 4, 485–495.

Torvik, P. J. and Bagley, R. L. (1984). On the appearance of the fractional derivatives in the behavior of real materials, *ASME J. Appl. Mech.* **51**, 294–298.

Torvik, P.J., Bagley, R.L. (1985). Fractional calculus in the transient analysis of viscoelastically damped structures, *AIAA J.* **23**, 918–925.

Trautenberg, E. A., Gebauer, K. and Sachs, A. (1982). Numerical simulation of wave propagation in viscoelastic media with non-reflecting boundary conditions, in [Mainardi (1982)], pp. 194–217.

Tricomi, F.G. (1940). Sul principio del ciclo chiuso del Volterra, *Atti Accad. Sci. Torino* **76**, 74–82.

Trinks, C. and Ruge, P. (2002). Treatment of dynamic systems with fractional derivatives without evaluating memory-integrals, **29**, 471–476.

Truesdell, C. (1972). *A First Course in Rational Mechanics*, John Hopkins University, Baltimore, MA.

Truesdell, C. and Toupin, R. A. (1960). *The Classical Field Theories of Mechanics*, in: Flügge, S. (Editor), *Encyclopedia of Physics*, Vol III/1, Springer-Verlag, Berlin.

Truesdell, C. and Noll, W. (1965). *The non-linear field theories of Mechanics*, in: Flügge, S. (Editor), *Encyclopedia of Physics*, Vol III/3, Springer-Verlag, Berlin.

Trujillo, J., Rivero, M., and Bonilla, B. (1999). On a Riemann-Liouville generalized Taylor's formula, *J. Math. Anal. Appl.*, **231**, 255–265.

Tschoegel, N.W. (1989). *The Phenomenological Theory of Linear Viscoelastic Behavior: an Introduction*, Springer Verlag, Berlin.

Tschoegel, N.W. (1997). Time dependence in material properties: an overview, *Mech. Time-Dependent Mater.* **1**, 1–31.

Tschoegel, N.W., Knauss, W. and Emri, I. (2002). Poisson's ratio in linear viscoelasticity: a critical review, *Mech. Time-Dependent Mater.* **6**, 3–51.

Turchetti, G. and Mainardi, F. (1976). Wave front expansions and Padè approximants for transient waves in linear dispersive media, in: Cabannes, H. (Editor), *Padè Approximants Method and its Applications* to Mechanics, Springer Verlag, Berlin, pp. 187–207. [Lecture Notes in Physics, Vol. 47]

Turchetti, G., Usero, D., and Vazquez, L.M. (2002). Hamiltonian systems with fractional time derivative, *Tamsui Oxford J. Math. Sci.* **18** No 1, 31–44.

Überall, H. (1978). Modal and surface wave resonances in acoustic-wave scattering from elastic objects and in elastic-wave scattering from cavities, in Miklowitz, J. and Achenbach, J.D. (Editors), *Modern Problems in Elastic Wave Propagation*, Wiley, New York, pp. 239–263.

Uchaikin, V.V. (1998). Anomalous transport equations and their applications to fractal walking, *Physica A* **255**, 65–92.

Uchaikin, V.V. (1999). Evolution equations foir Lévy stable processes, *Int. J. Theor. Physics* **38**, 2377–2388.

Uchaikin, V.V. (2000). Montroll-Weiss' problem, fractional equations and stable distributions, *Int. J. Theor. Physics* **39**, 2087–2105.

Uchaikin, V.V. (2002). Subordinated Lévy-Feldheim motion as a model of anomalous self-similar diffusion, *Physica A* **305**, 205–208.

Uchaikin, V.V. (2003). Relaxation processes and fractional differential equations, *Int. J. Theor. Physics* **42**, 121–134.

Uchaikin, V.V. (2008). *Method of Fractional Derivatives*, ArteShock-Press, Ulyanovsk [in Russian].

Uchaikin, V.V. and Sibatov, R.T. (2008). Fractional theory for transport in disordered semiconductors, *Comm. in Nonlinear Science and Numerical Simulations* **13**, 715–727.

Uchaikin, V.V., Sibatov, R.T. and Uchaikin, D. (2008). Memory regeneration phenomenon in dielectrics: the fractional derivative approach, *Physica Scripta* **T136**, 014002/1–6.

Uchaikin, V.V. and Zolotarev, V.M. (1999). *Chance and Stability. Stable Distributions and their Applications*, VSP, Utrecht.

Vainshtein, L.A. (1957). Group velocity of damped waves, *Sov. Phys. Techn. Phys.* **2**, 2420–2428.

Vainshtein, L.A. (1976). Propagation of pulses, *Sov. Phys. Usp.* **19**, 189–205.

Vallé, O. and Soares, M. (2004). *Airy Functions and Applications to Physics*, World Scientific, Singapore.

Van der Pol, B., and Bremmer, H. (1950). *Operational Calculus Based Upon the Two-Sided Laplace Integrals*, Cambridge University Press, Cambridge.

Van der Waerden, B.L. (1951). On the method of saddle points, *Appl. Sci. Res.* **82**, 33–45.

Van Groesen, E. and Mainardi, F. (1989). Energy propagation in dissipative systems, Part I: Centrovelocity for linear systems, *Wave Motion* **11**, 201–209.

Van Groesen, E. and Mainardi, F. (1990). Balance laws and centrovelocity in dissipative systems, *J. Math. Phys.* **30**, 2136–2140.

Vazquez, L. (2003). Fractional diffusion equations with internal degrees of freedom, *J. Comp. Math.* **21** No 4, 491–494.

Veber, V.K. (1983a). Asymptotic behavior of solutions of a linear system of differential equations of fractional order, *Stud. IntegroDifferential Equations* **16**, 19–125.

Veber, V.K. (1983b). Passivity of linear systems of differential equations with fractional derivative and quasi-asymptotic behavior of solutions, *Stud. IntegroDifferential Equations* **16**, 349–356.

Veber, V.K. (1985a). On the general theory of linear systems with fractional derivatives and constant coefficients in spaces of generalized functions, *Stud. IntegroDifferential Equations* **18**, 301–305.

Veber, V.K. (1985b). Linear equations with fractional derivatives and constant coefficients in spaces of generalized functions, *Stud. IntegroDifferential Equations* **18**, 306–312.

Voigt, W. (1892). Ieber innere Reibung fester korper, insbesondere der Metalle. *Annalen der Physik* **283**, 671–693.

Volterra, V. (1909a). Sulle equazioni integro-differenziali, *Atti della Reale Accademia dei Lincei*, Rendiconti Classe Sci. Fis. Mat e Nat. (Ser. V) **18**, 203–211.

Volterra, V. (1909b). Sulle equazioni integro-differenziali della teoria dell'elasticità, *Atti della Reale Accademia dei Lincei*, Rendiconti Classe Sci. Fis. Mat e Nat. (Ser. V) **18**, 295–301.

Volterra, V. (1913). Sui fenomeni ereditari, *Atti della Reale Accademia dei Lincei*, Rendiconti Classe Sci. Fis. Mat e Nat. (Ser. V) **22**, 529–539.

Volterra, V. (1913). *Lecons sur les Fonctions de Lignes*, Gauthier-villard, Paris.

Volterra, V. (1928). Sur la théorie mathématique des phénomènes héréditaires, *J. Math. Pures Appl.* **7**, 249–298.

Volterra, V. (1929). Alcuni osservazioni nei fenomeni ereditari, *Atti della Reale Accademia dei Lincei*, Rendiconti Classe Sci. Fis. Mat e Nat. (Ser. VI) **19**, 585–595.

Volterra, V. (1940). Energia nei fenomeni elastici ereditari, *Acta Pontificia Academia Scientifica*, **4**, 115–128.

Volterra, V. (1959). *Theory of Functionals and of Integral and Integro–differential Equations*, Dover, New York. [First published in 1930]

Von Hippel, A.R. (1962). *Dielectrics and Waves*, Wiley, New York.

Wall, H.S. (1948). *Analytic Theory of Continued Fractions*, Van Nostrand, New York.

Wang, C.C. (1965). The principle of fading memory, *Arch. Rat. Mech. Anal.* **18**, 343–366.

Wang, Y. (2002). A stable and efficient approach to inverse Q filtering, *Geophysics* **67**, 657–663.

Wang, Y. (2003). Quantifying the effectivness of stabilized inverse Q filtering, *Geophysics* **68**, 337–345.

Wang, Y. and Guo, J. (2004). Modified Kolsky model for seismic attenuation and dispersion, *J. Geophys. Eng.* **1**, 187–196.

Ward, I.M. (1983). *Mechanical Properties of Polymers*, Second Edition, Wiley, New York.

Warlus, S. and Ponton, A. (2009). A new interpretation for the dynamic behaviour of complex fluids at the sol-gel transition using the fractional calculus, *Reol. Acta* **48**, 51–58.

Watson, G.N. (1954). *A treatise of Bessel Functions*, Cambridge University Press, Cambridge.

Welch, S.W.J., Ropper, R.A.L. and Duren, Jr.R.G. (1999). Application of time-based fractional calculus methods to viscoelastic creep and stress relaxation of materials, *Mechanics of Time-Dependent Materials* **3** No 3, 279–303.

Weron, K. (1986). Relaxation in glassy materials from Lévy stable distributions, *Acta Physica Polonica* **A70**, 529–539.

Weron, K. and Kotulski, M. (1986). On the Cole-Cole relaxation function and related Mittag-Leffler distribution, *Physica A* **232**, 180–188.

Weron, K. and Jurlewicz, A. (2005) Scaling properties of the diffusion process underlying the Havriliak-Negami relaxation function, *Defects and Diffuson Forum* **237–240**, 1093–1100.

Weron, K., Jurlewicz, A. and Magdziarz, M. (2005). Havriliak-Negami response in the framework of the continuous-time random walk, *Acta Physica Polonica B* **36** No 5, 1855–1868.

Weron, A. and Weron, K. (1987). A statistical approach to relaxation in glassy materials, in Bauer, P. *et al.* (Editors), *Mathematical Statistics and Probability Theory*, Reidel, Dordrecht, pp. 245–254.

West, B.J., and Grigolini, P. (1997). Fractional diffusion and Lévy stable processes. *Phys. Rev. E* **55**, 99–106.

West, B.J., Grigolini, P., Metzler, R. and Nonnenmacher, T.F. (1997). Fractional diffusion and Lévy stable processes, *Phys. Rev. E* **55** No 1, 99–106.

West, B.J., Bologna, M. and Grigolini, P. (2003). *Physics of Fractal Operators*, Springer Verlag, New York. [Institute for Nonlinear Science] Chapter 7: Fractal Rheology, pp. 235–259.

Westerlund, S. (1991). Dead matter has memory!, *Phys. Scripta* **43**, 174–179.

White, J.E. and Walsh, D.J. (1972). Proposed attenuation-dispersion pair for seismic waves, *Geophysics* **37**, 456–461.

Widder, D.V. (1946). *The Laplace Transform*, Priceton University Press, Princeton.
Widder, D.V. (1971). *An Introduction to Transformation Theory*, Academic Press, New York.
Williams, G. and Watts, D.C. (1970). Non-symmetrical dielectric relaxation behavior arising from a simple empirical decay function, *Trans. Farady Soc.* **66**, 80–85.
Wineman, A.S. and Rajagopal, K.R. (2000). *Mechanical Response of Polymers, An Introduction*, Cambridge University Press, Cambridge.
Whitham, G.B. (1959). Some comments on wave propagation and shock wave structure with application to magnetohydrodynamics, *Comm. Pure Appl. Math.* **12**, 113–158.
Whitham, G.B. (1974). *Linear and Nonlinear Waves*, Wiley, New York.
Whittaker, E. T. and Watson, G. N. (1952). *A Course of Modern Analysis*, 4-th edn., Cambridge University Press, Cambridge [Reprinted 1990].
Wilson, D.K., 1992. Relaxation-matched modelling of propagation through porous media, including fractal pore structure. *J. Acoust. Soc. Amer.* **94**, 1136–1145.
Wiman, A. (1905a). Über den Fundamentalsatz der Theorie der Funkntionen $E_\alpha(x)$, *Acta Math.* **29**, 191–201.
Wiman, A. (1905b). Über die Nullstellen der Funkntionen $E_\alpha(x)$. *Acta Math.* **29**, 217–234.
Wintner, A. (1959). On Heaviside and Mittag-Leffler's generalizations of the exponential function, the symmetric stable distributions of Cauchy-Levy, and a property of the Γ-function, *J. de Mathématiques Pures et Appliquées* **38**, 165–182.
Wong, R. (1989). *Asymptotic Approximations of Integrals*, Academic Press, Boston, MA.
Wong, R. (Editor) (1990). *Asymptotics and Computational Analysis*, Marcel Dekker, New York.
Wong, R. and Zhao, Y.-Q. (1999a). Smoothing of Stokes' discontinuity for the generalized Bessel function, *Proc. R. Soc. London A* **455**, 1381–1400.
Wong, R. and Zhao, Y.-Q. (1999b). Smoothing of Stokes' discontinuity for the generalized Bessel function II, *Proc. R. Soc. London A* **455**, 3065–3084.
Wong, R. and Zhao, Y.-Q. (2002). Exponential asymptotics of the Mittag-Leffler function, *Constructive Approximation* **18**, 355–385.
Wright, E.M. (1933). On the coefficients of power series having exponential singularities, *Journal London Math. Soc.* **8**, 71–79.
Wright, E.M. (1935a). The asymptotic expansion of the generalized Bessel function, *Proc. London Math. Soc. (Ser. II)* **38**, 257–270.

Wright, E.M. (1935b). The asymptotic expansion of the generalized hypergeometric function, *Journal London Math. Soc.* **10**, 287–293.

Wright, E. M. (1940). The generalized Bessel function of order greater than one, *Quart. J. Math., Oxford Ser.* **11**, 36–48.

Wyss, W. (1986). Fractional diffusion equation, *J. Math. Phys.* **27**, 2782–2785.

Wyss, M and Wyss, W. (2001). Evolution, its fractional extension and generalization, *Fractional Calculus and Applied Analysis* **4** No 3, 273–284.

Xu, D. (1998). The long-time global behavior of time discretization for fractional order Volterra equations, *Calcolo* **35** No 2, 93–116.

Xu, M. and Tan, W. (2001). Theoretical analysis of the velocity field, stress field and vortex sheet of generalized second order fluid with fractional anomalous diffusion, *Science in China, Ser. A* **44** No 1, 1387–1399.

Xu, M. and Tan, W. (2003). Representation of the constitutive equation of viscoelastic materials by the generalized fractional element networks and its generalized solutions, *Science in China, Ser. G* **46** No 2, 145–157.

Xu, M. and Tan, W. (2006). Intermediate processes and critical phenomena: Theory, method and progress of fractional operators and their applications to modern mechanics, *Science in China, Ser. G* **49** No 3, 257–272.

Youla D.C, Castriota L.J and Carlin H.J. (1959). Bounded real scattering matrices and the foundations of linear passive network theory, *IEEE Trans. Circuit Theory* **CT-6**, 102–124.

Yu, R. and Zhang, H. (2006). New function of Mittag-Leffler type and its application in the fractional diffusion–wave equation, *Chaos, Solitons and Fractals* **30**, 946–955.

Yuan, L. and Agrawal, O.P. (2002). A numerical scheme for dynamic systems containing fractional derivatives, *Journal of Vibration and Acoustics* **124**, 321–324.

Yuen, D. A., Sabadini, R., Gasperini, P. and Boschi, E. (1986). On transient rheology and glacial isostasy, *J. Geophys. Res.*, **91**, 11420–11438.

Zahorski, S. (1982). Properties of transverse and longitudinal harmonic waves, in [Mainardi (1982)], pp. 41–64.

Zaslavsky, G.M. (1994). Fractional kinetic equation for Hamiltonian chaos. *Physica D* **76**, 11-0-122.

Zaslavsky, G.M. (1995). From Lévy flights to the fractional kinetic equation for dynamical chaos, in [Shlesingher *et al.* (1995)], pp. 216–238.

Zaslavsky, G.M. (2002). Chaos, fractional kinetics and anomalous transport, *Phys. Reports* **371**, 461–580.

Zaslavsky, G.M. (2005). *Hamiltonian Chaos and Fractional Dynamics*, Oxford University Press, Oxford.

Zelenev, V.M., Meshkov, S.I. and Rossikhin, Yu.A. (1970). Damped vibrations of hereditary - elastic systems with weakly singular kernels, *J. Appl. Mech. Tech. Physics* **11** No 2, 290–293.

Zemanian, A.H. (1972). *Realizability Theory for Continuous Linear Systems*, Academic Press, New York.

Zener, C. (1948). *Elasticity and Anelasticity of Metals*, University of Chicago Press, Chicago.

Zener, C. (1958). Anelasticity of metals, *Il Nuovo Cimento* (Ser. X) **7** Suppl. 2, 544–568.

Zhang, W. and Shimuzu, N. (1998). Numerical algorithm for dynamic problems involving fractional operators, *JSME Int. Jour. C* **41**, 364–370.

Zhang, W. and Shimuzu, N. (1999). Damping properties of the viscoelastic material described by fractional Kelvin-Voigt model, *JSME Int. Jour. C* **42**, 1–9.

Zhang, W. and Shimizu, N. (2001). FE formulation for the viscoelastic body modeled by fractional constitutive law, *Acta Mechanica Sinica* (English Series) **17** No 4 , 254–365.

Zimm, B. H. (1956). Dynamics of polymer molecules in dilute solution: viscoelasticity, flow birefringence and dielectric loss, *J. Chem. Phys.* **24**, 269–278.

Zolotarev, V.M. (1986). *One-dimensional Stable Distributions*, American Mathematical Society, Providence, R.I., USA.

Zolotarev, V.M. (1994). On representation of densities of stable laws by special functions. *Theory Probab. Appl.* **39**, 354–361.

Zolotarev, V.M. (1997). *Modern Theory of Summation of Random Variables*, VSP, Utrecht.

Index

Abel, N.H., 19
Bagley, R.L., 76
Brillouin, L., 98, 100
CISM, 75
Caputo, M., xi, 75, 76
Courant, R., 20
Dzherbashyan, M.M., 21, 233
Euler, L., 17
Fractional Calculus and Applied Analysis, 18
Fractional Dynamic Systems, 19
Gemant, A., 74
Gerasimov, A, 74
Gorenflo, R., xi, 232
Gross, B., 65
Hanyga, A., 54
Jeffreys, Sir H., 131
Journal of Fractional Calculus, 18
Kiryakova, V., 18
Leibniz, G.W., 17
Liouville, I, 19
Mainardi, F., 75, 76
Meshkov, S.I., 76
Mittag-Leffler, G.M., 211
Nishimoto, K, 18
Oldham, K.B., 17
Rabotnov, Yu.N., 74, 221
Riemann, B., 19
Ross, B., 17

Rossikhin, Yu.A., 76
Scott-Blair, G.W., 74
Sommerfeld, A., 98, 100
Spanier, I., 17
Stanković, B., 258
Strick, E., 54
Torvik, P.J., 76
Tricomi, F.G., 204, 235
Volterra, V., 54
Weyl, H, 19
Whitham, G.B., vii
Wright, E.M., 237
www.diogenes.bg/fcaa/, 19
www.fracalmo.org, 19

Abel integral equation, 230
Absorption coefficient, 87
Airy function, 120, 187
Anomalous diffusion, 259
Anomalous dispersion, 89, 102
Anti-Zener model, 33
Asymptotic expansion, 109
Asymptotic representation, ix
Attenuation coefficient, 78, 87, 96

Bernstein function, 42, 111
Bessel function, 95, 127, 128, 173, 175, 179, 182, 184, 239
Bessel-Clifford function, 239

Beta function, 165
Boltzmann superposition principle, 25, 26
Brillouin problem, 98
Brillouin representation, 101
Bromwich path, 80, 99, 126, 130, 132, 249
Bromwich representation, 80, 101, 109, 131
Buchen-Mainardi method, 116, 121
Burgers model, 38

Canonic forms, 35
Caputo fractional derivative, 6, 21, 61, 66, 73, 138, 151
Cauchy problem, 139, 140, 147, 148
Cauchy-Lorentz probability density, 254
Central limit theorem, 252
Characteristic function, 252
Cinderella function, 235
Combination rule, 30
Completely monotonic function, 42, 111, 216, 249
Complex compliance, 46, 51, 86
Complex factorial function, 156
Complex modulus, 46, 51, 86
Complex refraction index, 78, 81, 87, 90, 98
Complex wave number, 86
Compound Maxwell model, 35
Compound Voigt model, 35
Cornu spiral, 201
Correspondence principle, 60, 61
Creep compliance, 24, 30, 63, 79, 111, 113, 116, 139
Creep function, 42, 43, 65
Creep representation, 25, 79, 82, 110
Creep test, 24

Crossing-symmetry relationship, 86, 87

D'Alembert wave equation, 137
Dashpots, 30, 35, 48
Dawson integral, 197
Diffusion, 137
Diffusion equation, 84
Dispersion, ix, 85
Dispersion relation, 85, 87
Dissipation, ix, 85
Distortion-less case, 95
Dynamic functions, 23, 45, 46, 51

Energy dissipation, 48, 54
Energy storage, 48, 54
Energy velocity, 107
Entire function of exponential type, 109, 113
Erfi function, 197
Error function, 122, 133, 149, 191, 193, 225
Euelerian functions, 155
Euler series, 206
Euler-Mascheroni constant, 205
Exponential integral, 203, 207

Fast diffusion, 150
Feller-Takayasu diamond, 253
Forerunner, 106, 153
Fourier diffusion equation, 137
Fourier transform, viii, 1, 23, 46, 86, 141, 252–254, 257
Fox H-function, 233, 256
Fractional anti-Zener model, 62
Fractional Brownian motion, 259
Fractional calculus, vii, 1, 73
Fractional diffusive wave, 139
Fractional exponential function, 221
Fractional Maxwell model, 62, 67, 121, 123, 125

Fractional mechanical models, 61
Fractional Newton (Scott–Blair) model, 61, 67
Fractional operator equation, 63
Fractional oscillation, 230
Fractional relaxation, 11, 15, 60, 207, 230, 235
Fractional Voigt model, 61, 67
Fractional Zener model, 62, 63, 66, 72, 92
Frequency–spectral function, 43
Fresnel integral, 198, 200
Friedlander–Keller theorem, 118

Gamma function, 2, 6, 155
Gauss probability density, 148, 256
Gel'fand-Shilov function, 3
Green function, 80, 113, 139, 143, 147
Grey Brownian motion, 259
Gross method, 43
Group velocity, 78, 85, 88, 90, 96

Hankel function, 175
Hankel integral representation, 162
Hankel path, 160, 214, 249
Heaviside function, 24
Hilfer fractional derivative, 11
Hilfer function, 15
Hooke model, 31
Hurwitz polynomials, 37

Impact problem, ix, 127
Impact waves, 77, 82, 85, 109
Incomplete Gamma function, 171
Infinite divisibility, 254
Initial conditions, 40, 73
Internal friction, 50

Jeffreys problem, 131, 135

Klein-Gordon equation, 78, 94, 102
Krönig-Kramers relations, 88

Lévy stable distribution, 139, 149, 252, 256
Lévy-Smirnov probability density, 148
Ladder networks, 54
Laplace convolution, 3, 80
Laplace transform, viii, 1, 4, 9, 12, 23, 28, 40, 43, 46, 79, 86, 95, 99, 109, 110, 139, 141, 148, 186, 194, 207, 223, 231, 248, 251, 257
Lee-Kanter problem, 127, 128, 135
Linear viscoelasticity, vii, 23, 55, 73, 89
Liouville-Weyl fractional derivative, 15
Liouville-Weyl fractional integral, 15
Lorentz-Lorenz dispersion equation, 100
Loss compliance, 47
Loss modulus, 47
Loss tangent, 47, 51, 52, 66, 68, 90

Mainardi function, 258
Mainardi-Turchetti theorem, 113
Material functions, 23, 24, 35, 64, 109, 116
Maxwell model, 32, 84, 91, 106, 115, 125, 127, 135
Mechanical models, 30, 37, 49
Mellin transform, 257
Miller-Ross function, 220
Mittag-Leffler function, viii, 12, 21, 60, 65, 211, 214, 216, 227, 231, 233, 234, 246, 247, 249, 250
Modified exponential integral, 44, 204

Newton model, 31, 84
Normal dispersion, 89, 102

Operator equation, 37

Padè approximants, ix, 109, 133, 235
Phase coherence, 49
Phase shift, 46
Phase velocity, 78, 87, 88, 90, 96
Physical realizability, 37, 54
Pochhammer's symbol, 160
Positive real functions, 37
Pot, 59
Power-law creep, 58
Prabhakar function, 222
Probability density function, 147
Psi function, 169

Quality factor, 50
Queen function of fractional calculus, 235

Rabotnov function, 221
Rabotnov, Yu.N., 75
Rate of creep, 82, 110
Rate of relaxation, 82, 110
Ray–series method, 125
Reciprocity relation, 141, 144
Relaxation function, 42, 43, 65
Relaxation modulus, 24, 30, 63, 79, 111
Relaxation representation, 25, 79, 83, 110
Relaxation spectrum, 41
Relaxation test, 24
Relaxation time, 32–34, 80
Retardation spectrum, 41
Retardation time, 32–34, 80
Riemann–Liouville fractional integral, 2

Riemann-Liouville fractional derivative, 5, 6, 61, 73
Riemann-Liouville fractional integral, 230

Saddle–point approximation, ix, 126, 128, 131, 135
Saddle–point method, 109, 126, 127, 130, 132, 145
Scott–Blair model, 59
Self-similarity, 253
Semigroup property, 2
Signal velocity, 78, 90, 100, 102, 126
Signalling problem, 139, 140, 147, 148
Similarity variable, 137, 139, 142, 144
Slow diffusion, 150
Specific dissipation function, 50, 69, 90
Springs, 30, 35, 48
Standard Linear Solid, 33
Steady–state response, 98, 99
Steepest–descent path, 78, 101, 103, 106, 126, 128, 132
Stieltjes integral, 25
Stieltjes transform, 45
Stirling formula, 163
Storage compliance, 47
Storage modulus, 47
Strain creep, 25
Stress power, 47
Stress relaxation, 25
Stress–strain relation, 23, 30, 46

Telegraph equation, 84, 94, 97, 106
Thermal relaxation, 71
Thermoelastic coupling, 71
Time–spectral function, 41, 43, 64
Transient–state response, 99

Voigt model, 32, 84, 121, 123, 125

Wave propagation, vii, 48, 77, 137
Wave–front expansion, ix, 109, 111, 113, 116, 124, 135
Wave–front velocity, 81, 84, 90, 112, 126
Wave-mode solution, 85, 86, 99
Wright F-function, 143, 144, 240, 241, 258

Wright M-function, 143–145, 147, 150, 153, 240–243, 249, 251, 257–259
Wright function, viii, 120, 139, 149, 237–239, 241, 245, 248, 250, 258

Zener model, 33, 51, 84, 91, 115, 135